Assessment of Responsible Innovation

Responsible innovation encourages innovators to work together with stakeholders during the research and innovation process, to better align the outcomes of innovation with the values, needs and expectations of society. Assessing the benefits and costs of responsible innovation is crucial for furthering the responsible conduct of science, technology and innovation. However, until now there has only been limited academic work on responsible innovation assessment. This book fills this lacuna.

Assessment of Responsible Innovation: Methods and Practices presents tools for measuring, monitoring and reporting upon the responsible innovation process and the social, environmental, scientific and economic impacts of innovations. These tools help innovators to mitigate risk and to strengthen their strategic planning. This book aligns assessment tools and practices with the United Nations sustainable development goals (SDGs). The prospects as well as the limitations of various responsible innovation assessment approaches and tools are discussed, as well as their applicability in various industry contexts.

The book brings together leading scholars in the field to present the most comprehensive review of responsible innovation tools. It articulates the importance of assessment and value creation, the different metrics and monitoring systems that can be deployed and the reporting mechanisms, including the importance of effective communication.

Emad Yaghmaei is Senior Researcher at the Faculty of Technology, Policy and Management at Delft University of Technology, the Netherlands.

Ibo van de Poel is Antoni van Leeuwenhoek Professor in Ethics and Technology and Head of the Department of Values, Technology and Innovation at Delft University of Technology, the Netherlands.

Assessment of Responsible Innovation

Methods and Practices

Edited by
Emad Yaghmaei and Ibo van de Poel

Routledge
Taylor & Francis Group
LONDON AND NEW YORK

First published 2021
by Routledge
2 Park Square, Milton Park, Abingdon, Oxon OX14 4RN

and by Routledge
52 Vanderbilt Avenue, New York, NY 10017

Routledge is an imprint of the Taylor & Francis Group, an informa business

British Library Cataloguing-in-Publication Data
A catalogue record for this book is available from the British Library

Library of Congress Cataloging-in-Publication Data
A catalog record has been requested for this book

ISBN: 978-0-367-27975-2 (hbk)
ISBN: 978-0-429-29899-8 (ebk)

Typeset in Sabon
by Newgen Publishing UK

Contents

Figures

Tables

Abbreviations

AREA	anticipate, reflect, engage, act
BOP	bottom of the pyramid
CEN	European Committee for Standardisation
CEN/WS	European Committee for Standardisation (CEN) Workshop
CFP	corporate financial performance
CPS	Cyber-Physical-Systems
CSE	creative self-efficacy
CSI	corporate social innovation
CSP	corporate social performance
CSR	corporate social responsibility
CWA	CEN Workshop Agreement
DG RTD	Directorate-General for Research and Innovation
EC	European Commission
EFA	exploratory factor analysis
ELSI	ethical, legal and social impacts
EPSRC	Engineering and Physical Sciences Research Council
ERA	European research area
ESG	environmental social and corporate governance
EU	European Union
FAIR	findable, accessible, interoperable and reusable
FoP	front-of-pack
FSC	Forest Stewardship Council
GE	gender equality
GMOs	genetically modified organisms
GRI	Global Reporting Initiative
IoT	internet of things
IP	innovation potential
IRB	Internal Review Board
KMO	Kaiser–Meyer–Olkin
KPIs	key performance indicators
LCA	life cycle assessment
LoE	line of evidence
MARIE	MAinstreaming Responsible Innovation in European S3

MCDA	multi-criteria decision analysis
MDGs	millennium development goals
MoRRI	Monitoring the Evolution and Benefits of Responsible Research and Innovation
MSA	multi-stakeholder agreement
MUSE	meaningful, useful, sensible and effective
NGO	non-governmental organization
NLP	Natural Language Processing
OA	open access
OCEANIS	Open Community for Ethics in Autonomous and Intelligent Systems
OECD	Organisation for Economic Co-operation and Development
OPDCA	observe–plan–do–check–adjust
PDCA	plan–do–check–act/adjust
PE	public engagement
R&D	research and development
R&I	research and innovation
RFO	research-funding organization
RI	responsible innovation
RIL	Responsible Research and Innovation intensity level
RMA	reflexive monitoring in action
RPO	research-performing organization
RRI	Responsible Research and Innovation
SATORI	Stakeholders Acting Together on the Ethical Impact Assessment of Research and Innovation
SDGs	sustainable development goals
SEC	Securities and Exchange Commission
SEP	Standard Evaluation Protocol
SLSE	science, literacy and science education
SMART	specific, measurable, attainable, relevant and timely
SMEs	small and medium-sized enterprises
SRD	socially responsible design
SR TT	Societal Readiness Thinking Tool
StRIM	strategic responsible innovation management
SUPER MoRRI	Scientific Understanding and Provision of an Enhanced and Robust Monitoring system for RRI
SwafS	Science with and for Society
SWOT	strengths, weaknesses, opportunities and threats
TRL	technology readiness level
UN	United Nations
UNDP	United Nations Development Project
WBA	World Benchmarking Alliance
WFE	World Federation of Exchanges

Notes on the editors and contributors

Editors

Emad Yaghmaei is Senior Researcher at the Faculty of Technology, Policy and Management at Delft University of Technology, the Netherlands. His research interests cover governance of organizations and processes with focus on their ethical, social and governmental impacts. He is especially interested in designing and developing necessary policies and methods for implementation and evaluation of Responsible Innovation within an industrial context. His publications lie in the intersection of science, innovation, technology and society. His current work takes the outset in Responsible Research and Innovation (RRI) and its institutionalization within stakeholders across innovation ecosystems.

Ibo van de Poel is Antoni van Leeuwenhoek Professor in Ethics and Technology and head of the Department of Values, Technology and Innovation at the Faculty Technology, Policy and Management at Delft University of Technology in the Netherlands. He has published on engineering ethics, the moral acceptability of technological risks, design for values, responsible innovation, moral responsibility in research networks, ethics of newly emerging technologies and the idea of new technology as social experiment. He has recently received a European Research Council advanced grant on design for changing values: a theory of value change in sociotechnical systems.

Contributors

Vincent Blok is Associate Professor at the Philosophy Group at Wageningen University, the Netherlands. In 2005 he received his PhD in Philosophy at Leiden University with a specialization in Philosophy of Technology. Together with a research group of four postdocs and six PhD students, Blok is involved in several (European) research projects at the crossroads of business ethics, philosophy of technology and responsible innovation. His books include *Ernst Jünger's Philosophy of Technology: Heidegger* *and the Poetics of the Anthropocene* (Routledge, 2017) and *Heidegger's Concept of Philosophical Method: Innovating Philosophy in the Age of Global Warming* (Routledge, 2020). Blok's work has appeared in, amongst others, the *Business Ethics Quarterly*, *Business & Society*, *Journal of Responsible Innovation* and *Public Understanding of Science*. For more information about his current research and teaching, see www.vincentblok.nl

Elisabetta Borsella is Senior Analyst at Airi, the Italian Association for Industrial Research. She holds a degree in Physics cum Laude. She has expertise in laser interaction with matter and nanotechnologies. She had scientific responsibility for ENEA (Italian National Agency for New Technologies, Energy and Environment) on national and EU projects on nanotechnologies (until 2013). She was the coordinator of European Commission (EC) projects on nanotechnologies: NanoMat, IST-SINERGIA and BONSAI. Elisabetta has also published over 130 refereed journal articles, participated in international conferences and organized workshops in the field of nanomaterials. Since 2013 she has been involved in a number of Horizon 2020 (H2020) projects in the thematic areas of Responsible Research and Innovation (RRI), including Responsible Industry, PRISMA and caLIBRAte.

Joey van den Brink is Certification Manager at the non-profit organization B Lab Europe, the driving force behind the B Corp movement in Europe. Joey has been a scholar at Wageningen University and Research where he studied Environmental Economics with a focus on macroeconomic models, game theory and sustainable entrepreneurship. Through his professional career, Joey seeks to develop frameworks in which business as a force for good can thrive.

Donato Calace is the Director of Accounts and Innovation at Datamaran. He leads global innovation projects for the firm as well as the sales and account strategy. He leads on the engagement and cooperation with the academic world. In June 2016, Donato was awarded a PhD in the international program "The Economics and Management of Natural Resourcesg developed by the LUM Jean Monnet University (Italy), Megatrend University, Belgrade (Serbia), China Three Gorges University (CTGU – China), Louisiana Tech University (USA) and St Petersburg State Forest Technical University (SPbSFTU – Russia). Recently, he wrote the definition of materiality for the *Encyclopedia of the United Nations Sustainable Development Goals* (Springer).

Tung Tung Chan is a Researcher at the Centre for Science and Technology Studies (CWTS) of Leiden University in the Netherlands. Her research interests include advancing research intelligence applications, analyzing Responsible Research and Innovation (RRI) in practice and monitoring the development of open science in Europe. She has published several case studies for the European Commission Open Science Monitor, and co-created the Societal Readiness Thinking Tool for the NewHoRRIzon project. Her master thesis on China Scholarship Council (CSC)-funded PhDs in the Netherlands has informed ScienceGuide, and sparked discussions on the internationalization of doctoral education in the Netherlands via national television NOSop3. In 2018, she graduated from the Erasmus Mundus Joint Master's Degree in Research and Innovation in Higher Education (MARIHE). She obtained her bachelor's degree in Communication and Media at Erasmus University Rotterdam.

Cristina Covello is a foodservice consultant with an MBA from the University of Bath, UK. From furniture to food, she has worked on the creative side of multiple industries, transitioning from design and product development to marketing and strategy. She has a particular interest in the link between business strategy and responsible business and strongly believes in the old adage that businesses can "do well by doing good."

Adriana Farenga is Digital Marketing Executive at Datamaran. Adriana has a background in Philosophy and History of the Middle Age. She defended her PhD at University La Sapienza in Rome in 2016, and spent two years as a specialization student at the Institute of Philosophy of KU Leuven University, the Netherlands. Adriana is particularly interested in environmental sustainability and women's empowerment. She was part of the Communications team of the Institute for European Environmental Policy (IEEP), the non-governmental organization Global One and the National Institute for Health Research (NIHR). Also, since 2009 she has been working as a freelance journalist for the web-magazine Ambient&Ambienti.

Steven Flipse is Assistant Professor in Communication Design for Innovation at the Delft University of Technology, the Netherlands. He studies communication design in support of innovation practice. Steven's research focuses on interaction design to spur responsible innovation in practice.

Jilde Garst is a Postdoctoral Researcher in the Accounting, Auditing and Control department at the Erasmus School of Economics (Rotterdam, the Netherlands). After obtaining her MSc in Health Food Innovation Management (Maastricht University, 2011) and working as project manager for health-related non-governmental organizations, she conducted her PhD at Wageningen University in the area of responsible innovation and Corporate Social Responsibility. Her research interests are knowledge management, social and sustainable innovation, self-regulation and standards for social value creation and sustainability reporting. Her work has been published in *Business & Society*, amongst others.

Robert Gianni is a Political and Social Philosopher currently working at Maastricht University, the Netherlands. He has been working on the relation between technology and society with a particular focus on their regulations and intersections. He has analyzed some of the principles guiding democratic activities like responsibility and freedom. The current aim of his research is to critically scrutinize the different registers of justification of technological innovation and their repercussions for democratic systems. His overall interest is to understand the challenges faced by democracy with the emergence of digitalization and to explore the possibility of extending democratic approaches beyond the sphere of politics.

Agata Gurzawska is a Senior Research Analyst at Trilateral Research within the Policy, Ethics and Emerging Technologies (POET) team and a PhD candidate at the Department of Philosophy of the University of Twente, the Netherlands. Her areas of expertise and main research interests encompass: human rights and business, corporate social responsibility (CSR), strategic CSR, sustainability, ethics of technology, Responsible Innovation and new technologies in the business context. She has been engaged in a number of FP7 and Horizon 2020 projects, including SATORI, Responsible Industry, SHERPA, COPKIT and PREFET. Agata holds an MA in Law (University of Lodz) and an MA in Human Rights and Democratisation (European Inter-University Centre for Human Rights and Democratisation EMA). Driven by curiosity, ethical and human rights values, Agata wants to support, strengthen and create economic opportunities that are profitable for both businesses and society. She believes tailor-made technological innovations can create economic benefits and boost companies' social responsibility and sustainability.

Martin de Heaver is Managing Director of ORBIT (the Observatory for Responsible Research and Innovation in ICT) and has been commercializing both academic and industrial ICT research for over 20 years.

Carolyn Ten Holter is Communications Officer for the ORBIT project at the University of Oxford, UK, with a background in law, library science and marketing. She is undertaking a DPhil in responsible quantum technologies.

Konstantinos (Kostas) Iatridis is an Assistant Professor of Business and Society in the School of Management at the University of Bath, UK. He has worked with FTSE 100-listed companies on a range of corporate responsibility/sustainability consulting projects. Kostas' research explores the governance mechanisms companies use to address contemporary challenges for sustainable/responsible conduct. His research has been published in leading academic journals, including *Business Ethics Quarterly*, *Journal of Business Ethics* and *Journal of Environmental Management*.

Veikko Ikonen works as a Senior Scientist and project manager at VTT Technical Research Centre of Finland. His research focuses on the human-driven design of future technologies, applications and services, as well as ethical issues and the development of a Responsible Research and Innovation (RRI) approach. Veikko has experience in several EU projects, where user engagement and ethical issues for the product or application development had the central role. Veikko was previously a board member for Corporate Responsibility Network FIBS (Finnish Business Society) and is a member of the Ethics Committee of the Tampere region. He also acts as Research Integrity Adviser for VTT.

Edurne A. Inigo is a Postdoctoral Researcher at Wageningen University and Research, the Netherlands. Her current interests are system transitions to sustainability, responsible innovation and the circular economy. She obtained her PhD at Deusto Business School (DBS). She has also been a Visiting Researcher at Lappeenranta University of Technology. Prior to this, she studied at University of the Basque Country (LLM), London Metropolitan University (MA) and the London School of Economics and Political Science (Exec. Grad. Programme). She has managerial experience as a consultant and freelance in multiple projects, for the public and private sector, and has participated in European research projects. Her work has been published in *Industrial Marketing Management*, *Technological Forecasting and Social Change* and *Journal of Cleaner Production*, among others.

Katharina Jarmai is a Project Manager and Research Fellow at the Institute for Managing Sustainability, Vienna University of Economics and Business. She has coordinated the H2020-funded COMPASS project and co-developed the Responsible Innovation self-check tool for small and medium-sized enterprises (SMEs). She specializes in responsible innovation, innovation systems and social learning. She has additional qualifications in the field of moderation, group dynamics and organizational development.

Marina Jirotka is Professor of Human Centred Computing at the University of Oxford and co-principal investigator of ORBIT. She undertakes work focused on deepening societal comprehension of the impacts of technology and ameliorating negative effects by anticipating outcomes. She leads the human-centred computing group, an interdisciplinary research group that aims to understand the ways in which technology affects communication, collaboration and knowledge exchange within scientific, work and home settings.

Pim Klaassen works as Assistant Professor of Policy, Ethics and Communication in the Health and Life Sciences at the Athena Institute, Vrije Universiteit Amsterdam. The focus of his work is on the societal meaning of science and innovation, and on governance arrangements and (transdisciplinary) research methodologies conducive to Responsible Research and Innovation. He also holds a position as Senior Policy

Advisor and Coordinator of Safe-by-Design at the Dutch National Institute for Public Health and the Environment (RIVM).

Wouter van de Klippe is a part-time Junior Researcher. Wouter has a background in Science and Technology studies. He has a keen interest in working to ensure that science and technology are developed in ways that reflect the needs of broader society. He is interested in thinking critically about the sociopolitical nature of science and technology and is particularly interested in investigating ways to ensure that science and technology can work alongside and empower social movements. He works part-time on the SUPER MoRRI project while spending the rest of his week volunteering and working at organizations championing social justice issues.

Frank Kupper is Assistant Professor of Science Communication and Public Engagement at the Athena Institute, Vrije Universiteit Amsterdam. His work revolves around the study and design of science–society interactions and interfaces, with an emphasis on practices of sense and meaning making, the development of art- and design-based interventions and the facilitation of reflection, dialogue and change processes.

Maria João Maia is a Researcher at the Institute of Technology Assessment and Systems Analysis (ITAS) at the Karlsruhe Institute of Technology (KIT), in Germany. Her main areas of research include health technology assessment, the integration of RRI aspects in industry and the assessment of emerging technologies in healthcare. Maia has a background in health sciences and a PhD in Technology Assessment. After the PRISMA project where she coordinated a work package on the development and organization of stakeholder dialogues to assist industry to implement RRI in innovation processes within the framework of their corporate social responsibility policy, she is currently working on the ethical, legal and societal implications of intelligent prostheses and orthotics (INOPRO project); and neurotechnological innovations (FUTUREBODY project) as a path to steer research and development responsibly. Maia is an affiliated researcher at CICS.NOVA (Portugal) and a founding member of the Portuguese Technology Assessment Observatory.

Elvio Mantovani is the Scientific Director of Airi/Nanotec IT, the Committee for Nanotechnology and the other Key Enabling Technologies (KETs) set up by the Italian Association for Industrial Research to promote nanotechnologies and the other KETs in Italy. He has a doctorate in Chemistry from the University of Rome "La Sapienza" and is a Member of the Royal Society of Chemistry (UK). Before joining AIRI he worked both in academia, as researcher and lecturer, and in industry, at first as senior researcher and then with senior managerial positions. His present activities cover research and technology assessment and planning, technology transfer and international cooperation, with particular attention on responsible growth. In his capacity, Dr Mantovani has been/is involved, sometimes as coordinator, in all the cooperative EU projects (FP6, FP7, H2020), in which the Association has been/is involved.

Ingeborg Meijer is a Senior Researcher in research policy and evaluation with a focus on biomedicine and healthcare research and innovation. Her main goal is to make sure that society is able to use research. Working in industry, health research policy and consultancy on European policies, she realized that in academia there is still a lot to gain at the science–society interface. Through crossing disciplinary boundaries and collaboration with policy, funders, charities, patient organizations and European research and innovation partners, she aims to improve mutual understanding and sharing of responsibility. Her latest contributions are in European projects on Responsible Research and Innovation (RRI). Furthermore she pursues reconciliation between notions of research quality and impact on society.

Elena Mocchio is Head of Innovation Office. She holds a degree in Foreign Languages and Modern Literature from the Università Cattolica of Milan. She joined UNI in 2000, where she has assumed different posts to date. At first International Assistant in the Technical Department, in 2004 she became National and CEN/TC Technical Officer in different sectors, involved in the development of processes and tools for the technical department. In 2010 she joined the newly created External Relations, New Business and Innovation Directorate, a new structure for carrying out strategic lines and finding out new potential markets and stakeholders. She is in charge of the development of new projects, both at national and international level, in emerging and innovative fields.

Mika Nieminen is an Adjunct Professor who works as Principal Scientist at VTT, Technical Research Centre of Finland. His key areas of expertise are research and innovation organizations, policy and systems. Nieminen has published extensively, including books, articles, edited volumes and research reports. He has served as journal editor, conference organizer and member of various boards. He has also participated in policy and organization evaluations and development. His current research interests cover especially systemic innovations and transformations, development of impact assessment methods and Responsible Research and Innovation (RRI)-related questions.

Margherita Nulli is Project Officer for the ORBIT project at the University of Oxford. She previously worked as a researcher at the department of Experimental Psychology, with a focus on Cognitive Psychology.

Konstantina M. Pentaraki is currently an R&D Specialist at ELBISCO, a food production company. She holds two master's degrees related to Food Science from the Agricultural University of Athens and from Wageningen

University. She has great experience from internships in food production companies and from the National Agricultural Research Foundation in Greece. She is strongly motivated on innovation, sustainability and human nutrition issues. She was involved in the International Nonthermal Processing Workshop in November 2015 with a project related to sustainable innovation. She is now working on projects in cooperation with the Harokopio University and the Agricultural University of Athens regarding responsible food innovation.

Udo Pesch is Assistant Professor at Delft University of Technology. He is affiliated to the department of Values, Technology and Innovation of the faculty of Technology, Policy and Management. His disciplinary interests include responsible innovation, environmental politics, public policy and ethics. Recently published work includes articles on energy justice, citizens' initiatives in sustainable innovation and the role of emotions in participatory risk assessment. He has published a wide range of articles and book chapters, among others in *Social Studies of Science* (SSS), *Science, Technology & Human Values* (STHV), the *Journal of Responsible Innovation* (JRI), *Science and Engineering Ethics* (SEE) and *Technological Forecasting and Social Change* (TFSC).

Daniela Pimponi is Technology Analyst at Airi, Italian Association for Industrial Research. She holds a degree in Aeronautical Engineering and a PhD in Theoretical and Applied Mechanics. She has a specific background and research experience as a postdoc in Fluid Dynamics. She joined Airi in 2017, contributing to EU H2020 projects GoNano, Prisma, NanoRestArt and Gov4Nano. Her current interests are related to technology assessment and policy analysis, in particular on circular economy and RRI.

Ibo van de Poel is Antoni van Leeuwenhoek Professor in Ethics and Technology and head of the Department of Values, Technology and Innovation at the Faculty of Technology, Policy and Management at the Technical University Delft in the Netherlands. He has published on engineering ethics, the moral acceptability of technological risks, design for values, responsible innovation, moral responsibility in research networks, ethics of newly emerging technologies and the idea of new technology as social experiment. He has recently received a European Research Council advanced grant on design for changing values: a theory of value change in sociotechnical systems.

Andrea Porcari is Project Manager at Airi, Italian Association for Industrial Research, a network of companies and public research organizations aiming to promote industrial research and deployment of emerging and enabling technologies. He has a degree in Physics at the University of Milan (1998) and several years of experience in working in the ICT and semiconductor industry. He joined Airi in 2005, contributing to several EU, national and regional cooperative projects aiming to support and promote nanotechnologies and enabling technologies. His current interests are related to technology assessment and policy analysis, risk governance, Responsible Research and Innovation, co-creation processes, technology transfer and multi-stakeholder dialogues on emerging and enabling technologies.

Charlotte Portier has a background in international law and human rights law. As Manager of Sustainable Development at GRI, she focuses her work on the sustainable development goals and the contribution of the private sector, as well as the practice of corporate sustainability reporting. She started her career in Brussels, working for European networks, defending and advocating for the rights of persons with disabilities and social service providers. She holds a LLM in International Human Rights Law from Maastricht University.

Barbara Regeer is Associate Professor at the Athena Institute, Vrije Universiteit Amsterdam. Her research interests are in emerging innovative strategies for (sustainable) development, with a specific focus on the facilitation of multi-stakeholder processes, knowledge co-creation, social change and mutual learning between all actors involved, in areas such as food systems, agriculture, nature policy, mental health, youth care and disability mainstreaming.

Laurens Rook is Assistant Professor at the Technology, Policy, and Management Faculty of Delft University of Technology, the Netherlands. He received his PhD from the Rotterdam School of Management, and his bachelor and master's degrees from the University of Amsterdam, the Netherlands. His research focuses on creativity, online personalization and choice behavior, often in the setting of human–computer interaction. He has published articles in journals such as *Omega*, *International Journal of Human-Computer Studies* (IJHCS), *Creativity Research Journal* (CRJ), *Journal of Intelligent Information Systems* (JIIS), *Journal of Environmental Psychology* (JEP), and in conference proceedings of various IEEE and Association for Computing Machinery (ACM) programs.

Norma Schönherr is a Project Manager and Research Fellow at the Institute for Managing Sustainability, Vienna University of Economics and Business. She concentrates on the topics of Corporate Social Responsibility (CSR), Enterprise and Sustainable Development, as well as Empirical Social Research. Her expertise is characterized by many years of experience in evaluation and monitoring, as well as work on EU-funded research projects, mainly in the field of impact assessment of companies on global development and the analysis of CSR successes.

Jack Spaapen is an Independent Expert on research and innovation policy, in particular regarding questions about the societal impact of research.

He is a retired Senior Policy Advisor at the Royal Netherlands Academy of Arts and Sciences. He is very active at the national and European level in projects that focus on the arts, humanities and social sciences. He is the vice chair of the COST action ENRESSH, a large European project that works on promoting ASSH research and the evaluation of its impact on society. He is senior partner of the recently started H2020 project SHAPE-ID that targets the interdisciplinary integration of ASSH research into science, technology, engineering and math projects. He (co)chaired several international projects, for example the FP7 SIAMPI project on productive interactions between science and society (2009–2012), the Organisation for Economic Co-operation and Development (OECD) project on science advice (2015) and the EU project on Responsible Research and Innovation, RRI (2014–2015).

Bernd Carsten Stahl is Director of the Centre for Computing and Social Responsibility at De Montfort University and co-principal investigator of ORBIT. His interests include philosophical issues arising from the intersections of business, technology and information. This includes the ethics of ICT and critical approaches to information systems. He specializes in advising on the ethical issues arising from emerging technologies.

Roger Strand is a Professor and former Director at the Centre for the Study of the Sciences and the Humanities, University of Bergen, Norway. His research focuses on complexity and uncertainty at the interfaces between science and society. Strand served as Chair for the European Commission Expert Group on Indicators of Responsible Research and Innovation. During the latter years, many of his research contributions have been made within EU-funded (FP7 and Horizon 2020) projects, including TECHNOLIFE, EPINET, HEIRRI, SUPER-MoRRI and MAGIC.

Adele Tharani is a Research Fellow at the Institute for Managing Sustainability, Vienna University of Economics and Business. She has led the development of a responsible innovation self-check tool for small and medium-sized enterprises in the H2020-funded project COMPASS. She is an expert in impact measurement and management tools for corporate settings and works on topics related to corporate sustainability and corporate social responsibility.

Patricia Urban is a Research Assistant at the Institute for Managing Sustainability, Vienna University of Economics and Business. In her work, she focuses on responsible innovation and sustainability standards. She has a background in Socio-Ecological Economics as well as Social Anthropology.

Lisanne Urlings leads WBA's organization-wide research projects, including WBA's systems transformations framework and Theory of Change. She is the lead author of the report "Measuring what matters most – Seven systems transformations for benchmarking companies on the SDGs." She is involved in the development of various WBA benchmarks, particularly focusing on methodology development and translating research and input from stakeholders into clear metrics and actions for companies. Lisanne strongly believes business can and should drive and accelerate the positive change needed to put the world on a more sustainable and inclusive path and achieve the SDGs. She is passionate about research-based efforts that promote and drive this long-term change and impact. She holds an MSc in Global Business and Stakeholder Management from the RSM Erasmus University.

Robert Verburg is an Associate Professor in the area of technology and innovation management at Delft University of Technology, Netherlands. His research interests focus on leadership, managing people, entrepreneurial behavior, knowledge processes and innovation. He has published articles in management journals, such as the *Journal of Management* (JOM), *Group & Organization Management*, the *International Journal of Human Resource Management* (IJHRM), *Knowledge and Process Management* (KPM), *New Technology, Work and Employment* (NTWE), *Leadership Quarterly* (LQ), *European Management Journal* (EMJ) and the *International Journal of Project Management* (IJPM), where he is also an editorial board member. He is involved with both master and executive teaching in the area of innovation management and he is the Director of Studies for the Management of Technology programme at Delft University.

Lisa Verwoerd is a Researcher at the Athena Institute, Vrije Universiteit Amsterdam, where she focuses on evaluating and supporting transdisciplinary and reflexive research in the field of environmental policy. She also works at the PBL Netherlands Environmental Assessment Agency, where she supports the normalization of deliberative research methods within the organization.

Claudia Werker is an Associate Professor in Economics of Technology and Innovation at the Department of Technology, Policy and Management at Delft University of Technology as well as a visiting professor at the Research Area Technology, Innovation, Marketing and Entrepreneurship (TIME) at RWTH Aachen University. Claudia Werker received her PhD-degree in Economics (Dr. rer. pol.) from Freiberg University of Technology, Germany (summa cum laude), studying the creation and dissemination of innovation and knowledge in innovation systems ever since. Recently, she has been concentrating on the digital transformation in strategy, management and innovation policy, with a particular focus on how to combine scientific insights in these matters with practical application.

Acknowledgments

Emad Yaghmaei and Ibo van de Poel would like to thank individuals and organizations that contributed to this publication with inspiration in the recommendations and feedback they provided during the publication period. The editors also greatly appreciate the financial support provided by the PRISMA project, funded by the European Union's Horizon 2020 research and innovation program under grant agreement no. 710059. The opinions expressed in this article reflect only the authors' view and in no way reflect the European Commission's opinions.

Disclaimer

Introduction

Emad Yaghmaei and Ibo van de Poel

In this volume, we explore methods and practices for the assessment of Responsible Research and Innovation (RRI). RRI aims to encourage societal actors to work together during all phases of the research and innovation (R&I) process to better align R&I and its outcomes with the values, needs and expectations of society (Owen et al. 2013; Stilgoe et al. 2013; European Commission 2014). It explicitly calls on all stakeholders to solve societal challenges and encourages the development and implementation of innovations and technologies that address grand societal challenges. RRI allows leading stakeholders to demonstrate how they help to innovate responsibly, both by minimizing negative impacts and maximizing positive impacts on people and the planet. Understanding the benefits of responsible innovation is crucial to furthering inclusivity, collaboration and transparency in R&I systems. This needs to be built upon, and to deepen understanding of whether and how responsible innovation leads to societal, democratic, scientific and economic benefits, and to provide stakeholders with user-friendly yet advanced tools that help their efforts to improve the outcomes of R&I.

Assessing the benefits and costs of RRI is thus crucial for furthering the responsible conduct of science, technology and innovation. However, until now there has only been limited academic work on RRI assessment. This volume aims to overcome this lacuna by bringing together recent work on this topic. With RRI assessment, we here refer to all those practices, methods and approaches that either assess the effects of certain RRI activities or RRI actions or that assess to what extent existing practices, processes or outcomes of innovation meet certain RRI criteria.

Depending on the context and the approach chosen, RRI assessment can take a range of different shapes and be aimed at different objectives. One possible objective of RRI assessment is to gain insight and to learn in order to improve RRI efforts and their effects. Another objective may be to increase (mutual) accountability. RRI assessment may, for example, help to show that certain RRI (performance) criteria are met, that certain innovations meet specific standards of responsibility and care or to show that money on RRI initiatives is well spent. In these ways, RRI assessment may also be helpful to encourage or incentivize actors to take up RRI and to develop RRI initiatives.

Whereas many components of responsible innovation are well known, its concept is still to a limited extent applied in practice. To redress knowledge gaps about the benefits of responsible innovation in practice, this edited volume aims at assessing responsible innovation principles and aligning the assessment practices with the United Nations sustainable development goals (SDGs).The contributing authors describe several assessment strategies and methods to encourage different types of stakeholders to innovate responsibly and embed responsible innovation principles into their organizations and the operations of these organizations. We discuss the crucial factors for innovating responsibly, both in terms of what is already done and what could or should happen in the future. We discuss what the exact impact of those factors is on the innovating organization and why responsible innovation is important, and perhaps even inevitable.

In addition to the academic goal of bringing together scholarly work on RRI assessment, this volume has the more practical objective of helping to guide stakeholders on how they can align their strategies as well as monitor, measure and manage their contribution to responsible innovation. The edited volume provides different methods and practices for measuring, monitoring and reporting how well responsible innovation is done and what social, environmental, scientific and economic benefits are achieved by implementing responsible innovation principles. These assessment methods and practices are useful for mitigating risks, strengthening strategic planning and assisting stakeholders in increasing their contribution to responsible innovation. Stakeholders, depending on where they are on the course of implementing responsible innovation practices, can apply assessment methods and practices to increase their relevance and make optimal use of their established strengths.

More specifically, this book explores how responsible innovation can be assessed. This is achieved in the book by providing empirical evidence of the benefits of responsible innovation, by showing improvements in the monitoring system and by presenting proposed assessment schemes for responsible innovation where possible and useful. The edited volume is conducted with a focus on a range of different types of stakeholders and particularly is designed for use at organizational level, but may be applied at product, process, divisional or regional level as required.

Synopsis/overview of the contributions

The book consists of three parts. The first, and shortest, part presents some relevant reflections on responsible innovation, and sets the context for the topic. In the second part, we focus more specifically on responsible innovation in industry. This is a context in which assessment of responsible innovation is particularly important as many innovations originate in companies;

at the same time, until recently much of the literature on responsible innovation has mainly focused on universities, research institutes and the public sector. The third part forms the core of the book. It discusses several approaches and methods, either in use or being proposed, for assessment of responsible innovation.

The first part, "Reflections on responsible innovation," starts with a contribution by Robert Gianni that provides a general introduction into the notion of responsible innovation, or Responsible Research and Innovation (RRI), as it is often called in the context of the European funding programs. Gianni explains the scientific and democratic relevance of RRI and, in this light, presents an interpretation of the six keys to RRI hat have been proposed by the European Commission (EC): engagement, gender, open access, science education, ethics and governance.

The second chapter in Part I provides a reflection on the work of the EC workgroup on Policy Indicators for Responsible Research and Innovation by Roger Strand, who chaired this working group, and Jack Spaapen, the rapporteur of the working group. This reflection helps to set the stage for the later contributions, particularly in the third part on assessing RRI. It shows some of the challenges that come with assessing RRI, and defining policy indicators. For example, how can we do justice to the underlying "philosophical" or "intellectual" notion of RRI while at the same time coming up with a set of indicators that are practically applicable? Or, how do we develop a comprehensive notion of RRI while doing justice to existing operationalizations of RRI, for example in terms of the six keys of RRI defined by the EC?

The second part focuses on a specific context for RRI and RRI assessment, namely corporate settings, or "RRI in industry." The reason for including this specific focus in this book is twofold. One is that most of the RRI literature until recently has tended to focus on governmental actors, research funding and universities and has relatively neglected corporate settings, which are nevertheless one of the main contexts in which actual technical innovations are developed and brought to the market (Lubberink et al. 2017; Van de Poel et al. 2017). Second, the call for assessing, monitoring or measuring RRI seems particularly strong in this context. One reason might be that the profit motive is believed to be particularly strong for companies, so that without explicit attention for incentivizing RRI, for example through monitoring or assessing RRI performance, RRI will not get off the ground (cf. Gurzawska et al. 2017).

In Chapter 3 on strategic responsible innovation management (StRIM), Agata Gurzawska discusses how innovations in companies can be linked to corporate social responsibility (CSR). She notes that there is no strategic model yet that links CSR to corporate innovation. In proposing a model to do so, she also proposes a way in which RRI can become better integrated in companies.

Chapter 4, by Cristina Covello and Kostantinos Iatridis, discusses challenges and drivers of implementing responsible innovation in foodpreneurial small and medium-sized enterprises (SMEs). They discuss the case of 19 foodpreneurial SMEs operating in London and find such obstacles to RRI as lack of awareness, time, labor and money, while they find drivers such as personal values, and a concern for societal and environmental well-being. They focus on the inclusion dimension of RRI and find that SMEs practice inclusion through informal channels, and that trust and reciprocity are particularly important rather than that companies prioritize stakeholders in terms of salience based on legitimacy, power and urgency.

Andrea Porcari and colleagues in Chapter 5 discuss how RRI uptake in industry can be supported by a qualitative and multi-criteria approach to analyzing the costs and benefits of implementation of RRI. Their contribution is based on outcomes of the PRISMA project, funded by the EU, that piloted RRI in eight companies. The issue they address is that for companies it is often not evident what the benefits of RRI are, while the costs, in terms of time, labor and money, might seem considerable. Moreover, the costs would seem to be immediate while the benefits of RRI will only materialize after some time. Nevertheless, the pilot studies in the PRISMA project showed that it was possible to find specific RRI actions that are perceived as valuable by companies if one focuses on the specific company and the specific challenges that company faces. On the basis of this experience, they propose a more general approach that can help companies to identify possible RRI actions, and to make a first qualitative assessment of the expected benefits of such RRI actions, that is helpful in devising an RRI strategy at the company level.

In Chapter 6, Edurne A. Inigo and her colleagues discuss whether voluntary standards can support responsible innovation implementation and reporting in industry, with a focus on the European food sector. Voluntary standards can trigger reflection on values like health in food and support the translation of such values into more concrete product requirements that can guide the innovation process. Moreover, front-of-pack labels that signal that products meet certain standards can help to incentivize RRI in the food industry. However, voluntary standards also have their drawbacks; they may lack legitimacy because the process of standard formulation is often not fully transparent or not fully inclusive with respect to all the relevant stakeholders. Some of these drawbacks may be overcome by a larger role of the government, without necessarily switching to a more traditional approach of top-down regulation through legal rules.

The third part, on "Responsible innovation assessment," presents different approaches, methods and metrics that can be used to assess responsible innovation practices.

In Chapter 7, "Monitoring Responsible Research and Innovation in the European research area: the MoRRI project," Ingeborg Meijer and Wouter van de Klippe stress how indicators and metrics for RRI play a central role in monitoring systems of innovation practices. They discuss the benefits of RRI in four categories: societal, democratic, economic and scientific. While the primary output of the MoRRI project was the creation of a monitoring system highlighting changes at the national level, the successor project of MoRRI, SUPER MoRRI project, envisions an alternative future for monitoring and evaluation of RRI in which monitoring happens at the institutional level with local and context-dependent methods.

Following the monitoring and evaluation of RRI at the institutional level, Adele Tharani, Katharina Jarmai, Norma Schönherr and Patricia Urban introduce, in Chapter 8, a diagnostic self-assessment tool to support companies to adapt responsible innovation by drawing on insights from organizational learning. In the context of innovation, self-assessment can enhance organizational learning with a view to foster an organizational culture and innovation capabilities that in turn have a positive effect on innovation performance measures. They aim to enable learning by translating the concept of responsible innovation into concrete corporate practices and policies, ascribing them to specific business functions, so aligning responsible innovation with business realities.

In Chapter 9, Pim Klaassen, Lisa Verwoerd, Frank Kupper and Barbara Regeer present reflexive monitoring in action (RMA) as an interactive methodology for monitoring and assessing RRI that simultaneously facilitates enacting it. They argue that the RRI monitoring framework is most useful when adopted by actors who take part in R&I processes to carefully and systematically reflect on what "responsible" means in their given situation. The chapter illustrates the potential role that RMA can play to promote learning to act in a more responsible manner across researchers, innovators and organizations.

In Chapter 10, "A future-oriented evaluation and development model for Responsible Research and Innovation," Mika Nieminen and Veikko Ikonen examine a future-oriented evaluation model with a systemic view and multi-actor approach for integrating RRI into the innovation activity. A focal point of their model is co-creation, which helps to define values and operational goals for innovation processes and supports the assessment of the success of these. Similarly to previous chapters, the authors present the assessment process as a loop of continuous learning in which participating organizations and individuals learn from each other during the evaluation process and reflexively develop their activity. To assess the desirability and acceptability of an innovation, the authors propose a dialogic process of anticipating the impacts and outcomes of the innovation from the very beginning. They argue that judgments on the desirability and acceptability of an innovation are supported by creating shared visions for the innovation.

Claudia Werker, in Chapter 11, shows the need for a system approach to assess RRI. She addresses the opportunities and challenges emerging from the digital transformation which changes the playing field of RRI systems. Her chapter, "Assessing Responsible Research and Innovation (RRI) systems in the digital age," examines shared values for the innovation activity. She believes that it is crucial that stakeholders from all sectors, i.e. government, academia, industry and civic society, have a clear picture of their own values, how they change in time, what the values of other stakeholders in their RRI system look like and which opportunities and ways there are to develop shared values. Finding jointly acceptable directions and intensity of innovative activities, Werker suggests that we apply shared values as a cornerstone to assess the components and activities of RRI systems.

In Chapter 12, "RRI intensity: a proposed method of assessing the requirement for responsible innovation in ICT projects," Martin de Heaver, Marina Jirotka, Margherita Nulli, Bernd Carsten Stahl and Carolyn Ten Holter propose the concept of RRI intensity level (RIL) as a method of selecting what type and how much RRI activity is feasible, appropriate and effective. Through asking the question, "How can the level of RRI activity be aligned with the requirements of the innovation project?" the authors develop an approach for defining the desirable RIL based upon assessments of both the anticipated technology readiness level and the innovation potential of the research.

In Chapter 13, "The responsible side of innovation: towards the measurement of a new construct," Robert Verburg, Laurens Rook and Udo Pesch provide a measurement scale for assessing RRI of employees in business contexts. They focus on the creative behavior of individuals in organizations when it comes to innovation. This stimulates study of the responsibility of employees in international contexts and a comparison of employees from different levels and backgrounds against RRI principles. As such, the RRI measure could also be applied in practice as this will enable companies to compare different departments on their level of RRI. The measure could also feature as an outcome for training and development efforts in the area of RRI or be used for the assessment of employees or managers.

In the final chapter, Chapter 14, "RRI measurement and assessment: some pitfalls and a proposed way forward," Ibo van de Poel identifies potential pitfalls of measuring RRI and related practices of monitoring and reporting. Two main categories of pitfalls are discussed in this chapter, i.e. pitfalls in measuring RRI and pitfalls that arise from the strategic setting in which RRI assessment is applied. These two categories of pitfalls may reinforce each other and may escalate the danger of window-dressing rather than internalization of the underlying motives of RRI. Van de Poel argues that distinguishing more clearly between different rationales for doing RRI assessment helps to understand better the RRI measurement pitfalls. He distinguishes three such rationales – learning, accountability and incentivizing – and argues that these three rationales are by and large in tension with each other. He suggests that, if one wants to develop and deploy RRI assessment, one should make an explicit choice for one of the rationales rather than trying to serve all three at once.

Overview of best practices

Society's grand challenges cannot be solved by academic articles alone. We believe that a mixture of sharing successful practices along with research results may help leading organizations and institutions to foster responsible innovation practices. We offer thus a set of best practices that help organizations and institutions cut through the complexities of implementing responsible innovation and understand the benefits and positive impact of it for themselves and their stakeholders, communities and the environment.

The best practices are presented between the various chapters in Part III of the book. They provide various perspectives of how we can assess responsible innovation within stakeholders from different sectors, i.e. government, academia, industry and civil society.

Beginning with "The B Impact Assessment," Joey van den Brink from B Lab, a non-profit organization, presents the B Impact Assessment tool, which provides companies with the opportunity to evaluate their impact performance through a holistic framework that covers all aspects of the company's operations. The company's impact performance is captured in five key impact areas: governance, workers, community, environment and customers.

In best practice II, Tung Tung Chan and Ingeborg Meijer present the Societal Readiness Thinking Tool (SR TT), which aims at cultivating researchers' knowledge on RRI, and stimulating integration of RRI in their research project. The SR TT provides questions to guide researchers to reflect on the societal alignment of their work at critical stages in the project lifecycle.

Datamaran company applies natural language processing (NLP) in best practice III to quantify narrative disclosures in annual financial reports, sustainability reports, Securities and Exchange Commission filings, mandatory and voluntary regulatory documents, social media and online news. Donato Calace and Adriana Farenga present Datamaran's data-driven approach, by which companies can stay in charge of their materiality and risk processes and identify trends that are concealed in unstructured data.

Best practice IV, "PRISMA KPI analysis tool," provides a basis for discussion about RRI in corporate innovation contexts, and explores the links between RRI and innovation management-related key performance indicators (KPIs). Steven Flipse recommends collecting data in a qualitative manner, in conjunction with this tool, for projects at the beginning, middle and end of their runtime, in order to make sense of the scoring patterns.

Andrea Porcari and Elena Mocchio, in best practice V, introduce a standard for responsible innovation in industry. The standard under development (formally named as workshop CEN/WS[1] 105) aims to provide practical guidelines for organizations to approach and evaluate the most significant ethical, legal and social impacts (ELSI) expected by the development of their innovative products, and to identify an RRI vision and operative RRI actions (roadmap) to address them.

Assessing corporate performance on the SDGs, Lisanne Urlings, in best practice VI, shares the World Benchmarking Alliance (WBA) benchmark approach to drive the private sector's engagement in the SDGs. The WBA benchmark envisions a future where companies, investors, governments, civil society and individuals can quickly and easily compare businesses and motivate a "race to the top."

In the final best practice, Charlotte Portier describes how the Global Reporting Initiative (GRI) practical guide enables businesses to disclose SDG-related information and manage their impact on sustainable development. She argues that, by empowering business to report on their impact on sustainable development and the SDGs, GRI realizes its mission: to empower decisions that create social, environmental and economic benefits for everyone.

Note

1 A European Committee for Standardisation (CEN) Workshop (CEN/WS) is a flexible working platform open to the participation of any company, organization or interested expert for rapid elaboration of CEN Workshop Agreements (CWAs). A CWA is a technical document agreed by the participants of a CEN/WS, which can be the basis for future standardization works.

References

European Commission (2014). Rome Declaration on Responsible Research and Innovation in Europe. https://ec.europa.eu/research/swafs/pdf/rome_declaration_RRI_final_21_November.pdf. Accessed 21 January 2017.

Gurzawska, A., Mäkinen, M., & Brey, P. (2017). Implementation of Responsible Research and Innovation (RRI) Practices in Industry: Providing the Right Incentives. *Sustainability*, 9(10), 1759.

Lubberink, R., Blok, V., van Ophem, J., & Omta, O. (2017). Lessons for Responsible Innovation in the Business Context: A Systematic Literature Review of Responsible, Social and Sustainable Innovation Practices. *Sustainability*, 9(9), 721.

Owen, R., Bessant, J. R., & Heintz, M. (2013). *Responsible innovation: managing the responsible emergence of science and innovation in society*. Chichester: John Wiley.

Stilgoe, J., Owen, R., & Macnaghten, P. (2013). Developing a Framework for Responsible Innovation. *Research Policy*, 42(9), 1568–1580, doi: http://dx.doi.org/10.1016/j.respol.2013.05.008.

Van de Poel, I., Asveld, L., Flipse, S., Klaassen, P., Scholten, V., & Yaghmaei, E. (2017). Company Strategies for Responsible Research and Innovation (RRI): A Conceptual Model. *Sustainability*, 9(11), 2045.

Part I

Reflections on responsible innovation

1 Scientific and democratic relevance of RRI

Dimensions and relations

Robert Gianni

1.1 Introduction

Responsible Research and Innovation (RRI) has gained prominence in the academic debate in the last decade. Introduced in order to provide an operational framework for enabling communication between science and society, it has undergone a consistent number of analyses to understand its value and relevance for such operation (Burget et al. 2017). However, despite this large number of attempts, clarity about its efficacy is still far from being reached. Accordingly, initial enthusiasm has started to fade in favor of alternative, more specific regulatory measures. This chapter aims to evaluate the scientific and democratic relevance of RRI beyond skepticism about the acronym, in order to understand if it can still play a role in the dialogue between science and society.

In order to do so I will operate an analysis, on the basis of a pragmatist perspective, of the main features and challenges faced by science and democracy, in order to then extract their main objectives and principles. I will then introduce the evaluation phase with a brief analysis of the concept of responsibility so to offer a theoretical background that summarizes the dynamic nature of RRI. One of the puzzling aspects of RRI is in fact the difficulty to obtain a common understanding on what responsibility means and implies. This preliminary assumption appears often fragmented or controversial, making it difficult to obtain a homogeneous approach to RRI. In a following step, I will then be able to start the evaluation of the scientific and democratic relevance of RRI. In order not to lose the conceptual power of the principles embedded in the framework, instead of concentrating on the acronym, I will focus on the six keys (engagement, gender, open access, science education, ethics, governance) adopted by the European Commission (EC), which will reasonably last even beyond RRI. It will finally emerge that RRI has a high scientific and democratic relevance.

1.2 Science and society

When it comes to RRI, one of the most intuitive doubts is about the capacity of an ethical framework to increase or at least maintain the scientific relevance of research and innovation. Scientists and innovators might still believe that a pure technical process, one apparently devoid of any extra scientific consideration, can represent an efficient methodology to make their research progress. Even amongst those who are not explicitly supporters of technological determinism, it is possible to find caution about the value of a non-technical integration into the research process, at least at an earlier stage.

It would be hard to affirm that technology, and science more generally, does not have an impact on societies.[1] The intuition of Poincaré, demonstrated only much later by Edward Lorenz, that an infinitesimal variation has potentially infinite sets of consequences, has nowadays become evident also to non-specialists (Jasanoff 2016; Lorenz 1963; Turing 1950).[2] Besides, studies on complexity have demonstrated how linear models are not apt to explain effects springing from systems behavior (Bridgman 1927). The "butterfly effect" is exactly what poses scientists with challenges that are not only technical, but ethical and, in the end, political.

It is difficult not to recognize that current societies and thus scientific research are complex and have to deal with a growing "hyper-complexity" (Qvortrup 2003). According to Funtowicz and Ravetz (1993), science is now developed in a "post-normal" scenario. For Gorgoni (2018), the intensification of consequences in terms of both time and space has shifted foresight exercises from dealing with uncertainty to navigating in indeterminacy. The transition from a curiosity-driven method to a problem-oriented one, where knowledge and decisions are not detachable, together with the growth of complexity, has made it difficult to exclude values and interests from scientific research (Arnaldi & Bianchi 2016; Funtowicz & Ravetz 2008; Jasanoff 2004).

Following this path, some authors have pointed out that science not only has an indirect and unforeseeable impact on society, but is also a sociopolitical construct. Accordingly, the objectives of scientific research, or rather their topics, are always decided according to political decisions although these are often implicit (Jasanoff 2004, 2016; Wynne 1993).

From a more radical point of view, the relation between nature and society is a coproduction, determining that our representations are inseparable from the ways in which we live (Pellizzoni 2004).

Many authors have justified the necessity to include extra-technical aspects by enumerating recent cases that have generated outrage in public opinion. They have shown the potentially severe consequences that could occur when neglecting a broader perspective in the choice of scientific paths. Several negative examples, like genetically modified organisms, or other episodes connected to chemicals, have shown the limits of adopting a purely quantitative methodology for assessing the impact of a research product (Jasanoff 2016; Von Schomberg 2013, 2014). Besides, many of these events suggest that consequences are often considered to be bad not only in terms of safety, but according to hybrid, dynamic and sociotechnical aspects (Jasanoff 2016). The risks stemming from the bureaucratization of thought and the supremacy of technique over politics urge us to focus on humanistic and social ends (Horkheimer & Adorno 2002).

Technology and science more generally are artifacts that confront humans with new challenges and possibilities. Emerging technologies like artificial intelligence, genomics or synthetic biology are fascinating gamechangers in the future of humanity.[3] However, Dewey reminds us that scientific artifacts are not simply there in some kind of naturalistic or deterministic fashion; they are the results of interactions – socially conditioned phenomena resulting from accumulated culture (Dewey 1991, Ch. 3). To paraphrase the American philosopher, we do not naturally *need* them, but we *want* them (p. 106). Therefore, also the assessment of emerging technologies becomes more complex and needs to take into account a broader "range of values that humans care about when contemplating the future" (Jasanoff 2016, p. 58). Also for Funtowicz and Ravetz (1993), in a situation of post-normal science, and in order to deal with the major challenges arising in our societies, the epistemic dimension should be integrated with an axiological one, meaning that values should complement technical aspects.

Apart from the easily understandable reactions to negative episodes, though, it is important to put in evidence that science has a proactive power in increasing the well-being of humankind. In a recent analysis, Jasanoff has stated that technologies, as the current focus of science, are "devices with which modern societies explore and create potentially more liberating and meaningful designs for future living" (Jasanoff 2016, p. 242). In this respect, there is a growing amount of literature and evidence about the advantages of the integration of broader perspectives into science. Extending knowledge can bring benefits to scientific investigations because it increases the possibility of solving problems by including additional and new information (Pellizzoni 2004; Stilgoe et al. 2013). Although, as pointed out by Blok and Lemmens (2015), transparency claims can be seen as highly naïve or counterproductive (Blok 2018), it is also true that external information can but enrich the general level of knowledge needed to progress in scientific research and innovation.

It has been noticed that scientific research has consistently turned to a contextualization of knowledge production (Nowotny 2015; Pavie et al. 2014) which is often aimed at the "creation of wealth" (Krishna 2013). In this sense, although basic research still plays a prominent and prestigious role, we do witness a proliferation of small-scale laboratories and research group targeting specific, and often already existing, problems. This is probably due also to the fact that technologies mutate along with the societies in which they operate (Jasanoff 2016). Therefore, it has been shown that successful technologies are often those that adapt and respond to particular needs (Bijker et al. 1987).

A connected reason is the change in the production process, economic growth and the consequent advantages of a "productive" research and innovation process (Godin 2015). According to Blok and Lemmens, "most innovations take place in commercial or industrial settings" and products need to be profitable (Blok & Lemmens 2015, p. 20). Also research increasingly faces challenges arising from the necessity of meeting criteria of excellence that can be measured and that can produce growth. However, Mazzuccato has reminded us that, although this is easily overlooked, private successes often have public origins and therefore researchers and innovators should deal with these political origins (Mazzuccato 2015). What she points out is that often science (and innovation) cannot be totally implemented without a whole series of other inputs and measures often generated by institutional mechanisms, which then raise the question of the criteria for decision-making. The question of "framing" of research trajectory appears for many scholars to be inextricable from the ontology of products and processes (Gianni & Goujon 2018; Goffman 1974; Maesschalck 2017).

If we recap the different aspects of the question, we can see that nowadays science has to deal with an increasingly growing complexity, which requires innovative, experimental and broader methodologies. This has been practically urged by negative happenings, but also the potential advantages are seen as highly relevant.

The ways in which to implement such broadening are mostly targeting participation and deliberative processes in the wake of democratic models (Guston 2001). Although it might be argued that more innovative and bottom-up instruments are needed (Bucchi & Neresini 2008; Flipse 2012; Gottweis 2008; Pansera & Owen 2018), the aim is nevertheless to establish a discussion with a sufficiently extended range of different perspectives, which can then feed the trajectory of science research. The relevance of different angles concerns the quantity of agents potentially affected by the outcomes, as well as the whole "worlds of relevance" of the actors considered (Arnaldi & Bianchi 2016; Jasanoff 2016; Limoges 1993), and their influence in the decision-making process (Fung 2006; Gianni & Goujon 2018). The necessity to enlarge perspectives towards narratives is considered to be important to not reduce the inputs in the discussion to "rationalistic" ones, which would fall back into the same limited epistemic framework that they are supposed to abandon (Dewey 1991 [1927]; Funtowicz & Ravetz 1993; Ricoeur 2000; Wynn1993). However, it has been highlighted that the competence of participants to express their interests is also crucial. This means that they have to be able to adopt a mature view in order to be able to fully understand some of the technical aspects in question (Kitcher 2011). Furthermore, participants should be willing to accept a common set of rules for deliberating (Bohman & Regh 1997; Fishkin & Laslett 2003; Hajer & Wagenaar 2003; Reber 2016). This last indication highlights the necessity to promote bilateral relations between science and democracy. If the former is called to open the discussion on the overall objectives because of endemic and epidemic reasons, the tools pertaining to the latter should also be deployed with a greater scientific attitude.

1.3 Democracy: principles, models and methodologies

Evaluating the democratic relevance of RRI implies an overall judgment of the relation between science and democracy and their possible intersection.

In this regard, it is important to operate a brief examination of the key features of the concept of democracy in order to understand what are the driving principles and its main objectives. In this way we will then be able to compare them with the objectives and tools embedded in RRI.

Democracy is a very complex concept with several ramifications and subgroups. According to Sartori, sticking to a definition or to the etymology of the word would be an error that would disserve rather than serve democracy (Sartori 1987, p. 13). Democracy is a historical model that changes in a perpetual tension between its ideal features and actual reality. The understanding of the demos has changed throughout the centuries according to moral, epistemic and political assumptions. Besides, the increase in complexity, due to the extension and intensification of interactions, has generated the need for increasingly sophisticated and continuously evolving forms (Galli 2011). In this sense, it is difficult to talk of one model of democracy, but rather of an ideal assuming different shapes.

All scholars dealing directly or indirectly with the concept of democracy share the belief that this political model aims to preserve and promote the moral and juridical ideals that found in the French Revolution their most symbolic expression: freedom, equality and solidarity. However, not all agree on the different weight of each principle, or on their relation with the other two.

Some authors have stressed the crucial role of equality (and that of defense from evil power) (Galli 2011). Sartori upholds equality, although a multidimensional one, as the protecting principle against power abuses (Sartori 1987). For Dahl, democracy should rest on the principle of intrinsic equality (Dahl 1989). Other authors, although less directly concerned by the principle and more about the methods through which to attain democratic processes, are also discussing new forms of democracy on the normative basis of equality (Manin 2012). The concept of equality itself is also used in different ways, depending on whether it concerns fundamental rights (Dahl 1989), applied ones, or whether it addresses the actual material conditions of an equal access to the public sphere (Honneth 2014; Sen 1999).

A similar, but perhaps more complex, destiny has characterized the priority given to the concept of freedom. Several scholars have endorsed the distinction proposed by Berlin between a negative and positive understanding of freedom (Berlin 2002). Northern American authors distinguish between freedom as protection from hindrances, and liberty as its institutional implementation (Pitkin 1988). Some philosophers have instead developed a more sophisticated structure for justifying the conceptual priority of freedom over equality (Bourgeois 2017; Honneth 2014; Kitcher 2011). These authors share a multilayered understanding of the concept which includes social structures of deliberation. This shared understanding appears to be a solid and updated development of the two meanings of individual freedom as self-determination and self-realization, as delineated respectively by Kant and Herder (Gianni 2018; Hegel 1991 [1821]; Honneth 2014; Kant 1997).

The third ideal, solidarity, has characterized the public debate in a more concrete way, due to its more tangible nature. At the theoretical level we often find it under the vests of redistributive justice, but it is nevertheless considered to be inextricable from a just democratic process (Carrabregu 2016; Dews 1992; Rodotà 2014; Sartori 1987). At a first glance, solidarity is a much more controversial point and it is not conceived as a presupposition but rather as an objective or a balancing force (Dahl 1989).

Although we find different understandings of these principles, democracy is for all of the cited authors the only viable alternative to enforce them.

The appropriate tools that could serve the purpose emerge from the conceptual priority given to one principle and to the diagnosis of its actual enforcement. Accordingly, between the extremes of direct democracy on the one hand, and technocratic (Audard 2009), guardianship (Dahl 1989), or authoritarian democracy (Sartori 1987) on the other hand, we can find a rich galaxy of proposals. Amongst the "theories of democracy in the plural," Holden organizes his analysis around five clusters: radical, new radical, pluralist, elitist and liberal-democratic theory (Holden 1999). Dahl distinguishes between Madisonian, populistic and polyarchal democracy (Dahl 1989). Sartori's account rests on four typologies: electoral, participatory, referendum and competitive, where the "major contrast is rendered in terms of participatory versus competitive theories" (Sartori 1987, p. 14).

Consequently, also the tools activated within these frameworks (and across them) are of various nature. According to a specific perspective, but also to a determined historical moment, the tools have undergone a diachronic and synchronic development (Sintomer 2011). If, on the one hand, some features of democracy have evolved, contributing to its refinement, others, like the mechanism of drawing lots, have gone through a less linear progress (Sintomer 2011). Within the discussion over one conceptual tool, the indications have stressed different aspects. Tools for enhancing direct forms of democracy, for instance, have been encouraged according to refined hybrid formats so to address underrepresentation and disaffection (Manin 2012; Sintomer 2011; Vandamme 2018) or as provocative attempts to encourage a public debate on the subject of democracy itself (Van Reybrouck 2014).

This conceptual universe is fueled by different perspectives being composed by philosophers (Honneth), political scientists (e.g. Sintomer, Dahl and Manin) and historians (Van Reybrouck). These approaches differ in their methodology, objective and normative depth. However, there are some features that crosscut most of these analyses on democracy. Some of the ideal criteria that identify a democratic process are commonly identified with participation, transparency, responsiveness, inclusiveness, deliberation and, on a more formal level, balance. According to Sartori and Dahl, these features belong to the ideal (Sartori 1987) or normative (Dahl 1989) sphere of democracy, and they serve as regulatory criteria for their implementation in real or descriptive democracies. The two authors are clear in recognizing the sufferance of democracy in its actual forms but they blame it on the discrepancy between intention and facts. Despite the mistrust that we could have in democracy, they encourage implementation of those criteria to shorten the gap between norms and facts. The distance between the objectives and the reality of the model is what most probably concurs to generate what Galli calls a "discomfort" about democracy (Galli 2011).

It is important to consider these criteria because they are all expressions of what we might conceive as our current reference ideal of democracy. Today's dominant understanding of democracy is one that enables all individuals to express their preferences through mechanisms of deliberation on the basis of mutual respect (Dewey 1991; Reber 2016). Reasons, language, communication and deliberation are not only buzz words which we encounter in political theory or rhetorical slogans, but substantial forms through which citizens can realize their full potential as individuals (Goodin 2012; Honneth 2014). As reminded by Sartori, "Democracy is ... a characteristically flexible system. Above all, or first, democracy is a procedure for processing whatever a society demands, whatever comes up through 'voice' and with voice" (Sartori 1987, p. 77).

We could list several attempts to improve participation and deliberation in democratic processes. They are often experimental because they must deal with the challenges arising from the complexification of the public. The indication to create a community or a "public" (Dewey 1991) in order to discuss general concerns is in itself a challenge to which many scholars are trying to offer an answer (Fagotto & Fung 2006; Goodin 2012; Hajer & Wagenaar 2003; Jacquet 2018). However, the basic assumptions rest on the cognitive certainty that, through means of knowledge formation and communication exchanges, democratic processes will help to improve freedom, equality and solidarity.

After having understood the current challenges for scientific investigations and for democratic processes, we can now evaluate if RRI is relevant for the objectives and methodologies required by current scientific and democratic models.

1.4 Responsible research and innovation features

If we want to assess the democratic and scientific relevance of RRI, it is important to first agree on one understanding of RRI. Recent literature shows in fact that such an attempt proves to be harder than one could expect ten years after its introduction in the public debate.

Although we might consider its conceptual history fairly dense since its first occurrences in 2008 and 2009 at the academic and national level

(Robinson 2009), and 2012 at the European level (Pearson 2018), RRI still seems to be far from reaching a clear and shared understanding. Despite the admirable quantity of contributions published so far[4] (Burget et al. 2017; Goos & Lindner forthcoming) and notwithstanding the fortune that, although patchworked, it has encountered at the policy level, RRI still represents for many commentators and practitioners more of a riddle than a solution. Furthermore, the initial enthusiasm detectable in academic contributions and policy investments has recently started to fade, although the reasons for such decline are not homogenous.[5]

According to some recent analyses, one of the potential motives can be traced to a general absence of harmony between policy-makers and academics but also within such clusters (Klassen et al. 2018; Wickson & Forsberg 2015). Pearson has also listed the discrepancies between national funding schemes and that of the EC one, highlighting how such divergence can generate an overall confusion for researchers in the implementation of RRI (Pearson 2018). In the UK, for instance, RRI is often identified with the AREA scheme (anticipate, reflect, engage, act),[6] while the EC has proposed six thematic keys for its implementation (engagement, gender, science education, open access, ethics, governance).[7]

From another angle, the relation between the academic debate and the policy-makers has suffered from a similar situation. Klassen et al. (2018) have highlighted the lack of communication between the academic debate and the policy-makers at the EC level, with regard to the understanding of RRI. Accordingly, on the one hand, a large part of the academic debate mostly draws on the objectives set in Von Schomberg's definition (Burget et al. 2017) and discusses Stilgoe et al.'s (2013) methodology. On the other hand, the EC has not explicitly integrated academic features in their framework, promoting instead a soft implementation of six thematic keys. Goos and Lindner have noticed "a degree of incongruence between these academic debates and their manifestations at the EC policy level" (Goos & Lindner forthcoming, p. 7).

The proliferation of definitions that has characterized the development of RRI, although they might be coupled according to their objectives (van den Hoven 2013; van Oudheusden 2014), has most probably also played a slowing-down role in the take-up of common strategies for its actualization.

At a practical level, it emerges that the actual integration of RRI within some research domains has not been successful, although also here the reasons are varied. At project level within the EC funding framework, research results have shown a general inconsistency between overall objectives and common strategies for implementing RRI. If in some program lines RRI has been central, in some it has been overlooked, and others have focused, directly or indirectly, only on some of its features, but not on the notion itself.[8]

In general, this depiction shows that it is at best difficult to agree on a conception of RRI which we could use to assess its scientific and democratic relevance.

However, despite this methodological complexity, mostly due to a solid epistemic and politic plurality (Reber 2016), it is important to choose one conceptual reference in order to evaluate the scientific and democratic relevance of RRI.

In order to determine our choice, we first need to propose an analysis of the main driver of RRI, responsibility, so as to understand also other aspects connected to the framework that are implicitly driving its tools.

1.4.1 Responsibility: a multi-level concept responding to historical challenges

Responsibility is a concept that has grown with the development of agents' interactions amongst themselves and with their external environment. If the very first modern occurrences of the term are strictly connected to a political dimension (McKeon 1957; Rosanvallon 2015), its more refined developments are to be found in the juridical and moral realm (Foddai 2005; Ricoeur 2000; van de Poel 2011).

With regard to the juridical sphere, we can detect a solid deontological tradition, building on Kant's moral and legal doctrines, which finds in legal positivism (and neo-positivism) an important recent doctrine (Hart 2008; Kelsen 2005 [1934]; Paulson 1992). According to this line of thought, responsibility is the legal duty to follow a certain number of rules, laws or principles in order to be active members of a given society. It is in many respects retroactive or focused on the relation that agents hold with their past actions. It usually has a "negative" connotation because it concentrates on the restrictions or sanctions that an individual would encounter in case of fault and is probably still nowadays the most common understanding of it (Moore 1984).

Originally entrenched in the framework of criminal law, this delineation started to be questioned with the development of the production process and the promulgation of civil laws at the beginning of the twentieth century (Ewald 1986). The accent previously placed on the author of an action as accountable and thus punishable was then shifted to the victim in the position to claim for compensation (Ricoeur 2000). This alternative perspective represented a crucial turning point because it concentrated not on the guilty person but on the victim and the damages she could suffer. This ended up overshadowing the connection of an action to an actor and the related notions of fault and punishment in favor of anonymous forms of compensation – usually economic ones (Ricoeur 2000). Driven by the increasing rise of risks and societal claims for security and solidarity, this new understanding aimed at dealing with the emerging impact of technology in societies, which criminal laws could not sufficiently address. This passage has been defined as the one "from individual management of fault to a socialized management of risk" (Engle 1993, p. 16).

However, it has been pointed out that this specific understanding of responsibility embeds severe side effects if detached from the notion of imputation and accountability. Ricoeur has shown that, with the extension of the sphere of risks in terms of space and time, the loss of identifiable fault and punishment of the responsible actor can lead to the paradoxical effect of placing actions under the sign of fatalism, which is the opposite of responsibility (Abbagnano 1957; Ricoeur 2000, p. 26). If fault cannot be traced back to an accountable person, then the connection between agents and actions vanishes, generating, as was the case when these laws were promulgated, a system of insurance where there can be a total loss of individual responsibility (Ewald 1986).

If the juridical understanding of responsibility as accountability has slowly shrunk in the philosophical debate, discussions on the moral notion of responsibility have instead proliferated (Pellé & Reber 2015; Vincent et al. 2011). The first modern occurrences of the moral traits of responsibility can be detected in Kant's moral philosophy, when the German philosopher defined the conditions for attributing an action to an agent and all the moral implications deriving from them. However, also moral responsibility faced the changes of societal interactions already encountered by legal scholars. The main shift, as we have seen, was the one passing from a single individual and her capacity to be held accountable and punished for her actions, to the damages potentially affecting someone external, a vulnerable figure. Then, the opening to the external effects of actions rather than the internal logic for acting generated a withdrawal from the criminal aspects of responsibility and also concurred with a proliferation of its moral features. In moral philosophy this has raised a very complex debate about the means and criteria, but most of all about the object of responsibility. If in civil law the question of detecting the victim is already difficult, in moral philosophy defining sufferance and vulnerability is even more complex (Ricoeur 2000).

Moral responsibility has fundamentally been addressed by two main perspectives. On the one hand, following a Kantian approach, we find scholars analyzing it through the intentions of an agent (Kant 1997). On the other hand, in line with different schools of thought, responsibility is assessed according to the consequences of an action (Pettit 1997). It has been pointed out that the main problem with the former concerns the disconnection between intended and actual effects of an action, which, by underestimating the complex chain of causes and consequences, can disregard the side of the victim and the related sufferance (Williams 1984). The latter, focused on calculating consequences, instead does not often consider the importance of subjective and non-rational values (Sen & Williams 1982). Furthermore, consequentialism appears to be facing several difficulties in dealing with the challenges connected to complex societies, and the extension, in terms of time and space, of risks (Ricoeur 2000). In fact, from when and until which point, it might be argued, can we consider ourselves responsible for the consequences of our actions? Following Ricoeur, we can recognize the dilemma we are facing when considering that "simply neglecting the side-effects of an action would render it dishonest, but an unlimited responsibility would make action impossible" (Ricoeur 2000, p. 33).

It is not by chance then that more recently a third stream has tried to address this dilemma between intentionality and externalities by encouraging an existential imperative of responsibility conceived as care (Grinbaum & Groves 2013; Jonas 1984). Given the uncertainty about the future and the increasingly growing influence of technologies in life as such, responsibility has started to address the necessity to take care of our "environment" in similar ways as we would take care of our beloved ones (Pavie 2018; Pellé 2018; Tronto 1993). The emerging "imperative" of responsibility stems from the fact that damages are already tragically present, and that consequences of our actions cannot be circumscribed. This perspective is radically future-oriented and should be implemented according to contextual ways, so as to be able to cover those gray areas where laws or objective principles have proved to be insufficient. The point is not to address the relation between human beings but that to the natural environment that makes human existence possible. In this sense it can be seen as a total effort which must be deployed in every action. As highlighted by Derrida, responsibility exceeds what is known because otherwise it would be the implementation of a technical-cognitivist device, nothing more than the mechanical deployment of a theorem (Derrida 2008; Foddai 2005).

More recent contributions concerning the role of responsibility in science have also recognized the valuable contribution of this existential meaning of the term (Adam & Groves 2011; Levinas 1998 [1974]; Stilgoe et al. 2013). This stream has sometimes intertwined with virtue ethics, a theoretical framework where an agent is called to develop a character able to make her do, not only what is right, but also what is good (MacIntyre 2007; Pesch 2015; Pettit 2015).

However, it has been noticed that also this last notion, when taken alone, does not resolve the potential infinity of consequences stemming from an action and the impossibility for an individual to take them all in charge (Ricoeur 2000; Spämann 1975). Besides, given the increasingly plural nature of societies, it is debatable what a virtuous behavior might mean. Although they are powerful forces able to permeate the gray areas of human actions, ethics of care and virtue ethics suffer from indeterminacy, which makes them unsuitable for assuming an objective regulatory stance.

If all these different understandings encounter some major limitation, we could be wondering what meaning of responsibility can be appropriate to offer a regulatory role for research and innovation.

Furthermore, as we have seen, it might be argued that it becomes increasingly difficult to distinguish between individual and collective responsibility given that actions and consequences are tangled amongst different actors. I believe that we can find a solution to this problem if we take a look at some previous conceptual attempts to solve this question.

The theoretical and practical polysemy characterizing the concept of responsibility has been puzzling commentators for a long time. It is difficult to settle the question with regard to the ultimate sense of responsibility. What seems to reconcile the different authors who have analyzed its historical path, though, is that we can find an evolution integrating the constant necessity to guarantee the correct functioning of social relations with the historical changes in agency (Foddai 2005; Genard 1999; Levi-Bruhl 1884; Villey 1977). Responsibility, wrote Lévi-Bruhl in 1884, is a fragile notion owing its multiple senses to the superposition of elements of different periods. Also for Villey, although his understanding differs from that of Lévi-Bruhl, responsibility is the fruit of a secular evolution due to a series of changes and needs for a social, political and juridical character. "Similar to a geological formation," takes up Foddai, the concept of responsibility "reveals in each of its layers the context of ideas and structures of thought in which it has been forged" (Foddai 2005, p. 7).

It has been noticed that there is often a discrepancy between theory and action when it comes to responsibility. What Bradley called a vulgar understanding of responsibility differed, in his view, from the doctrines of the philosophers (Bradley 2018). Arnaldi and Bianchi, following Weber, distinguish a theoretical plan where responsibility is traversed by a theoretical clash, and the practical one, where the two sides are often applied in a hybrid way (Arnaldi & Gorgoni 2016). Jean-Louis Genard in his reconstruction also concludes that, when it comes to a polysemic concept like responsibility, theory often separates what is conjoined in practical affairs (Genard 1999).

Accordingly, many scholars have suggested the necessity to adopt an integrative and complementary conception of responsibility (Gianni 2016; Grunwald 2018). Foddai believes that an analysis with regard to its historical evolution does not necessarily impede the formulation of a conception of responsibility that enshrines its different meanings (Foddai 2005). Rosanvallon augurs a new foundation of political responsibility on the basis of its two main senses, accountability as well as its role towards future outcomes (Rosanvallon 2015). Sartori also refuses a reduction of political responsibility to the former, endorsing the strong role of the latter for a good democratic practice (Sartori 1987).

Spämann had already suggested that responsibility should be an arbitration between the controllable and foreseeable effects of an action and its unlimited consequences (Spämann 1975). Also Ricoeur, following a Hegelian scheme, calls for a "moral judgment in some specific circumstance" on the basis of an ethics "careful to preserve a distance between imputability, solidarity and shared risks" (Gianni 2018; Ricoeur 2000, pp. 34–35).

What most of the commentators are then concluding is that the uptake of responsibility should be conceived in a complementary and equidistant way because all the meanings of the concept contribute to defining its overall sense.

This integrative and contextual approach appears to be in line with our current models of democracy that see responsibility as a device able to not only protect current societies, but also produce better futures (Dahl 1989;

Sartori 1987). It is important to highlight this positive and proactive side of responsibility that departs from the solid ground of hard regulations to create better futures. As suggested by Sartori, policy-makers are responsible in a double sense: they are responsible in front of the people *of* what they have done, but also *for* what they are going to do.

However, there is still a small passage that is missing, although it is implicit in a holistic, or ethical-political understanding of responsibility.

Responsibility is becoming even more complex with the extension of power that humanity is assuming through technology. However, it has been highlighted that, although the concepts may overlap, power is not identical with freedom (Sartori 1987). The latter can only be exercised, and in a responsible way, through a series of institutional conditions which define their possibilities and appropriateness (Böckenförde 2010; Gianni 2018; Honneth 2014; Seillan 2016).

This argument, which might appear abstract at a first glance, helps us in understanding that an actor cannot be assumed to be responsible if she has not been guaranteed the necessary level of freedom. Without entering the debate about the social pathologies caused by a short-circuit between social pressures about responsibility and actual individual capacities (Ehrenberg 1998), it is worth recognizing that actors, researchers for instance, need to follow a whole set of rules and drivers that frame their investigations. So, if researchers cannot be seen to be fully responsible for their choices, it is because they are entrenched in complex systems, or "orders of recognition" (Ferry 1991) and their freedom to choose what they would consider right cannot be fully exercised (Beck 1992; Gianni 2016; Stilgoe et al. 2013).

In general, I believe it is important to highlight this polyhedric and multilevel nature of the concept of responsibility, which can be implemented in different ways and which offers a high degree of flexibility. At the same time, these implementations are always framed by a discourse and specific institutional measures that render them not only effective but possible. A multi-level but integrative understanding of responsibility can represent an improvement in terms of both efficiency and justice for researchers and innovators. If it is true that actors cannot operate outside a given regulatory framework the stress on care or virtue can still offer a more contextual and subjective way to enforce responsible practices maintaining the process efficient. Without the necessity to overburden scientists or innovators with additional and strong regulations, it is left to them to translate a responsible approach into actions. At the same time, a multi-level conception will be stronger in addressing the level of responsibility demanded from actors and its actual feasibility, avoiding processes of "over-responsibilization." A risk that we already incur is the continuous and disproportionate appeal to responsible behaviors that create a short-circuit between what an actor is called to do and what she is actually capable of doing (Ehrenberg 1998). In this way the overall objective and importance of responsible practices become at best unbearable.

Therefore, we have assumed that responsibility is an integrative, multidimensional and contextual regulator of different perspectives, called to be responsive to the challenges of our societies (Pellizzoni 2004). We now need to evaluate to what extent current institutional mechanisms of RRI enable such an understanding.

1.5 Six keys and their relevance for science and democracy

In this part I will propose a reading of the criteria adopted by the EC to develop responsible practices of research and innovation, more specifically of the six keys.

The reasons to choose the EC's keys are mainly twofold. The first one is that the six keys represent the methodological tools adopted by the main public funding institution in Europe, and therefore inevitably affect the framing of research and innovation. It could be argued that RRI has also been developed according to different criteria (Owen et al. 2013). However, the preference assigned to the keys does not exclude the adoption of "actions" in those thematic areas. Klassen et al. have encouraged such an integration between the academic debate and policy-oriented measures (Klassen et al. 2018). Besides, as has been pointed out by Stilgoe et al., research and innovation have to go through a whole series of steps before they can actually be introduced in society (Stilgoe et al. 2013). Similarly, responsibility requires institutional measures in order to be realistically enacted. Therefore, institutional measures are always necessary for their implementation.

As recently stated by the Commissioner Carlos Moedas quoting Jean Monnet, nothing is possible without the citizens and nothing is sustainable without the institutions.[9]

The second reason lies in the fact that, despite a surrounding skepticism about the future of RRI, the keys will most probably continue to play a role in the ethical development of research and innovation.

If we accept this methodological aspect we can agree that it is more efficient to work or assess the existing institutional measures because they appear effective and they are not necessarily limited to the acronym of RRI.

When RRI was adopted by the EC, former Commissioner Georghean-Quinn provided the research community with six main thematic areas through which RRI could and should be implemented. Engagement, gender, open access, science education, ethics and governance are the key references to follow, in order to assess a responsible approach to research and innovation.

1.5.1 Engagement

Engagement is defined by the EC as:

> Engagement of all societal actors – researchers, industry, policy -makers and civil society – and their joint participation in the research and

> innovation process, in accordance with the value of inclusiveness, as reflected in the Charter of Fundamental Rights of the European Union.
> (EC 2012)

Being responsible then entails that:

> the societal challenges are framed on the basis of widely representative social, economic and ethical concerns and common principles. Moreover, mutual learning and agreed practices are needed to develop joint solutions to societal problems and opportunities, and to pre-empt possible public value failures of future innovation.
> (EC 2012)

This definition encourages an integrative framework of the different perspectives in society and it addresses the problem of framing by opening to the variety of interests at stake. It relies on common measures and fosters deliberative processes. Furthermore, the explicit reference to the Charter of Fundamental Rights of the European Union provides a thin, although important, normative reference to the debate on which RRI can rest. In this regard, it appears to be implementing a notion of responsibility that is wide, inclusive and contextually flexible. Furthermore, participatory practices are the main tools suggested to increase both scientific investigations and democratic processes. It is important though that engagement is conceived in a meaningful way, overcoming the risks of power imbalance and window-dressing (Fung 2006). Therefore, the basic presupposition for the translation of responsible approaches needs to complemented by further specifications.

1.5.2 Gender

Gender can be seen as a first additional specification of engagement, defining not only the necessity to establish a dialogue, but also delineating basic ways of avoiding discriminations amongst participants. The EC believes that it is crucial to address inequality in the representation of women and men in different processes. Gender aspects should not only target a quantitative increase of female representation, but also a deeper sensibility to the contents of research and innovation. Gender is meant to represent a dimension where actors can reflect on the biases based on biological assumptions that affect scientific research and its products.

The different quantitative and qualitative aspects of gender are difficult to disentangle. It might be argued that quantitative satisfaction of gender aspects in one process could also decrease the necessity to deepen the discussion about its qualitative integration. Gender discussion could limit its scope on particular aspects, overlooking the preconditions that enact general forms of discrimination as economic assets, political power and cultural authority (Connell, 2005). Or, to paraphrase Haraway, science should not ignore the plurality of the world's interpretations (Haraway 1991).

However, quantitative evidence of social discrimination can trigger a discussion about its reasons and actions of implementation (Wojniak 2017). If it is true that "the under-representation of women continues to characterize participation in science & technology" (EC 2016), this fact can be a strong motivation for change. Tangible, measurable products of science, either technologies or articles,[10] can represent an entry stage for climbing back up the technology ladder to its paradigm stage (Bessant 2013).

Gender entails a double role within the framework of RRI. On the one hand, it defines the necessity to be inclusive, while on the other hand it suggests that a quantitative opening of the process must be accompanied by qualitative aspects. It appears to be strongly valuable for science and democracy. This key in fact can support scientific investigation by providing knowledge about gender differences and needs. It can also increase democratic legitimacy by overcoming forms of discrimination and enhancing equal participation.

1.5.3 Open access

Open access can be seen as complementary to science education. In order to increase the level of knowledge, researchers, innovators and policy-makers should commit to providing access to the results of their research. Sharing does not only increase the overall level of knowledge but also favors a more transparent management of research and innovation funding.

Open access is one of the two areas that are meant to support the active and passive participation of a broad range of actors in the development of scientific processes. The EC officially conceives it as "giving free online access to the results of publicly-funded research" (EC 2012), but its extension can be widened according to the specific context. It can be agreed that an easier access to sources of information can favor knowledge building, which is crucial to support a conscious and just decision. In fact, it is not difficult to conceive the features of open access as those of transparency, a quest that has become stronger than ever, and that represent a cornerstone of contemporary democracies (Hollyer et al. 2011; Shapiro 2012).[11] Science can benefit from the increase of openness if it is conceived also at the political level, defining the choices and the trajectory that scientists are called to make and the institutional conditions they need to deal with. It strengthens the democratic process by the necessary degree of justification that transparent decisions are required to provide at different stages of the process (Weil et al. 2013).

1.5.4 Science education

Science education aims at increasing the amount of knowledge as well as its quality. Through means of education, the objective is to make society able to understand science and "fully participate and take responsibility in the research and innovation process."

Science education is a crucial step if we want to complete the passage between information and knowledge. As stated by Sartori,

> knowledge implies a mental control over the information that is in no way supplied by the information itself. Given an equal amount of information, a person may either be competent or remain incompetent, depending on whether she perceives correctly which means are appropriate to what ends, and thereby which consequences derive from which decisions or action.
>
> (Sartori 1987, p. 117)

Deficit of understanding is for the public a critical aspect, undermining the possibility of responsible practices (Gianni 2018; Wynne et al. 2007). As clearly highlighted by juridical responsibility, the basic prerequisite for accountability and liability is the epistemic capacity of understanding external inputs (Hart 2008).

The information acquired through open access should then be preceded and integrated by an education to understand and to filter information in order to transform it into knowledge. Therefore, we can conceive this not only as education of future generations, but also as the overall capacity building that any participation process should foresee if eager to integrate competent opinions into its debate (Kitcher 2011). This key is the enabler of an emancipatory process in the wake of our current models of deliberative democracies. Rendering a larger range of the population able to understand scientific data can also help overcome fears and mistrust about science, and can offer a relevant source of additional information for researchers.

1.5.5 Ethics

Ethics stands for the necessity of doing the right thinking in the research and innovation process. The EC clearly highlights that ethical thinking should not be reduced to existing mandatory legal aspects but should be implemented by respecting fundamental rights and highest ethical standards. The clear normative reference to the Charter of Fundamental Rights expresses the objective of gaining "acceptability" through societal relevance.

However, in order to be consistent with the analyses of the other five keys, it would be more sensible to consider a more complex t version of this term. Although ethics is one of those controversial terms the meaning of which can be very broad (Reber 2016), the EC's definition of ethics, and its relation with the other keys, seems to reflect a pragmatist interpretation of the term. Accordingly, ethics, rather than being a set of predefined rules impeding progress, is the integration of objective rights and subjective values into a common and contextual framework (Adorno 1993, 2005; Dewey 1991; Gianni 2018; Ritter 1982).

Following this line, ethical processes aim at overcoming unilateral or unjustified perspectives by considering social interactions in a holistic way. This fifth key concerns the analysis and integration of different values and norms into the design and development of research and innovation. This interpretation matches the abovementioned understanding of responsibility that integrates the different levels of regulatory aspects. Ethics requires that different cultural values and interests are taken to the fore according to different levels of regulations. It offers a multi-level, refined tool to improve pluralism in democratic societies. In this sense it is reasonable to conceive it as a pragmatist approach that distances itself from both applied ethics and traditional philosophy by embracing a dynamic perspective (Keulartz et al. 2004).

1.5.6 *Governance*

The description of the last key, governance, is particularly relevant for the purpose of our analysis and for understanding the sense of the keys. "Policy-makers also have a responsibility to prevent harmful or unethical developments in research and innovation" (https://ec.europa.eu/research/swafs/pdf/pub_rri/KI0214595ENC.pdf).

Governance in fact should be conceived as an umbrella term expressing the necessity that policy-makers develop models of RRI that integrate all other keys.

According to Jessop, governance is an "important mean to overcome the division between rulers and the ruled in representative regimes, and to secure the input and commitment of an increasingly wide range of stakeholders in policy formulation and implementation" (Jessop 2002, p. 3). According to this definition and to the delineation made by the EC, governance within the RRI framework seems to be responding to these objectives. Governance is considered to be more efficient because it crosscuts different social dimensions and can be implemented in a faster way with respect to government (Kuhlmann & Ordonez-Matamoros 2017). Lee and Petts suggest that governance in the RRI framework should also be conceived according to the upscale of science policies, becoming a multi-level process and thus developing a set of soft-law instruments besides the hard-law ones (Goos & Lindner forthcoming; Lee & Petts 2013).

The focus on soft regulations which RRI seems to encourage can be considered a weakening factor because it relies on the willingness of actors to commit, implying a concrete level of arbitrariness, which can undermine its regulatory force (Wickson & Forsberg 2015). At the same time, such openness can represent a potential strength because of the capacity to be adapted to different contexts by a larger number of actors (Gorgoni 2018).

Although some of the "tools" entailed in the governance of RRI might not represent a novelty (Zwart et al. 2014), the way in which they are combined represents an original and more refined attempt to offer a multi-level and contextual regulatory framework (Fisher & Rip 2013; Gianni & Goujon 2018; Goos & Lindner forthcoming; Grunwald 2011, 2016, 2018).

Governance, within this framework, is characterized by the importance allocated to a broader assessment of implementation and societal impact; a more proactive attitude towards inclusiveness and competition; and an accentuated responsiveness to the future, which opens up space for a debate about the goals of research and innovation.

All these aspects are clearly detectable by the adoption of an ethical understanding of responsibility (fifth key), which implies a more democratic and proactive understanding of technical processes.

1.6 The implementation role of the six keys for science and democracy

After having operated an analysis of the EC's six keys which enables us to implement them at different levels, it is important to go a step further and to highlight two additional aspects about their operationalization.

The first aspect concerns the keys taken singularly. Although we have proposed an interpretation of each key based on an ideal understanding, this perspective is not blind to the criticisms that each of them might be incurring.

Commentators have highlighted two main issues. The first one concerns the potential window-dressing function they could play, and the second one focuses on the difficulty implementing certain keys in specific research fields. For instance, some authors have highlighted the risk for engagement to be instrumentalized in different manners (Callon et al. 2009; Fung 2006; Honneth 1991; Stilgoe et al. 2013). Also other keys like gender might raise some issues with regard to the relation between a thin (quantitative) and thick (qualitative) integration of the concept (Connell 2005). Open access understood as transparency has been questioned as it might clash with the necessary levels of confidentiality (Blok 2018; Blok & Lemmens 2015).

Although these criticisms sound legitimate, I believe there are two more aspects that we need to consider to help us overcome these limitations. The first one is that these shortcomings do not always apply, and therefore we should not be dismissing the keys *a priori*. Several aspects of research are or should be public, especially their outcomes, and should go under scrutiny in order to improve them. Besides, the fact that key concepts encounter several barriers should not be a reason to dismiss them, but rather a motivation to seek for appropriate or alternative forms of implementation.

Several authors have also suggested that, when implementation does not appear evident or easy, it might be sensible to shift up the reflection to a meta-level (Stahl et al. 2013; Van den Hoven et al. 2012) or to a second order (Gianni & Goujon 2018; Maesschalck 2017) in order to question the framing that drives the overall scientific process. A meta-level reflection can help in addressing the apparent clash between normative drivers and the practical obstacles pertaining to a specific field because it suggests a reflection on the broader meaning of a normative concept, so to find alternative ways to promote its uptake. For instance, open access can also be conceived as a form of transparency not on the methodology or on specific data, but on the objectives that a certain methodology should produce. In fields where gender ratio might be seen as achieved, other forms of gender-biased mechanisms or product might be scrutinized. In other words, it is more sensible to understand the keys not as indications to perform a clear and specific assessment, like a box-ticking exercise, but rather as overall dimensions, the meaning of which needs to be explored in order to improve the ethical stance of a research process.

A second-order reflexivity is meant instead to protect normative objectives from instrumentalization. If engagement, for instance, can be criticized because it appears not effective in many cases, this weakness should not always be identified with the principle as such, but with the ways in which is implemented. In this sense, a second-order reflexivity can deepen the knowledge about the overall process so as to unveil potential gaps between supposed objectives and actual working mechanisms. By questioning the framework, practitioners can better understand the main objectives at the heart of the keys, assess an objective with the measures put in place, and eventually question the mismatch between the two aspects without the necessity for getting rid of the concept as such.

An example of implementation of the six keys taken singularly has been carried out in a recent EC research project, BODEGA, dealing with human factors and their development after the introduction of automatization in border controls. The project conducted field studies in different border check-points observing the usual procedures put in place to check passenger flows entering and exiting the Schengen area. By applying the six keys, the project was able to provide a whole set of recommendations for improving the border control process without leaking sensitive information or undermining security needs. One example concerns uniforms for border guards which are designed for men and can become ineffective or disturbing when used by a woman. As clothing and uniforms might have been designed before women were entitled to enter police corps, this suggests that what is missing are simply broader participatory structures that can bring to the attention some diverse issues. What has also emerged is that the taboo surrounding certain research fields or practitioners is often detrimental because it prevents a broader analysis that would not undermine their main scope.[12]

The second aspect about their operationalization focuses on the relation between the keys. An approach that I believe has not been discussed extensively but that I think is indeed part of the EC's intentions is to conceive the keys in a relation of complementarity.[13] Such coupling, that we could also interpret in a lexical way, can be justified by a factor that it's not directly linked to the fact that they are part of one strategic document.

The keys in fact entail a logic and practical complementarity in the framework of a deliberative democracy. It is difficult to think of a sincere participatory process that prescinds from the appropriate competences. Also, an open discussion about the future that neglects gender aspects might raise some perplexity. The examples could be several, but what they suggest is that each key is necessary in order to produce a responsible approach which can be inclusive not only of different actors, but also of underrepresented issues. This is corroborated by the fact that the keys tend to merge an epistemic stance with a value-based one, answering the challenges of post-normal science (Funtowicz & Ravetz 1993). It might be correctly argued that the absence of one of the keys from an analysis does not necessarily produce irresponsible approaches. However, I believe that there would be a great difference in terms of a democratic approach if one of the keys were completely dismissed.

Despite the possible and reasonable objections that this holistic approach could raise, I believe that this hypothesis could represent a fruitful attempt in the cross-fertilization process between science and democracy. Besides, although the keys and the underlying understanding of better futures have also been questioned (De Saille & Medvecky 2016), I believe it would be more fruitful to improve the general understanding of RRI through mechanisms that are in place and that are meant to favor such criticisms. It is most probably through an increment of engaging practices based on aware and ethical stances that we might hope to refine the understanding of RRI.

The ideal character and the challenges of this perspective should not, in my opinion, discourage experimentation. RRI and democracy both being soft-regulatory frameworks covering different fields, they will always suffer from the distance between ideal features and their actual implementation, but this does not imply that we should not try to fill this gap. As Dewey has repeatedly suggested, democracy should not be reduced to a fixed political model but rather understood as a method to be applied in all social interactions (Dewey 1991).

To summarize, the six keys proposed by the EC are instrumental to creating participatory processes in which a wide range of actors can interact to deliberate about intended futures while considering diverse values and interests. This perspective is supported by a multi-level understanding of responsibility and by the democratic principles embedded in the keys. The conception of democracy they enact is one where subjects should be put in the situation to self-determine their ends on the basis of moral and epistemic pluralism.

At this point of our analysis where we have adopted a reference model of RRI, let us briefly recap the evaluation of its scientific and democratic relevance.

Considering scientific relevance, we can recall how this six-key-based RRI framework integrates an epistemic moment (technical knowledge) with the inclusion of actors carrying different interests, values and information (pluralism). It appears that scientific accuracy can but benefit from such broadening of perspectives. The criticism directed at the appropriateness of some keys for an efficient research and innovation process has also proven to be misdirected. The argument that certain drivers like transparency and openness cannot be fully implemented in research (Blok & Lemmens 2015) can in fact be overcome if we accept that these and other normative tools, when implemented, can be used at different levels (Gianni & Goujon 2018).

If we believe in the necessity and advantages of an opening of scientific research objectives, we can then judge RRI and its keys as a useful methodological tool to improve this process. The keys as a responsible approach to research and innovation support the engagement of a wider range of conscious actors' interests and values into the delineation of futures. Furthermore, they include the necessary measures to implement effectively and rightly such exchange between scientists and citizens. When correctly and robustly implemented, RRI can represent an effective tool to help science progress in its social contribution by integrating social values and interests into its trajectory.

Therefore, following this perspective, it is plausible to deduce that RRI is in principle of high relevance to the development of scientific investigation.

On the basis of the conception of a dynamic, active and deliberative model of democracy, like the one proposed by Dewey, it is not difficult to see how RRI can be judged in line with it and is thus relevant. RRI tries to bring democratic practices into science by introducing some basic tools through which citizens can express their views in an equal and free manner. Furthermore, RRI aims to reduce inequalities by empowering citizens to actively participate in technology trajectories. The possibility of developing a reflexive deliberation about ethical aspects and debate on additional measures to be integrated to technical processes appears nowadays not only as a managerial opportunity but also as a democratic necessity.

I do not think that it is misleading to suggest that the keys are tools that innovatively try to put citizens in the position to exert their "humanity," i.e. to enter the discourse about science trajectory through means of equality and freedom steered by solidarity. By integrating an epistemic dimension with a value-based one, RRI is an attempt to implement a participatory democratic model in science – one that enables citizens to express their values and preferences about objectives in a reasonable and competent way. Therefore, it implies an understanding of research and innovation as fields that should be subjected to the principles of democracy. This appears in line with the radicalization of democracy promoted by pragmatists (Dewey 1991).

It avoids the risk of a purely expert-based approach without losing the reasonable ground on which to initiate a dialogue. It offers an institutional, democratic and thus balanced structure in which actors are called to interact and imagine their horizons. It does so by recognizing and respecting the plurality of perspectives at stake. RRI acknowledges plurality as a challenge and aims to generate futures that are the conscious fruits of this diversity.

As stated by Galli, if even democracy does not have a center any more, it is not possible that it does not have an end (Galli 2011).

1.7 Conclusions

We have analyzed the main features and challenges of scientific and democratic processes to understand what could be relevant for their respective futures. We have understood that they share the need to integrate broader perspectives into their decision-making process. At the same time, the chapter tried to highlight that the quest for a democratization of science should not be seen as an intrusion on scientific rigor but rather as a mutual deliberation about the objectives of science. I have then highlighted that RRI is an operational framework, with the EC's six keys functioning as regulatory criteria in order to increase the cross-fertilization between science and democracy. According to this perspective RRI appears highly relevant from a scientific and democratic point of view because it entails a series of tools that actualize the necessity of enabling citizens and scientists to communicate with each other.

However, this does not mean that RRI is necessarily going to keep its promises if the different actors are not going to commonly implement these indications. It is useful to remember that one of the main significances of the original concept of responsibility under Roman law was exactly the one of guaranteeing consistency between promises and facts (Benveniste 1969). Only if its features will be implemented in an institutionally substantive way will we avoid mistrust and disaffect about responsibility, and will we be able to shorten the gap between science and society.

Acknowledgments

This research was supported by the Province of Limburg, The Netherlands, under grant number SAS-2019-00061, and by the European Commission under grant number 741402.

Notes

1 I will talk here of science in general, although the focus is on research and innovation which are often different in their scope. The differences and overlaps will emerge in the tendency to increase the contextual response of science in general.
2 The popularity of the butterfly effect is witnessed by the enormous number of different contributions in different divulgation sectors.
3 For an interesting project about this, see: www.sienna-project.eu
4 Burget et al. reviewed 235 articles mentioning RRI, but they could not include larger contributions and deliverable projects, which would increase this number substantially.
5 For an analysis at the policy level, see NewHoRRIzon project, D.4.1, https://newhorrizon.eu/wp-content/uploads/2019/03/D-4.1-Diagnosis-RRI-in-Societal-Challenges-1.pdf

6 https://epsrc.ukri.org/index.cfm/research/framework/
7 https://ec.europa.eu/research/swafs/pdf/pub_public_engagement/responsible-research-and-innovation-leaflet_en.pdf
8 https://newhorrizon.eu https://ec.europa.eu/research/evaluations/index.cfm?pg=h2020evaluation
9 https://ec.europa.eu/commission/commissioners/2014–2019/moedas/announcements/ep-high-level-conference-eu-research-and-innovation-our-daily-life_en
10 According to the SHE Figures report:

> "In the period spanning from 2010 to 2013, the propensity to integrate a *gender dimension in research content* measured in scientific articles in the EU-28 ranged from virtually zero in agricultural sciences, engineering and technology, and natural sciences to over 6% in the social sciences"
> (EC 2016).

11 See also http://www.oecd.org/about/secretary-general/opennessandtransparency-pillarsfordemocracytrustandprogress.htm
12 The remarkable results obtained suggest that it is an optimistic path to explore: https://bodega-project.eu
13 For a general discussion about the keys, see Spaapen et al. (2015).

References

Abbagnano, N. (1957). Il giudizio di responsabilità nella morale e nel diritto. *Rivista di Filosofia*, 1, pp. 30–45.

Adam, B. and Groves, C. (2011). Futures tended: care and future-oriented responsibility. *Bulletin of Science, Technology and Society*, 31(1), pp. 17–27.

Adorno, T. W. (1993 [1963]). *Hegel: Three Studies*. Cambridge, MA: MIT Press.

Adorno, T. W. (2005 [1951]). *Minima Moralia: Reflections on Damaged Life*. London: Verso.

Arnaldi, S. and Bianchi, L. (2016). *Responsibility in Science and Technology*. Wiesbaden: Springer.

Arnaldi, S. and Gorgoni, G. (2016). Turning the tide or surfing the wave? Responsible Research and Innovation, fundamental rights and neoliberal virtues. *Life Sciences, Society and Policy*, 12(6), pp. 1–19.

Audard, C. (2009). *Qu'est-ce que le Libéralisme ? Éthique, Politique, Société*. Paris: Gallimard.

Beck, U. (1992). *Risk Society: Towards a New Modernity*. London: Sage Publications.

Benveniste, E. (1969). *Le Vocabulaire des Institutions Indo-Européennes*. Paris: Les Éditions de Minuit.

Berlin, I. (2002 [1969]). *Liberty: Incorporating Four Essays on Liberty*. Oxford: Oxford University Press.

Bessant, J. (2013). Innovation in the Twenty-First Century. In: R. Owen, J. Bessant and M. Heintz (eds.), *Responsible Innovation. Managing the Responsible Emergence of Science and Innovation in Society*. Hoboken, NJ: John Wiley, pp. 1–25.

Bijker, W. E., Hughes, T. P. and Pinch, T. J. (eds.), (1987). *The Social Construction of Technological Systems: New Directions in the Sociology and History of Technology*. Cambridge, MA: MIT Press.

Blok, V. and Lemmens, P. (2015). The Emerging Concept of Responsible Innovation. Three Reasons Why It Is Questionable and Calls for a Radical Transformation of the Concept of Innovation. In: B.-J. Koops, J., Oosterlaken, I. Romijn, et al. (eds.),

Responsible Innovation 2: Concepts, Approaches,and Applications. Dordrecht: Springer, pp. 19–35.

Blok, V., Lubberink, R., van den Belt, H., Ritzer, S., van der Kruk, H. and G. Danen (2018). Challenging the ideal of transparency as a process and as an output variable of Responsible Innovation. The case of 'The Circle'. In: R. Gianni, J. Pearson, J. Reber. (eds.), *Responsible Research and Innovation. From Concept to Practices*. Oxford: Routledge, pp. 225–244.

Böckenförde, E. W. (2010). *Diritto e Secolarizzazione. Dallo Stato Moderno all'Europa Unita*. Rome: Laterza.

Bohman, J. and Regh, W. (eds.), (1997). *Deliberative Democracy*. Cambridge, MA: MIT Press.

Bourgeois, B. (2017). *Sept Questions Politique du Jour*. Paris: Vrin.

Bradley, F. H. (2018). *Ethical Studies*. London: Forgotten Books.

Bridgman, P. (1927). *The Logic of Modern Physics*. New York: MacMillan.

Bucchi, M. and Neresini, F. (2008). Science and Public Participation. In: E. Hackett, O. Amsterdamska and M. Linch (eds.), *Handbook of Science and Technology Studies*. Cambridge, MA: MIT Press, pp. 449–473.

Burget, M., Bardone, E. and Pedaste, M. (2017). Definitions and conceptual dimensions of responsible research and innovation: a literature review. *Science and Engineering Ethics*, 23(1), pp. 1–19.

Callon, M., Lascoumes, P., Barthe, Y. (2009). *Acting in an Uncertain World: An Essay on Technical Democracy*. Cambridge, MA: MIT Press.

Carrabregu, G. (2016). Habermas on solidarity: an immanent critique. *Costellations*, 23(4), pp. 507–522.

Connell, R. W. (2005), Change among the gatekeepers: men, masculinities, and gender equality in the global arena. *Signs*, 30(3), pp. 1801–1825.

Dahl, R. A. (1989). *Democracy and Its Critics*. New Haven, CT: Yale University Press.

Derrida, J. (2008). *The Gift of the Death*. Chicago, IL: Chicago University Press.

De Saille, S. and Medvecky, F. (2016). Innovation for a steady state: a case for responsible stagnation, *Economy and Society*, 45(1), pp. 1–23.

Dewey, J. (1991 [1927]). *The Public and Its Problems*. Athens, OH: Swallow Press / Ohio University Press.

Dews, P. (ed.), (1992). *Autonomy and Solidarity. Interviews with Jürgen Habermas*. London: Verso.

Ehrenberg, A. (1998). *The Fatigue of Being Oneself: Depression and Society*. Paris: Odile Jacob.

Engle, L. (1993). Vers une nouvelle approche de la responsabilité: le droit français face à la dérive américaine. *Esprit*, 192(6), pp. 5–31.

European Commission (2012). *Responsible Research and Innovation. Europe's Ability to Respond to Societal Challenges*. Directorate-General for Research and Innovation, Brussels. Available at: https://ec.europa.eu/research/swafs/pdf/pub_public_engagement/responsible-research-and-innovation-leaflet_en.pdf

European Commission (2016). *She Figures 2015*. Directorate-General for Research and Innovation, Brussels. Available at: https://ec.europa.eu/research/swafs/pdf/pub_gender_equality/she_figures_2015-final.pdf

Ewald, F. (1986). *L'Etat Providence*. Paris: Grasset et Fasquelle.

Fagotto, E. and Fung, A. (2006). Empowered participation in minneapolis: the neighborhood revitalization program. *International Journal of Urban and Regional Research*, 30(3), pp. 638–655.

Ferry, J. M. (1991). *Les Puissances de l'Experiénce*. Paris: Cerf.
Fisher, E. and Rip, A. (2013). Responsible Innovation: Multi-Level Dynamics and Soft Intervention Practices. In: R. Owen, J. Bessant, and M. Heintz (eds.), *Responsible Innovation. Managing the Responsible Emergence of Science and Innovation in Society*. Hoboken, NJ: John Wiley, pp. 165–184.
Fishkin, J. and Laslett, P. (eds.), (2003). *Debating Deliberative Democracy*. London: Blackwell.
Flipse, S. M. (2012). *Enhancing Socially Responsible Innovation in Industry. Practical Use for Considerations of Social and Ethical Aspects in Industrial Life Science and Technology*. Delft University. Available at: https://pdfs.semanticscholar.org/9979/9d2d7e436ccd7d7fa0b595ae106515f4cf3c.pdf.
Foddai, M. A. (2005). *Sulle Tracce della Responsabilità. Idee e Norme dell'Agire Responsabile*. Turin: Giappichelli Editore.
Fung, A. (2006). Varieties of participation in complex governance. *Public Administration Review*, 66(1), pp. 66–75.
Funtowicz, S. O. and Ravetz J. R. (1993). Science for the post-normal age. *Futures*, 25(7), pp. 739–755.
Funtowicz, S. O., Ravetz, J. R. (2008). Beyond complex systems: emergent complexity and social solidarity. In: D. Waltner-Toews, J. J. Kay and N-M. Lister (eds.), *The Ecosystem Approach: Complexity, Uncertainty, and Managing for Sustainability*. New York: Columbia University Press, pp. 309–322.
Galli, C. (2011). *Il Disagio della Democrazia*. Turin: Einaudi.
Genard, J.-L. (1999). *Grammaire de la Responsabilité*. Paris: CERF.
Gianni, R. (2016). *Freedom and Responsibility. The Ethical Realm of RRI*. London: ISTE/Wiley.
Gianni, R. (2018). The Discourse of Responsibility. A Social Perspective. In: R. Gianni, J. Pearson and J. Reber (eds.), *Responsible Research and Innovation. From Concept to Practices*. Oxford: Routledge, pp. 11–34.
Gianni, R. and Goujon, P. (2018). What are the Conditions for the Ethical Implementation of RRI? Responsible Governance and Second-Order Reflexivity. In: R. Gianni, J. Pearson and J. Reber (eds.), *Responsible Research and Innovation. From Concept to Practices*. Oxford: Routledge, pp. 172–207.
Godin, B. (2015). *Innovation Contested – The Idea of Innovation over the Centuries*. London: Routledge.
Goffman, E. (1974). *Frame Analysis: An Essay on the Organization of Experience*. New York: Harper and Row.
Goodin, R. (2012). *Innovating Democracy. Democratic Theory and Practice After the Deliberative Turn*. Oxford: Oxford University Press.
Goos, K. and Lindner, R. (Forthcoming). Genealogies of the concept of RRI – from fragmented notions towards commonalities and disparities. To be published.
Gorgoni, G. (2018). Responsible Research and Innovation and the governance of human enhancement. *Nano Ethics*, 12(3), pp. 257–267.
Gottweis, H. (2008). Participation and the new governance of life. *BioSocieties*, 3(3), pp. 265–286.
Grinbaum, A. and Groves, C. (2013). What is 'Responsible' about Responsible Innovation? Understanding the Ethical Issues. In: R. Owen, J. Bessant and M. Heintz (eds.), *Responsible Innovation. Managing the Responsible Emergence of Science and Innovation in Society*. Hoboken, NJ: John Wiley, pp. 119–142.

Grunwald, A. (2011). Responsible innovation: bringing together technology assessment, applied ethics, and STS research. *Enterprise and Work Innovation Studies,* 7(IET), pp. 9–31.
Grunwald, A. (2016). *The Hermeneutic Side of Responsible Research and Innovation.* London: ISTE/Wiley.
Grunwald, A. (2018). Responsibility Beyond Consequentialism. The EEE Approach to Responsibility in the Face of Epistemic Constraints. In: R. Gianni, J. Pearson and J. Reber (eds.), *Responsible Research and Innovation. From Concept to Practices.* Oxford: Routledge, pp. 35–49.
Guston, D. H. (2001). Integrity, responsibility, and democracy in science. *SciPolicy. A Journal of Science and Health Policy,* 1(2), pp. 168–189.
Hajer, M. and Wagenaar, H. (eds.), (2003). *Deliberative Policy Analysis. Understanding Governance in the Network Society.* New York: Cambridge University Press.
Haraway, D. (1991). *A Cyborg Manifesto.* New York: Routledge.
Hart, H. L. A. (2008 [1968]). *Punishment and Responsibility. Essays in the Philosophy of Law.* Oxford: Oxford University Press.
Hegel, G. W. F. (1991 [1821]). *Elements of the Philosophy of Right.* Cambridge: Cambridge University Press.
Holden, B. (ed.), (1999). *Global Democracy. Key Debates.* Oxford: Routledge.
Hollyer, J. R., Rosendorff, B. P. and Vreeland, J. R. (2011). Democracy and transparency. *The Journal of Politics,* 73(4), pp. 1191–1205.
Honneth, A. (1991). *The Critique of Power: Reflective Stages in a Critical Social Theory.* Cambridge, MA: MIT Press.
Honneth, A. (2014). *Freedom's Right. The Social Foundations of Democratic Life.* Cambridge: Polity Press.
Horkheimer, M. and Adorno, T. W. (2002 [1947]). *Dialectic of Enlightenment.* Stanford, CA: Stanford University Press.
Jacquet, V. (2018). The role and the future of deliberative mini-publics: a citizen perspective. *Political Studies,* 67(3), pp. 639–657.
Jasanoff, S. (2004). *States of Knowledge. The Co-production of Science and Social Order.* London: Routledge.
Jasanoff, S. (2016). *The Ethics of Invention. Technology and the Human Future.* New York: Norton.
Jessop, B. (2002). *Governance and Metagovernance: On Reflexivity, Requisite Variety, and Requisite Irony.* Available at: www.lancaster.ac.uk/fass/resources/sociology-online-papers/papers/jessop-governance-and-metagovernance.pdf
Jonas, H. (1984). *The Imperative of Responsibility: In Search of Ethics for the Technological Age.* Chicago, IL: University of Chicago Press.
Kant, I. (1997 [1788]). *Critique of Practical Reason.* Cambridge: Cambridge University Press.
Kelsen, H. (2005 [1934]). *Pure Theory of Law.* Berkeley, CA: California University Press.
Keulartz, J., Schermer, M., Korthals, M. and Swierstra, T. (2004). Ethics in technological culture: a programmatic proposal for a pragmatist approach. *Science, Technology, & Human Values,* 29(1), pp. 3–29.
Kitcher, P. (2011). *Science in a Democratic Society.* New York: Prometheus Books.
Klassen, P., Rijnen, M., Vermulen, S., Kupper, F. and Broerse, J. (2018). Technocracy vs Experimental Learning in RRI. On Making the Most of RRI's Interpretative

Flexibility. In: R. Gianni, J. Pearson and J. Reber (eds.), *Responsible Research and Innovation. From Concept to Practices*. Oxford: Routledge, pp. 77–98.

Krishna, V. (2013). Changing Social Relations between Science and Society: Contemporary Challenges. Available at: https://halshs.archives-ouvertes.fr/halshs-00910707/document

Kuhlmann, S. and Ordonez-Matamoros, G. (eds.), (2017). *Research Handbook on Innovation Governance for Emerging Economies. Towards Better Models*. Cheltenham: Edward Elgar.

Lee, R. G. and Petts, J. (2013). Adaptive Governance for Responsible Innovation. In: R. Owen, J. Bessant and M. Heintz (eds.), *Responsible Innovation. Managing the Responsible Emergence of Science and Innovation in Society*. Hoboken, NJ: John Wiley, pp. 143–164.

Lévi-Bruhl, L. (1884). *L'Idée de la Responsabilité*. Paris: Bourloton. Available at: https://archive.org/details/lidederesponsab00lvgoog/page/n15

Levinas, E. (1998 [1974]). *Otherwise Than Being or Beyond Essence*. Pittsburgh, PA: Duquesne University Press.

Limoges, C. (1993). Expert knowledge and decision-making in controversy contexts. *Public Understanding of Science*, 2(4), pp. 417–442.

Lorenz, E. (1963). Deterministic non-periodic flow. *Journal of the Atmospheric Sciences*, 20, pp. 131–141. Available at: https://journals.ametsoc.org/doi/pdf/10.1175/1520-0469%281963%29020%3C0130%3ADNF%3E2.0.CO%3B2

MacIntyre, A. (2007). *After Virtue: A Study in Moral Theory*. Notre Dame, IN: Notre Dame University Press.

Maesschalck, M. (2017). *Reflexive Governance for Research and Innovative Knowledge*. London: ISTE/Wiley.

Manin, B. (2012). *Principes du Gouvernement Représentatif*. Paris: Flammarion.

Mazzuccato, M. (2015). *The Entrepreneurial State. Debunking Public vs Private Sector Myths*. New York: Public Affairs.

McKeon, R. (1957). The development and significance of the concept of responsibility. *Revue Internationale de Philosophie*, 1, pp. 3–32.

Moore, M. S. (1984). *Law and Psychiatry. Rethinking the Relationship*. Cambridge: Cambridge University Press.

Nowotny, H. (2015). *The Cunning of Uncertainty*. Oxford: Polity.

Owen, R., Bessant, J. and Heintz, M. (eds.), (2013). *Responsible Innovation. Managing the Responsible Emergence of Science and Innovation in Society*. Hoboken, NJ: John Wiley.

Pansera, M. and Owen, R. (2018). *Innovation and Development: The Politics at the Bottom of the Pyramid*. London: ISTE/Wiley.

Paulson, L. S. (1992). The neo-Kantian dimension of Kelsen's Pure Theory of Law. *Oxford Legal Studies*, 12(3), pp. 311–332.

Pavie, X. (2018). From Responsible-Innovation to Innovation-Care. Beyond Constraints, a Holistic Approach of Innovation. In: R. Gianni, J. Pearson and J. Reber (eds.), *Responsible Research and Innovation. From Concept to Practices*. Oxford: Routledge, pp. 245–267.

Pavie, X., Scholten, V. and Carthy, D. (2014). *Responsible Innovation. From Concept to Practice*. Singapore: World Scientific.

Pearson, J. (2018). Ever Deeper Research and Innovation Governance? Assessing the Uptake of RRI in Member States' Research and Innovation Programmes.

In: R. Gianni, J. Pearson and J. Reber (eds.), *Responsible Research and Innovation. From Concept to Practices*. Oxford: Routledge, pp. 99–128.

Pellé, S. (2018). Responsibility in Research and Innovation. The Potential of Care Ethics. In: R. Gianni, J. Pearson and J. Reber (eds.), *Responsible Research and Innovation. From Concept to Practices*. Oxford: Routledge, pp. 268–285.

Pellé, S. and Reber, B. (2015). Responsible Innovation in the light of moral responsibility. *Journal on Chain and Network Science*, 15(2), pp. 107–117.

Pellizzoni, L. (2004). Responsibility and environmental governance. *Environmental Politics*, 13(3), pp. 541–565.

Pesch, U. (2015). Engineers and active responsibility. *Science and Engineering Ethics*, 21(4), pp. 925–939.

Pettit, P. (1997). The Consequentialist Perspective. In: M. Baron, P. Pettit and M. Slote (eds.), *Three Methods of Ethics: A Debate*. Oxford: Blackwell, pp. 92–174.

Pettit, P. (2015). *The Robust Demands of the Good. Ethics with Attachment, Virtue and Respect*. Oxford: Oxford University Press.

Pitkin, H. F. (1988). Are freedom and liberty twins? *Political Theory*, 16(4), pp. 523–552.

Qvortrup L. (2003). *The Hypercomplex Society*. New York: Peter Lang.

Reber, B. (2016). *Precautionary Principle, Pluralism and Deliberation*. London: ISTE/ Wiley.

Ricoeur, P. (2000). *The Just*. Chicago, IL: Chicago University Press.

Ritter, J. (1982). *Hegel and the French Revolution. Essays on the Philosophy of Right*. Cambridge, MA: MIT Press.

Robinson, D. (2009). Co-evolutionary scenarios: an application to prospecting futures of the responsible development of nanotechnology. *Technological Forecasting and Social Change*, 76, pp. 1222–1239.

Rodotà, S. (2014). *Solidarietà. Un'Utopia Necessaria*. Rome: Laterza.

Rosanvallon, P. (2015). *Le Bon Gouvernement*. Paris: Seuil.

Sartori, G. (1987). *The Theory of Democracy Revisited*. Chatam, NJ: Chatam House.

Seillan, H. (2016). *Danger et precaution. Le roman des mots*. Paris: Les Belles Lettres-Manitoba.

Sen, A. (1999). *Development as Freedom*. Oxford: Oxford University Press.

Sen, A. and Williams, B. (eds.), (1982). *Utilitarianism and Beyond*. Cambridge: Cambridge University Press.

Shapiro, I. (2012). *The Moral Foundations of Politics*. New Haven, CT: Yale University Press.

Sintomer, Y. (2011). *Petite Histoire de l'Expérimentation Démocratique*. Paris: La Découverte.

Spaapen, J. (Rapporteur); Strand, R. (Chair); Bauer, M. W., Hogan, E., Revuelta, G. and Stagl, S. (Members); and Paula, L. and Guimarães Pereira, Â. (Contributors) (2015). *Indicators for Promoting and Monitoring Responsible Research and Innovation. Report from the Expert Group on Policy Indicators for Responsible Research and Innovation*. European Commission. Available at http://ec.europa.eu/research/swafs/pdf/pub_rri/rri_ indicators_final_version.pdf

Spämann, R. (1975). Nebenwirkungen als moralisches Problem. *Philosophisches Jahrbuch*, 82(2), 323.

Stahl, B., Eden, G. and Jirotka, M. (2013). Responsible Research and Innovation in Information and Communication Technology: Identifying and Engaging with the Ethical Implications of ICTs. In: R. Owen, J. Bessant and M. Heintz (eds.), *Responsible Innovation. Managing the Responsible Emergence of Science and Innovation in Society.* Hoboken, NJ: John Wiley, pp. 119–218.

Stilgoe, J., Owen, R. and Macnaghten, P. (2013). Developing a framework for responsible innovation. *Research Policy*, 42(9), pp. 1568–1580.

Tronto, J. (1993). *Moral Boundaries. A Political Argument for an Ethic of Care.* London: Routledge.

Turing, A. M. (1950). Computing machinery and intelligence. *Mind,* 49, pp. 433–460.

Vandamme P.- E. (2018). Le tirage au sort est-il compatible avec l'élection? *Revue Francaise de Science Politique*, 68(5), pp. 873–894.

:van den Hoven, J. (2013). Value Sensitive Design and Responsible Innovation. In: R. Owen, J. Bessant and M. Heintz (eds.), *Responsible Innovation. Managing the Responsible Emergence of Science and Innovation in Society.* Hoboken, NJ: John Wiley, pp. 75–83.

van den Hoven, M.J., Lokhorst, G. and van de Poel, I. (2012). Engineering and the problem of moral overload. *Science and Engineering Ethics,* 18(1), pp. 1–13.

van de Poel, I. (2011). The Relation Between Forward-Looking and Backward-Looking Responsibility. In: N. Vi, I. van de Poel and J. Van den Hoven (eds.), *Moral Responsibility, Beyond Free Will and Determinism.* Dordrecht: Springer, pp. 37–52.

van Oudheusden, M. (2014). Where are the politics in responsible innovation? European governance, technology assessments, and beyond. *Journal of Responsible Innovation*, 1(1), pp. 67–86.

Van Reybrouck, D. (2014). *Contre les élections.* Arles: Actes Sud.

Villey, M. (1977). Esquisse historique sur le mot responsable. *Archives de philosophie du droit*, XXII, pp. 45–58.

Vincent, N., van de Poel, I. and Van den Hoven J. (eds.), (2011) *Moral Responsibility, Beyond Free Will and Determinism.* Dordrecht: Springer.

Von Schomberg, R. (2013). A Vision of Responsible Research and Innovation. In: R. Owen, J. Bessant and M. Heintz (eds.), *Responsible Innovation. Managing the Responsible Emergence of Science and Innovation in Society.* Hoboken, NJ: John Wiley, pp. 51–74.

Von Schomberg, R. (2014). The Quest for the 'Right' Impacts of Scienceand Technology: A Framework for Responsible Research and Innovation. In: J. Van den Hoven, N. Doorn, T. Swierstra, et al. (eds.), *Responsible Innovation 1. Innovative Solutions for Global Issues.* Dordrecht: Springer Science and Business Media.

Weil, D., Graham, M. and Fung, A. (2013). Targeting transparency. *Science*, 340, pp. 1410–1411.

Wickson, F. and Forsberg, E. M. (2015). Standardising responsibility? The significance of interstitial spaces. *Science and Engineering Ethics*, 21(5), pp. 1159–1180.

Williams, B. (1984). *Moral Luck.* Albany, NY: Suny Press.

Wojinak, J. (2017). Responsible Research and Innovation. How to put gender equality into practice? *Zarzadzanie Publiczne*, 38(2), pp. 163–176.

Wynne, B. (1993). Public uptake of science: a case for institutional reflexivity. *Public Understanding of Science*, 2(4), pp. 321–337.

Wynne, B., Felt, U., Callon, M., Gonçalves, M. E., Jasanoff, S., Jepsen, M., Joly, P. B., Konopasek, Z., May, S., Neubauer, C., Rip, A., Siune, K., Stirling, A. and

Tallacchini, M. (2007). Taking European Knowledge Society Seriously. *Report of the Expert Group on Science and Governance to the Science, Economy and Society Directorate*. Directorate-General for Research. European Commission, Brussels.

Zwart, H., Landeweerd, L. and Van Rooij, A. (2014). Adapt or perish? Assessing the recent shift in the European research funding arena from 'ELSA' to 'RRI'. *Life Sciences, Society and Policy*, 10(11), doi: 10.1186/s40504-014-0011-x

2 Locomotive breath? *Post festum* reflections on the EC Expert Group on Policy Indicators for Responsible Research and Innovation

Roger Strand and Jack Spaapen

2.1 Introduction: the alien phenomenon of RRI

The work of historians and archaeologists is very hard. With great effort they search archives, buildings, monuments and excavation sites for traces from the past. Worse, somehow they need to make sense of these traces of alien practices, cultures and institutions, inscribed with forgotten signs and languages.

Out of the many endeavours and events of the 21st century with which the historians of the future might struggle, this chapter will provide an early account of a particular episode in the development of the governance of scientific research and (in particular, technological) development. The episode on our mind is that which was called "Responsible Research and Innovation" – RRI – and more specifically, the attempts at making RRI a powerful concept in research and innovation (R&I) policy in the European Union (EU). One of these attempts was the creation of an expert group at the European level, called the Expert Group on Policy Indicators for Responsible Research and Innovation. The group worked intensively throughout 2014 and their final report was published in 2015 (Strand et al., 2015). This chapter is not at all a neutral and disinterested account of the work of this expert group; in fact, the authors were strongly involved in the work, as "rapporteur" (Jack Spaapen) and "chair" (Roger Strand), respectively.[1] Rather, we provide a first-hand narrative with some reflections on the results and the (lack of) translation of these results into policy and governance, at least to date, mid-2019. We have chosen to do so not only to enrich the archives for the historians of the future but also because we believe these reflections to be relevant for all those who work to develop the governance of R&I. From the perspective of 2019, it seems that although the concept of RRI might not survive for long – this we cannot know – the underlying structural challenges it was made to address will not go away any time soon. In other words, the future might be in the past when it comes to the basic ideas of RRI.

The blunt way of stating these structural challenges is to liken science to a runaway train, hijacked by policy and industry. At the beginning of the 21st century, many countries and regions held economic policies that emphasized the role of science in relation to larger innovative processes. In many countries programmes were set up to connect scientific developments closer with industrial sectors in society and/or with wider societal challenges. In the Netherlands, the Top sector policy may serve as an example. Through this policy, nine economic sectors were identified that were deemed of vital importance for the Dutch economy, and a funding scheme was developed for intense collaboration between academic researchers and industry or public organisations (Government, s. a.). At the EU level, large parts of subsequent framework programmes were devoted to connecting science and industry and society in similar ways, culminating in the H2020 framework programme where, apart from the European Research Council section, all other parts of the budget are reserved for industry and the Grand Societal Challenges. Whereas scientific development had been taken as a hallmark of societal progress for at least two centuries (Rommetveit et al., 2013), it was mainly after the end of World War I (1918) that science became seen as an engine for economic growth (Godin, 2006). An important expression of this political imaginary was Vannevar Bush's (1945) report "Science: The Endless Frontier". Geographical frontiers, such as the Pacific Coast, had been conquered, thanks to explorers like Lewis and Clark, and other frontiers had to be given up at the entrance into post-war, post-colonial times. Bush's formula for conquering the "endless frontier" was that generous support to basic science would lead to technological inventions, new consumer products and therefore sustained employment and economic growth. At the turn of the 20th century, the economic policies of industrialised and wealthy nations emphasised not only being generous to science but also actively pushing research towards industrial innovation, increasing innovation rates and thereby keeping or gaining competitive advantage over other nations. In this political climate, Schumpeter's idea of "creative destruction" had been accepted and transformed into a maxim of accelerating and "disruptive" innovation, which was seen by governments as inherently desirable (notwithstanding the social side-effects of market disruptions).

The desire for the New was slowly contrasted with concerns about the New. Nuclear technology became controversial after the bombing of Hiroshima and Nagasaki, and later stayed controversial for different reasons after disasters with reactors such as in Three Miles Island (1979), Chernobyl (1986) and Fukushima (2011). More generally, concerns started to grow over the role of Science in other societal issues such as the defence industry, in industrial pollution, and from the 1980s, in engineering of life with the aid of molecular biology and biotechnology. Initial concerns were voiced outside of the political institutions and arenas and often labelled as anti-progress (religious, Luddite, irrational, pre-modern, New Age etc.). From around the 1980s, however, concerns also found expression in civil society which could be voiced inside the political institutions, notably as "ethics", "technology assessment", "risk assessment" and "public engagement" with science.

These different expressions co-existed and grew in force towards the end of the 20th century (Greenpeace and other non-governmental organisations concerned with the environment or the food industry or health issues) and became in many European countries part of the regular debate about the role of science in society, although in other countries remained more marginal in the policy discourse. They were somewhat supported, though, in a few national contexts and also through the minor "Science-and-Society" and "Science-in-Society" programmes within the EU framework programmes.

In 2011, the term "Responsible Research and Innovation" was introduced by philosopher and EU DG RTD (Directorate-General for Research and Innovation) policy officer René von Schomberg as he was summing up results from Science-in-Society research in FP7, the seventh EU Framework Programme for Research and Innovation. He defined RRI as:

> *a transparent, interactive process by which societal actors and innovators become mutually responsive to each other with a view on the (ethical) acceptability, sustainability and societal desirability of the innovation process and its marketable products.*
>
> (von Schomberg, 2011, p. 9)

The term was soon adopted by DG RTD into its policy. By political intervention it was even made into a "cross-cutting principle" of the eighth framework programme, the so-called "Horizon 2020". In von Schomberg's initial mention, the potential of science and technology to cause harm was explicit:

> *we are confronted with the Collingridge dilemma, implying that ethical issues could be easily addressed early on during technology design and development whereas in this initial stage the development of the technology is difficult to predict. Once the social and ethical consequences become clearer, the development of technology is often far advanced and its trajectory is difficult to change.*
>
> (von Schomberg, 2011, p. 8)

Von Schomberg thus introduced the reference to Collingridge (1980) and his dilemma of the social control of technology into EU policy. Elsewhere, von Schomberg argued for the need for a new paradigm of governance. Faced with the accelerating runaway train of science and technology development, law, ethics and the market were all too slow and weak and they arrived late to the scene of the accident. Scholars such as von Schomberg argued that the new paradigm – whether called RRI or something else – could be created by joining the theoretical and practical forces of the various strands of science critiques – philosophy, ethics, feminist critique, studies of risks and hazards, participatory technology assessment and many others. The new paradigm was envisaged as "soft governance": not strict laws – how can you regulate that which still is not discovered or even imagined? – but rather (more or less) voluntary self-governance by scientists and technologists who would learn to deeply reflect on the prospects of doing good as well as harm to society and the world and who would also deeply engage in deliberation with citizens, not least those who might suffer from the potential harms. The paradigm was imagined in terms of more or less constructivist philosophies of science and technology. Science was not seen as a fully deterministic process that merely discovers the Truth about Nature, but rather as a human, social and cultural practice that could be shaped and influenced by politically motivated decisions. Helga Nowotny, Peter Scott and Michael Gibbons (2001) coined the term "socially robust knowledge", referring to their earlier work on how the social contract of science had changed by the closer intertwining of science, industry and government. Socially robust knowledge, in their definition, would be knowledge that combined epistemic and social values – being scientifically sound, but also tested in societal practice. A similar concept of "post-normal science" was previously introduced by Funtowicz and Ravetz (1993).

When the European Commission (2012) adopted RRI as a policy principle, the result showed little evidence of regard to the Collingridge dilemma and the constructivist philosophies. Instead, it translated the intellectual traditions that formed the basis of RRI into operational tasks, or "keys" as they became known. Accordingly, the policy leaflets and explanations from the European Commission described RRI as something that could be verified by a box-ticking exercise:

1. Were there ethical issues, and were they taken care of in an appropriate manner?
2. Were there gender issues, and were they taken care of in an appropriate manner?
3. Did the R&I project engage with civil society?
4. Did it also educate the public to improve the quality of the social dialogue?
5. Did it provide open access of its results to the public?
6. Was there an appropriately responsible governance of the R&I project? (This "key" actually held the potential to redirect the attention to von Schomberg's point and was mostly left undefined or even omitted.)

This was often called the "keys'" approach to RRI – popular within DG RTD and by extension relevant to the many clients (including ourselves) who relied on the wealthy patronage of the Directorate-General to fund their research projects on and around RRI during the latter phase of FP7 and subsequently Horizon 2020. Still, within the intellectual RRI community that formed, von Schomberg's definition and similar approaches, such as the "AREA framework" endorsed by the UK Engineering and Physical Science Research Council, were the ones considered in line with the original rationale of at least trying to deal with the runaway train. What developed was a sort of subversive humanism, usually the mildest sort, in which the formal deliverables of FP7 and Horizon 2020 RRI projects complied with the five- or six-keys approach and delivered results on them, while the academic outputs – written and oral – took a freer stance towards the keys.

Sometimes the subversive humanism also included attempts at simultaneously embracing the key approach and the philosophical origin of RRI and somehow integrating them or expanding the former with the latter. The Expert Group on Policy Indicators for Responsible Research and Innovation was an example of such an attempt. Its mandate, as specified by DG RTD, was "to help the Commission identify existing indicators and to propose new indicators that can measure impacts of RRI activities in qualitative and quantitative terms" (Strand et al., 2015, p. 15). The terms of reference defined the aim of RRI as to "improve the societal alignment of R&I, in particular by improving public engagement, science education, gender equality, open access to scientific information, ethics and governance", that is, in terms of the six keys. Furthermore, the mandate defined the need for indicators in terms of an underlying agenda of evidence-based policy-making, demanding that indicators be "SMART" (an acronym which in this particular context meant "specific, measurable, attainable, relevant and timely"). In other words, the mandate was quite far from von Schomberg's vision of philosophically informed self-governance among researchers who sought reflection and deliberation in civil society. Our group went to the task with the same mild brand of subversive humanism as could be observed in many of the RRI/Science-with-and-for-Society projects conducted in the same period. We did propose indicators, and they did address the six keys. However, our indicators were hardly SMART in the above sense and in few cases quantitative or easily measurable for a policy directorate. Indeed, we politely opposed the agenda of evidence-based, top-down command-and-control style of governance and rather proposed a large suite of indicators together with an approach to selecting among them that re-introduced the ideal of self-governance. In that sense, our approach was "smart" in the street-wise sense of the word; we acknowledged the contextual character of R&I by presenting a toolbox of indicators, mostly qualitative, from which the community of researchers and stakeholders could choose those relevant for their project. The context of health research is completely different from the context of global warming or migration or electromagnetic radiation by mobile telephone masts. And it is the community of researchers and stakeholders that is best to choose the most relevant indicators. Our Expert Group therefore chose a slightly more subversive strategy than the concurrent\ Monitoring the Evolution and Benefits of Responsible Research and Innovation (MoRRI) project, which actually produced measurable indicators that apparently could work in a command-and-control type of governance system. In doing so, we could be seen to find inspiration in the report of a previous expert report within the same area, called "Taking European Knowledge Society Seriously" (Felt et al., 2007). That report could hardly be seen to answer the operational questions it was asked and hence got little direct policy impact in the short term. Its depth of analysis, however, had made it somewhat of a classic within the academic community and still a source of inspiration a decade after it was produced.

2.2 The results from the Expert Group: a suite of indicators for a network approach

Even outside the logic of control-and-command, the question of monitoring and devising indicators can make good sense. Also in bottom-up self-governance it is reasonable to ask if a certain action or activity was successful and met the purpose it was intended to fulfil. To reflect upon one's criteria for success is worthwhile irrespective of model of governance. In the case of RRI, there have been various proposals for RRI indicators. The outcomes of the MoRRI project are described elsewhere in a wealth of documentation.[2] Wickson and Carew (2014) were not obliged to stay with the six keys and presented an indicator approach closer to the academic literature on RRI.

As mentioned above, the Expert Group chose to try to integrate the "philosophical" or "intellectual" concept of RRI, based on von Schomberg's definition, with that of the six keys. The former was given priority in the sense that it gave direction to the choice of monitoring framework and the content of indicators. For instance, the Expert Group interpreted "ethics" as a key of RRI in terms of those issues and aspects of RRI (philosophically defined) that pertained to ethics or happened to be actualised in practices and institutions with the ethics label. Following this argument, the ethical acceptability of R&I outcomes was seen as the main ethics dimension of RRI, whereas the typical issues of medical research ethics such as informed consent fell outside the scope. How such issues (such as informed consent or more generally ethical clearance) were dealt with institutionally, however, and by which mechanisms, could be an RRI ethics issue. Similar arguments were carried out for the other keys.

Giving the philosophical concept of RRI priority also implied the need to deal creatively with the idea of a monitoring framework to make it compatible with the appropriate understanding of governance: not command-and-control but an appreciation of the role of new and broader dialogues between science and civil society. The approach chosen was largely inspired by the concept of productive interactions developed in the FP7 SIAMPI project, referring to societal change being the consequence of interactions between researchers and stakeholders in society, defined as a dynamic network around a societal issue like, for example, global warming, renewable energy or migration (Spaape and van Drooge, 2011).[3] This was stated in the executive summary of the Expert Group report as follows:

> *This means that the group sees RRI from a network perspective, consisting of stakeholders jointly working on a set of principles guided by the RRI keys. When monitoring RRI policy, the prime focus therefore*

> *should be on the development of RRI agendas in these networks. In other words, the focus of monitoring and indicators should be on the governance of RRI in these networks, and in particular on what kind of RRI policies are developed (best practices). Secondly, one should discuss the kind of indicators best fitting research and innovation (R & I) practices and goals within these networks, and then decide upon a limited set of indicators (because things have to be manageable). Such indicators are thus highly contextual, meaning that there will be no one list for all.*
>
> (Strand et al., 2015, p. 5)

The strategy of the Expert Group was accordingly to propose a large set of more or less practicable indicators that could give ideas and inspiration not only to the European Commission but to anyone who might be interested in monitoring the success of RRI activities in a learning process.

2.2.1 *What to indicate?*

Another design issue, perhaps a less controversial one, was the question of what the indicators were supposed to indicate. RRI activities and initiatives could be seen as a means to an end, namely to change R&I practices so that they would be more "responsible" in the RRI sense. In principle, one might have a theory or a body of evidence that already specified what such means would have to be in order to reach that end. As a matter of fact, one did not. The causality between RRI means and RRI outcomes was not established. On the other hand, measuring the degree of responsibility of R&I practices was not a trivial exercise, either. The Expert Group opted for a hybrid model:

> *To be responsible in general and in the specific terms of RRI includes three dimensions: performance, perception and key actors. RRI performance depends both on the processes that promote RRI activities and on the effects that these processes have: outcome. Acting responsibly defines who we are – we are acting in a certain manner (performance). But responsibility also includes the key element of perception – to be seen to act responsibly. The weight given to process, outcome and perception indicators for a given issue depends upon the nature of that issue. Again, what we offer here is a generic and flexible framework that in itself can be applied responsibly as well as irresponsibly.*
>
> (Strand et al., 2015, p. 16)

2.2.2 *What about substantive values such as sustainability and social justice?*

In addition, there was the issue of the comprehensiveness (or rather the lack thereof) of the six keys. In line with the strategy of "mild subversive humanism", the Expert Group kept the keys but presented an argument for extending the framework with two additional keys: sustainability and social justice/inclusion. Although the argument never gained any traction, at least not until the writing of this chapter, we took some pride in how it was crafted. It was presented as a corollary of the Europe 2020 strategy of the EU, an overarching policy of which actually Horizon 2020 was an act of implementation. The slogan of Europe 2020 was "smart, inclusive and sustainable growth" (Strand et al., 2015, p. 11). The Expert Group argued that, if R&I was to give growth, then inclusive and sustainable R&I was needed to yield inclusive and sustainable growth. While perhaps not an entirely valid syllogism, the argument was expressed within the orthodox EU policy discourse at the time; within what a Rawlsian would have called the comprehensive normative doctrine.

The resulting framework could be presented as a matrix (Table 2.1).

The remaining task was to populate the cells. The Expert Group did so with creativity, with little regard for the statistics that were already sitting there in DG RTD, in part because we expected the MoRRI project to take a more pragmatic approach, and in part because the policy winds inside DG RTD were blowing away from, or even against, RRI towards the end of 2014. As it became ever less likely that the indicators would actually be adopted by DG RTD in their daily work, it appeared ever more important to present something that could inspire RRI governance "out there", in national research funding organisations and in self-governance in research-performing organisations. Again, the ethics indicators may serve as an example of the creative approach. The following were the proposed indicators for the process dimension of the RRI ethics key (excerpts from report) (Strand et al., 2015, pp. 34–35):

- *Documentation of institutional attention to normative tensions related to research integrity policies and actions. (This may manifest itself in a*

Table 2.1 The indicator framework of the Expert Group report

Criteria	*Performance indicators*		*Perception indicators*	*Key actors*
	Process indicators	*Outcome indicators*		
Public engagement				
Gender equality				
Science education				
Open access				
Ethics				
Governance				
Sustainability				
Social justice/inclusion				

Source: Reproduced from Strand et al. (2015).

number of ways, e.g. activities such as open meetings or seminar series on ethics and research integrity and the inclusion of such discussions in research strategy papers, research proposals.) This is a qualitative indicator that will have to be based on strategic sampling of information – process indicator.

- *Documented change in R&I priorities (research or research funding) attributable to multi-stakeholder and/or transdisciplinary processes of appraisal of societal relevance and ethical acceptability (presence/frequency; qualitative descriptions; best practices) – process indicator.*
- *Presence of mechanisms for multi-stakeholder and/or transdisciplinary processes of appraisal of societal relevance and ethical acceptability (presence/frequency; qualitative descriptions; best practices) – process indicator.*
- *The formal and actual scope of the ethics review/IRB [Internal Review Board] clearance (e.g. whether a committee or a reviewer is free to identify any type of ethical issue or concern or has to limit the analysis to a predefined set, and the degree to which this freedom is enacted) – process indicator.*

The first entry of this list illustrates again how the "philosophical" concept of RRI informed the design. Research integrity as such might be an issue of research ethics but, in this understanding, not of RRI. Rather, the RRI issue is whether the R&I actors and practices are aware of the tensions between the theory and practice of research integrity, which again was seen to be related to the tensions between Mertonian norms and a conception of science as a fully autonomous, ivory tower-type institution on one hand, and the social contract of science that was envisaged by the more constructivist or at least coproductionist views on science that informed RRI thought, on the other. To be concrete, installing and applying ethics guidelines for co-authorship is a research integrity issue. To reflect and act upon the tension between such ethics guidelines and the real-world pressures and demands for strategic authorship in a highly competitive research field is an RRI issue. Even just to understand what this indicator referred to, not to speak of how to measure it, demanded insight into these theoretical issues. But why should such insight not be demanded? Why would it be appropriate to allow people to govern science if they have no theoretical understanding of the process of how science is organised and carried out?

2.3 What does it mean to take a "network approach"?

The preceding paragraph ended with two rhetorical, tendentious questions. Of course we stated them because we believe that it is *inappropriate* that science is governed by individuals and institutions who lack a theoretical understanding of science. Such an understanding cannot be gained in a short while or by a short text; a proper learning process would require the study of STS scholars such as Sheila Jasanoff, Bruno Latour, Donna Haraway and Brian Wynne (to mention a few) in addition to the already-mentioned Nowotny, Gibbons, Funtowicz, Ravetz, Godin and others. It is only fair, however, that the authors of this text try to indicate what we mean by a "network approach" and a "coproductionist perspective" to those readers to whom such concepts are alien. We shall do that by providing our own, possibly somewhat idiosyncratic, outlook on the governance of science at the time of writing.

At the beginning of the 21st century, in countries with a functional research sector, institutes operating within the academic sphere (universities, academy institutes and other) also perform their research in the context of issues and challenges in policy or society at large. The networks that are formed by these wider communities are as a rule more diverse and often also more temporary than traditional academic networks. Both researchers and policy makers have to review their perspective on this position, both in terms of how they interact with the broader environment (for example, with regard to their R&I agenda) and in terms of how they assess the success of new forms of collaboration. The networks are characterised by a variety of scientific stakeholders (various disciplines) and stakeholders from society, be it industry, government or society at large. Somehow, all these different backgrounds, interests and work practices have to be attuned in new arrangements in which goals, performance and monitoring and evaluation have to be elaborated.

Research that aims at helping innovate society in a responsible way, whether this is through new ideas, technologies or new forms of organisation, new insights or processes, or otherwise, is thus mostly the outcome of an intricate process in which a multitude of actors operate, with participants coming into the collaboration with different perspectives, interests and expectations. The mix of partners differs per sector and per subject, and so will the weight of different RRI keys. Sometimes there is a small enduring network of established actors, with a high level of coordination, for example in public health research where researchers collaborate with specific governmental bodies, hospitals, general practitioners and patient organisations, and sometimes industry (Spaapen and van Drooge, 2011, p. 214). Typically, the research agendas in such networks are developed in close collaboration with the relevant stakeholders who have more or less the same goal. It is relatively easy to agree about which RRI factors are relevant and how they should be applied. In other cases, this is not so. Networks can be much more open and dynamic, with stakeholders coming in and going out, and having rather different goals in mind. One example is the case of energy transition, where goals and development paths can lie far apart (nuclear energy versus renewable energy) and stakeholders feel less connected to each other (say, the car industry and Greenpeace). But also in those cases, RRI should be part of the agenda development. As we write, in 2018, *automated cars* are emerging as an issue of growing importance and potential for controversy. We shall take that issue as an example of a diverse network filled with different interests.[4]

The automobile industry in Europe and the USA is starting to invest in the development of electrical and self-driving cars, from Ford's $1 billion investment in an artificial-intelligence company to Uber's 2016 purchase of self-driving truck company Otto. Tesla, Nissan and BMW all say they will have fully driverless cars by 2021. And a report from the Boston Consulting Group suggests that by 2030 more than 5 million conventional vehicles could be replaced by automated ones in the USA alone (Lang et al., 2017).

Many European governments have set up policies to ban fossil fuel cars in the next two decades or less. Obviously, a main driver for this development is the necessity to reduce the level of CO_2 reduction in line with the Paris agreements. But the new transportation system will have a major influence on many other aspects of society. It will be a relief to drivers who have to commute every day from home to work or elsewhere, and then it will most likely influence the behavioural patterns of these drivers. To predict what will happen is however very difficult since, as it turns out, the environmental impact of automated vehicles is extremely uncertain. One report from the American Department of Energy found that automated vehicles could reduce fuel consumption for passenger cars by as much as 90%, or increase it by more than 200% (Worland, 2017).

The wide range of potential outcomes is the result of a long list of variables about how a future with automated cars will take shape. These variables partly lie in the scientific and technological options that will be developed, and partly in the behaviour and interests of a multitude of societal stakeholders, ranging from industry to governments to consumer organisations to the general public. Many researchers expect that automated cars will lead to a sharp increase in the average miles travelled by a given vehicle, thanks to the human factor. Key barriers to hopping in a car – fatigue, age or intoxication, to name a few – will disappear, they expect, and car owners will be free to travel further and more frequently. Workers may choose to live even further away from the office, opting to sleep in the car or use that time to squeeze in a workout. And, once in the city, car owners might instruct their vehicle to drive around in circles rather than pay for parking. And there is more. Beyond changed driving patterns, simple technology advances will reduce the environmental toll of automated cars. Most importantly, engineers say that the largely accident-free vehicles can eliminate safety equipment, such as antilock brakes and airbags, that has increased the weight – and fuel consumption – of vehicles. Automated cars can also travel closer together, allowing them to take advantage of aerodynamics. Trucking fleets are already trying to take advantage of this fuel-saving measure.

With all the insecurities the question is how to (try to) steer – *govern* – this important development forward. Who are the key players and how will they relate to each other, and will they develop, and carry out, an R&I

agenda in a responsible way? Researchers alone cannot do this. They have sought to model how humans might respond to automated driving using surveys, driving data and lab experiments, but ultimately the sheer number of choices and assumptions involved in transportation has made reaching concrete conclusions about driving behaviour difficult. Regulators cannot do it. While they represent the obvious way to protect against the potential environmental downsides of automated vehicles, they will not know what the most efficient way is to influence drivers' behaviour as long as they are uncertain about technical and social developments. The automobile industry cannot make the transition without changing their business model that historically has been founded on fossil fuel cars. Clearly, the network around this issue is much wider (for example, how will enough electricity be produced without negative effects on the environment, and by whom?), and the actors in this network will be confronted with RRI issues on a more or less permanent basis.

To sum up: there are many uncertainties in the scientific and societal developments of automated cars and road systems. The future will be decided via a mix of technical solutions, human behaviour and political and business decisions. Many RRI keys play an important role in many of the discussions between the different stakeholders: how to engage civil society, educating the public to strengthen social dialogue, ethical and gender issues. But perhaps most of all the question how to govern such a process in a way that is inclusive and socially just. While scientists and policy makers say reducing emission figures will be key to addressing man-made climate change, there is a wide range of potential outcomes and this is the result of a long list of technical, political and human variables about how a future with automated cars will take shape.

We see an area where there seems to be a lot of consensus about the necessity of reducing the emission of CO_2 as a prime cause of global warming and climate change. But at the same time, there is a lot of uncertainty how to reach those goals in a responsible way in a societal network with lots of different interests and uncertainties. The main challenge is to develop a process of development that is based on only responsible governance. Arguably, this is more difficult to manage in the latter case than in the case of public health, mentioned before, where we see a more or less stable network with partners close to each other.

2.4 RRI: a sinking ship?

Whether such a responsible development will be possible in the USA of the early 21st century remains to be seen. For those who still believe in so-called European values, as professed in the EU Charter on Fundamental Rights, including the present authors, it seems that RRI ought to be part and parcel of European development. Listening to policy makers in Brussels, or reading the research policies at the entry into the ninth Framework Programme for Research and Innovation, however, RRI seems to be a sinking ship, if not already at the bottom of the Atlantic ocean. Still, if Europe wants to be different from the rest of the world, more democratic, more social, more ethical, etc., then it is difficult to see how one could avoid the values and principles informing RRI in one way or another as part of the development of R&I in a responsible society. To do this effectively, a new thinking about agenda setting and evaluation of network activities is necessary. In short, this should be based on the combination of a theory of change and a convincing narrative to explain both the main goal of R&I and the ways how to achieve these goals.

Evaluation of R&I in the 21st century is therefore both demanding and exciting. It is exciting because the pace of scientific research speeds up by the year thanks to the growth of science and global collaboration. Adding to that is the influence of digitisation, enabling new ways of cooperation, and collecting data and exchanging them globally, the growing need for inter- and transdisciplinary collaboration (especially between human and technological scientific fields as machines and human bodies are becoming ever more connected; see Strand and Kaiser, 2015)) and the worldwide challenges that are in desperate need of scientific input (climate change, energy transition, genetic engineering, migration and political crises). One could say that science is finally becoming a truly international trade.

It is demanding too, because a broad variety of stakeholders (from policy, industry and wider society) is pervading university life at an ever faster pace. Academic researchers are required to think about the relevance of their work for those stakeholders, and need to engage in a negotiation process with them. There is no escape here, because more and more research funding is allocated in ways that require societal impact, both at the national and European level. Europe spent about 80 billion euros in the current framework programme H2020 and two of the three main parts – the Grand Societal Challenges and industrial leadership) directly promote collaboration between science and society. At the national level governments design large programmes to stimulate collaboration between science and society, like, for example, in the Netherlands the Top Sector policy and the Science Agenda (Dutch National Research Agenda, s. a.). In this, demands for societal impact and public engagement come together, the Dutch National Research Agenda being the result of a nationwide public consultation about what would be the most urgent questions that science could help solve in society. Solutions will have to come from collaboration between researchers and variegated stakeholders in society. To assess these kinds of large multi-partner network activities, the Dutch have a comprehensive evaluation method available, the Standard Evaluation Protocol (SEP).

2.5 … or a Flying Dutchman?

The national evaluation framework for publicly funded research in the Netherlands may serve as an example of comprehensive evaluation which

Table 2.2 Overall framework

	Scientific quality	*Relevance to society*
Demonstrable output		
Demonstrable use		
Demonstrable recognition		

includes RRI-like elements. Operating since 2003, the SEP is now in its third version, running from 2015 to 2021. The philosophy behind the SEP is that: (1) evaluators should balance the assessment of scientific quality and societal relevance; and (2) SEP offers an overall framework for all scientific fields, but within that all fields should be able to use indicators that are appropriate for the research practices of a particular field. In other words, there is room for a bottom-up development of indicators as long as they fit into the overall framework. The overall framework consists of two main criteria, scientific quality and societal relevance, with three indicator categories for both: output, use and recognition (Table 2.2).

Next to these two main evaluation criteria, SEP includes a number of other criteria that are part of the overall assessment framework. These are viability, PhD programmes, diversity and research integrity. With the exception of PhD programmes, all of these criteria overlap with the RRI keys set by the EU. Viability is meant as a management criterion which includes governance issues, diversity includes gender issues and research integrity includes ethical issues.

Research entities to be evaluated have to produce a self-evaluation report that regards all of these elements, preferably via a narrative highlighting the connection between the scientific goals and their relevance to society. Indicators may be used to underpin this narrative, but, as mentioned above, the SEP does not prescribe which indicators to use. There are two important ideas behind this: one is that there is not one set of indicators which is useful for all fields, and the other is that the research community should be allowed to take responsibility by choosing how to represent its research production and the ways it interacts with stakeholders in society. The main difference with other evaluation systems is that, in the SEP, the responsibility is placed with the community of researchers and stakeholders to enable self-governance.

RRI, in von Schomberg's (2011) definition, fits in this approach of comprehensive evaluation. As mentioned above, some of its main keys have been introduced in the national evaluation process, SEP 2015–2021, like diversity/gender issues and ethics/integrity.

To assess research in this broader context, rather different evaluation regimes than the current ones are necessary. Innovative approaches can be found in a limited number of countries, for example the Scandinavian countries, the UK, Canada, Australia and the Netherlands. Building on the possibilities that SEP 2015–2021 offers, the humanities faculties have developed a novel approach which is based on the *narrative* as a central element in the self-evaluation report. The idea is simple: tell the world why it is important that you are doing the research you are doing. To do this in a convincing way for a wider audience, you need to clarify the connection between the scientific goals and the possible impact results can have on society. And you need to show how you envisage engaging stakeholders and how you collaborate with them. To be short, research is part of a bigger innovation process, and its role in that process has to be shown.

In principle, this is a qualitative approach, but the narrative should be supported by concrete evidence of activities in the realms of science and society. The purpose of this new kind of evaluation is different from traditional evaluations: it is not competitive and ranking-oriented (who publishes the most), but oriented towards collaboration and mutual learning: how can we best bring together necessary knowledge and expertise from science and society to solve major problems in the world?[5]

2.6 Off track and into the sea

We can imagine how the historians and archaeologists of the future might conclude how governance of science emerged as a particularly difficult societal challenge in the 20th and 21st centuries, characterised by high but unclear stakes, partial understanding, disputed criteria and sometimes outright confusion. Metaphors have their benefits and risks when understanding is partial. Metaphors can help to give a first, tentative grasp of the situation but may also direct the attention too much and effectively limit it. "Governance" is already, by its etymology a nautical metaphor, indicating that science is a ship to be navigated. It is not, if at all, on the sea, it is a huge fleet, and in a sense it is not too wrong to think of the marginal attempts of imposing RRI as top-down policy concept as the attempt to launch another little ship into that fleet. Playing on that metaphor, it was little wonder that it sank, and better prospects of success if it could become a network approach Flying Dutchman, that is, a ghost that might haunt, fascinate and inspire the minds of the thousands of sailors and fellow travellers who together constitute the collective governance of science of the future. Whether that ghost will carry the name RRI is not necessarily as relevant. The underlying issues of difficult collective choices will remain.

Indeed, in order for human historians and archaeologists to exist in the future, the governance of science will have had to improve. Initially in this chapter, we likened science to a runaway train, hijacked by policy and industry. This is an appropriate metaphor in order to emphasise the seriousness of the situation. Indeed, just by compiling a list of some of the products of science and engineering in the 20th century – nuclear technologies, plutonium, chemical and biological weapons, surveillance technologies, genetic engineering of bacteria and virus, increasing CO_2 levels, microplastics, organic pollutants, algorithms that know nothing but pretend to know everything about us, etc. – it seems quite surprising that human civilisation did not collapse already in that century (or is the downfall happening now with some delay?). How much more can it take of irresponsible technological development?

The problem with the metaphor of the runaway train, however, is that it does little but consolidate a sense of disempowerment, as in the Jethro Tull rock song that gave the title to this book chapter, *Locomotive Breath*.

To create support for expansive research funding, 20th- and 21st-century policy makers crafted their policies within the discourse of technological optimism and scientific determinism: science is a train, its destination is Utopia and we should simply decide to go faster. As the Utopia becomes dystopic, optimism flips into pessimism and despair, as in the song lyric.

However, science is not a train, there are no rails and there is no predetermined trajectory. To admit that is the first step towards a more proper governance. The next step is to admit that there is no captain but rather a fuzzy, diverse, self-organised network of sailors and travellers with sometimes rather different visions of what can or should be done (think of the Chinese researcher He Jiankui and his genetically designed twin babies: Regalado, 2018). We hypothesise that the future historians who encounter this text will find these claims to be self-evident, otherwise they won't be there to encounter it. For them, then, the alien character of the early 21st century was that it was a time when these simple insights were controversial and could only to a very little extent be adopted into governmental policies and governance practices.

Notes

1 Our observations and reflections are entirely our own and should not in any way be taken to represent the views of other members of the mentioned expert group or individual fonctionnaires of the European Commission, nor seen to express the official views of the European Commission at any point in time. One author (Strand) has performed the work with this text as part of the project SuperMoRRI. This project has received funding from the European Union's Horizon2020 Research and Innovation Programme under grant agreement no. 824671. We are obliged to clarify that the Funding Agency cannot be held responsible for any use that may be made of the information the text contains.

2 All MoRRI deliverables can be accessed from http://www.morri-project.eu. The full indicator report, called Deliverable 4.3, can be found at http://www.technopolis-group.com/wp-content/uploads/2018/02/D4.3_Revised_20022018_clean.pdf

3 The FP7-funded SIAMPI project developed indicators to assess the social impact of research. More information can be found on the project website: www.siampi.eu

4 This example is adapted from an article by Justin Worland in *Time* of 16 November 2017, Why self-driving cars might not lead to a huge drop in fuel consumption.

5 The website for this approach was developed for the humanities, but may serve as inspiration for all fields: www.qrih.nl/en

References

Bush, V. (1945). *Science: The Endless Frontier*. United States Office of Scientific Research and Development, U.S. Government print office, Washington, DC.

Collingridge, D. (1980). *The Social Control of Technology*. Pinter, London.

Dutch National Research Agenda (s. a.) *Dutch National Research Agenda*. Retrieved from: https://wetenschapsagenda.nl/?lang=en.

European Commission (2012). *Responsible Research and Innovation: Europe's ability to respond to societal challenges*. DG Research and Innovation. DOI:10.2777/95935

Felt, U., Wynne, B., Callon, M., Gonçalves, M., Jasanoff, S., Jepsen, M., Joly, P., Konopasek, Z., May, S., Neubauer, C., Rip, A., Siune, K., Stirling, A. and Tallacchini, M. (2007) Taking European knowledge society seriously. Report of the Expert Group on Science and Governance to the Science, Economy and Society Directorate, Directorate-General for Research, European Commission (EUR 22700). Retrieved from https://ec.europa.eu/research/science-society/document_library/pdf_06/european-knowledge-society_en.pdf.

Funtowicz, S. and Ravetz, J. (1993). Science for the post-normal age. *Futures*, 25: 739–755.

Godin, B. (2006). The linear model of innovation. The historical construction of an analytical framework. *Science, Technology, & Human Values*, 31 (6): 639–667. DOI: 10.1177/0162243906291865.

Government (s. a.) *Encouraging innovation*. Retrieved from www.government.nl/topics/enterprise-and-innovation/encouraging-innovation.

Lang, N., Rüßmann, Chua, J. and Doubara, X. (2017) *Making Autonomous Vehicles a Reality. Lessons from Boston and Beyond*. The Boston Consulting Group, Boston, MA.

Nowotny, H., Scott, P. and Gibbons, M. (2001). *Re-thinking Science*. Polity, Cambridge.

Regalado, A. (2018, November 25). Exclusive: Chinese scientists are creating CRISPR babies. *MIT Technology Review*. Retrieved from www.technologyreview.com

Rommetveit, K., Strand, R., Fjelland, R. and Funtowicz, S. (2013). *What can history teach us about the prospects of a European Research Area?* (EUR 26120). Retrieved from https://ec.europa.eu/jrc/en

Spaapen, J. and van Drooge, L. (2011). Introducing 'productive interactions' in social impact assessment. *Research Evaluation*, 20 (3): 211–218. DOI: 10.3152/095820211X12941371876742.

Strand, R. and Kaiser, M. (2015). *Ethical Challenges Report on Ethical Issues Raised by Emerging Sciences and Technologies*. Council of Europe, Committee on Bioethics, Strasbourg.

Strand, R., Spaapen, J., Bauer, M. W., Hogan, E., Revuelta, G., Stagl, S., Paula, L. and Pereira, Â. G. (2015). Indicators for Promoting and Monitoring Responsible Research and Innovation. Report from the Expert Group on Policy Indicators for Responsible Research and Innovation. (EUR 26866). Retrieved from http://ec.europa.eu/research/swafs/pdf/pub_rri/rri_indicators_final_version.pdf

von Schomberg, R. (2011). Introduction. In R. von Schomberg (Ed.), *Towards Responsible Research and Innovation in the Information and Communication Technologies and Security Technologies Fields* (pp. 7–15). Publications Office of the European Union, Luxembourg.

Wickson, F. and Carew, A. L. (2014). Quality criteria and indicators for responsible research and innovation: learning from transdisciplinarity. *Journal of Responsible Innovation*, 1 (3): 254–273. DOI: 10.1080/23299460.2014.963004.

Worland, J. (2017, November 16). Why self-driving cars might not lead to a huge drop in fuel consumption. *Time*. Retrieved from www.Time.com

Part II

RRI in companies

3 Strategic responsible innovation management (StRIM)

A new approach to responsible corporate innovation through strategic CSR

Agata Gurzawska

3.1 Introduction

Over recent decades, companies have operated in complex business environments. On the one hand companies are facing fierce competition and need to achieve and manage innovation to maintain a competitive advantage and improve their financial performance and market share. On the other hand, businesses need to fulfil their legal obligations and seek to improve their corporate social responsibility (CSR) for multiple stakeholders in order to gain greater social acceptance of their activities. Despite the popularity of both innovation management and CSR most companies do not necessarily connect them and manage them strategically (Husted and Allen 2007; Husted, Allen et al. 2015). We lack a broader holistic perspective on the complex connection between innovation and CSR (McWilliams and Siegel 2006; Yin and Jamali 2016). In other words, we do not have a strategic and generic business model connecting innovation and CSR, through which innovation practices can create business value and positive societal and environmental change. The disconnectedness of these two aspects of corporate activities may lead to missed business opportunities, avoidable financial losses for a company in the long run, as well as negative economic, societal and environmental impacts. Therefore, the challenge is to innovate in a responsible way. An effective strategic approach to responsible corporate innovation would have sustainable outcomes for both business and society. I argue that intertwining innovation and CSR may bring opportunities for both business and society. In order to bring sustainable economic, societal and environmental outcomes, companies should have a strategic approach to innovation and CSR management. This chapter places the assessment of responsible innovation (including measuring, monitoring and reporting) in a broader picture of a business strategy. This analysis is crucial to understand why, when and how the evaluation and control of responsible innovation should be done in order to mitigate risks and strengthen strategic planning.

The remainder of this study is as follows: Section 3.2 defines the concepts of innovation and CSR, explores their relation to business strategy and addresses current discussions providing the theoretical understanding based on the existing literature; Section 3.3 explores the connection between innovation and CSR with a reference to the concepts of Responsible Research and Innovation (RRI), CSR-driven innovation, innovation-driven CSR and the multi-stakeholder approach; Section 3.4 derives from the strategic management literature and formulates a new approach called strategic responsible innovation management (StRIM); and Section 3.5 summarizes the findings.

3.2 Innovation, CSR and their relation to business strategy

3.2.1 Innovation

Innovation in the business context can be broadly defined as the successful application of new ideas (Dodgson, Gann et al. 2013). Various definitions of innovation exist; however this study takes a holistic approach to innovation, characterised as 'the implementation of a new or significantly improved product (good or service), or process, a new marketing method, or a new organisational method in business practices, workplace organisation or external relations' (OECD 2005). Therefore, four types of innovation can be differentiated that encompass a wide range of changes in companies' activities: product innovations, process innovations, organisational innovations and marketing innovations (OECD 2005). The crucial characteristic of innovation is that it is not a single action, a single idea or the invention of a single new device, but rather an integrated process involving various steps (Trott 2008; Conway and Steward 2009). Innovation emerges from various sources and has a multiplicity of influences, e.g. new regulations, new technological standard, collaborative partners or potential funding opportunity (Dodgson, Gann et al. 2013). However, there are two main sources of innovation, namely market pull and technology push (Bennett and Cooper 1981). Innovation originating from market pull comes from consumers' needs and identifying new market opportunities or a segment of an existing market that has been neglected (Whittington 2001; Baker 2014), whereas technology push comes from scientists and engineers, thereby scientific discovery or the availability of new technology leads to the development of a product (Bennett and Cooper 1981). Innovation is a crucial competence because it enables a company to adapt to the dynamically changing needs of the marketplace and is pivotal to the profitability and long-term survival of any company (Hauser, Tellis et al. 2006).

As far as innovation is said to be a key driver of economic development (Hanekamp 2007), it may challenge ethical values and human rights. One of Kranzberg's laws states that 'technology is neither good nor bad; nor is it neutral' (Kranzberg 1986). Recent controversies around misuse of Facebook users' data by Cambridge Analytica and potential implications for the US presidential elections in 2016 or the Social Credit System introduced in China (to score its citizens based on various data, including social media information) (Reuters 2018) raise legitimate concerns as to whether innovation, science and technology can be left to operate autonomously in the market without societal guidance and regulation. These considerations raise questions about companies' responsibilities for their innovation activities with respect to the environment, stakeholders and society. This issue is often discussed in relation to the more widely known concept of CSR.

3.2.2 Corporate social responsibility

The debate on whether companies have responsibilities to society beyond making profits has created an enormous amount of interest and controversy over the last 60 years. Nevertheless, companies more often recognise that they do have responsibilities that go beyond immediate shareholders and making profits (Crane and Matten 2016). Companies operate in, and interact with, society and the environment and serve customers in one or more countries. They fulfil their responsibilities by applying the well-established concept of CSR, which is a popular business management concept. Many companies have put some sort of CSR policy in place or have made explicit CSR communications (Maignan and Ralston 2002).

The most well-known definition of corporate responsibilities is Carroll's (1979, 1991) definition of CSR which identifies four categories of responsibilities: economic, legal, ethical and discretionary/philanthropic (Carroll 1979, 1991). Dahlsrud (2008), in a comprehensive review of CSR definitions, identifies five dimensions of CSR, specifically, environmental, social, economic, stakeholder and voluntariness. Garriga and Melé (2004) suggest there are four groups of CSR theories, namely instrumental, political, integrative and ethical theories. Basu and Palazzo (2008) propose that business approaches to study of CSR can be divided into three categories: (1) stakeholder-driven (pressures from external stakeholders); (2) performance-driven (the effectiveness of CSR actions in terms of their purpose by the organisation and their impact on the outside world); and (3) motivation-driven (the reasons why organisations undertake CSR) (Basu and Palazzo 2008). From the managerial perspective, CSR can be defined as:

> the process by which managers within an organisation think about and discuss their relationship with stakeholders as well as their roles in relation to the common good, along with the behavioural dispositions with respect to the fulfilment and achievement of those roles and relationships.
> (Basu and Palazzo 2008)

There are many ways to think about CSR, but broadly speaking, CSR refers to responsibility, hence duties and obligations or motivation and opportunities of the companies towards society (Gurzawska et al. 2015).

CSR derives from organisational sensemaking and dwells as an intrinsic part of a company's character (Basu and Palazzo 2008).

3.2.3 Strategy and strategic management

In today's complex business world, corporate success depends on crafting and implementing effective business strategies (De Kluyver and Pearce 2006). According to Mintzberg's and McHugh's 'grass-roots model' (1985), a strategy consists of intended and focused plans, on the one hand, and of emergent and often highly unexpected developments, on the other hand (Mintzberg and McHugh 1985). Rapid changes in the competitive environment require 'crafting long-term vision for an organisation while maintaining a degree of flexibility about how to get there and creating a portfolio of options for adapting to change' (De Kluyver and Pearce 2006). Therefore, strategic management can be defined as the field dealing with 'the major intended and emergent initiatives taken by general managers on behalf of owners, involving utilisation of resources, to enhance the performance of companies in their external environments' (Nag, Hambrick et al. 2007). Husted and Allen (2000) incorporate these definitions, and construe strategy as plans and actions taken to create unique resources and capabilities that leverage organisational routines and that are the source of sustainable competitive advantage and superior performance. From a practice point of view, strategy is about creating tools for managers to decide about how the company is going to win in the future (Husted and Allen 2010).

The purpose of strategic management is to bring about the conditions under which the organisation is able to create value. Value can be created through either the development of new products, processes, organisational innovations and marketing innovations or through the creation of entirely new markets (Husted and Allen 2007). In other words, companies use strategies to provide a competitive advantage or avoid a competitive disadvantage (Powell 2001). Husted and Allen (2010) argue that corporate strategy aims to achieve superior performance and both economic and social objectives of the company (gain economic and social competitive advantage or avoid economic or social disadvantage) (Husted and Allen 2010). They perceive strategy as a rational process that seeks two valuable and rational ends – economic value creation and social value creation (Husted and Allen 2010).

3.2.3.1 Innovation and strategy

There is a deep connection between corporate and innovation strategy in many business sectors (Berkhout 2013). Innovation is regarded as the focal point of a company's strategy (Milling and Stumpfe 2000). Innovation influences various aspects of companies' economic performance and corporate competitiveness, ranging from effects on sales and market share to changes in productivity and efficiency (OECD 2005). Nevertheless, benefits from innovation vastly depend upon how well it is managed. Therefore, in recent years, the field of managing and shaping innovation has attracted considerable attention from academics, policy makers and business practitioners (Conway and Steward 2009). The definition of innovation implies a process that involves the elements of strategic management, and therefore it should also reflect this systemic, strategic approach to innovation. According to this approach, the company is a system of interrelated and interdependent parts (White and Bruton 2010) that involves a framework of inputs, transformations, outputs (Muller, Välikangas et al. 2005; De Weerd-Nederhof 2007; Davila, Epstein et al. 2012) and feedback along the entire process (White and Bruton 2010). Figure 3.1 presents the system approach to innovation management.

Innovation is managed strategically through creating supportive structures, practices and processes which additionally requires that companies are in harmony with the contextual conditions in which they operate (Dodgson, Gann et al. 2013). Innovation is an organisation-wide concern that affects all units in a company, its organisational structure, people, processes, procedures and systems (White and Bruton 2010). Many

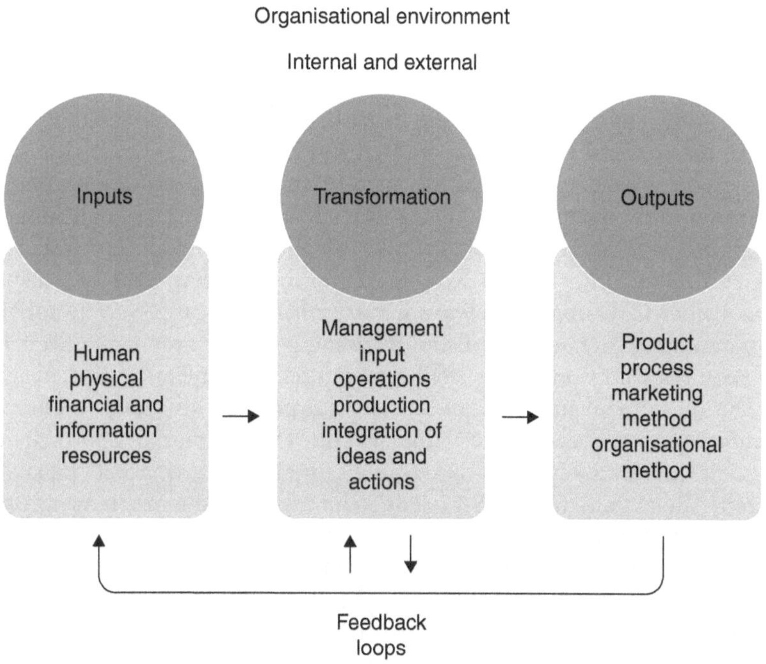

Figure 3.1 A system approach to innovation management. (Adapted from De Weerd-Nederhof; White and Bruton, 2010; and Davila et al. 2012; Muller et al. 2005.)

businesses that engage in innovation have separate research and development (R&D) divisions, innovation managers or innovation boards, which are often a driving force for their success.

3.2.3.2 CSR and strategy

Over recent decades CSR has shifted from the margins to the mainstream of business practice. The concept and scope of CSR have also evolved, from mere philanthropic action to so-called strategic CSR (Baron 2001; Lantos 2001; Gugler and Shi 2009), where corporate responsibility is managed in a systematic and intentional way. As emphasised by Galbreath (2006), 'CSR is ultimately a strategic issue, one that cannot be separated from a firm's overall strategy' (Galbreath 2006). According to the strategic CSR literature, CSR is strategic when it is integrated into companies' core business operation and, therefore, as an important part of corporate competitive strategy as it yields substantial business-related benefits to the company (Burke and Logsdon 1996). In other words strategic CSR is:

> a business strategy that is integrated with core business objective and core competencies of the company, and from the outside is designed to create business value and positive social change, and is embedded in a day-to-day business culture and operations.
>
> (McElhaney 2009)

Strategic CSR engagement is at the heart of the core business model and is brought into central value creation (Midttun 2009).

Many companies choose a CSR agenda and engage in 'doing good' by developing and engaging in projects that meet societal and environmental obligations and objectives, irrespective of a company's interest (Husted, Allen et al. 2015). Porter and Kramer (2006) assert that businesses usually only consider CSR in generic ways instead of in ways appropriate for their future strategies. In contrast to simply 'doing good', a strategic approach to CSR requires that companies deliberately create, implement and measure strategic investment and outcomes of social projects that seek a competitive advantage and economic value (Porter 1985; Liedtka 2000). In this way social projects replicate the strategic intention of 'market' behaviour (Husted, Allen et al. 2015). CSR needs to be considered more in terms of the opportunities it provides to the business and a fundamental value creation driver, rather that a reactive-defensive strategy and a support function to reduce risks and costs for shareholders (Midttun 2009; von Weltzien Høivik and Shankar 2011). Lantos (2001) differentiates strategic CSR from ethical CSR (moral responsibility to any individuals or groups where a company might inflict actual or potential injury) and altruistic CSR (being a 'good corporate citizen' by 'giving back' to society, regardless of whether or not this will benefit the company itself). He emphasises that strategic CSR

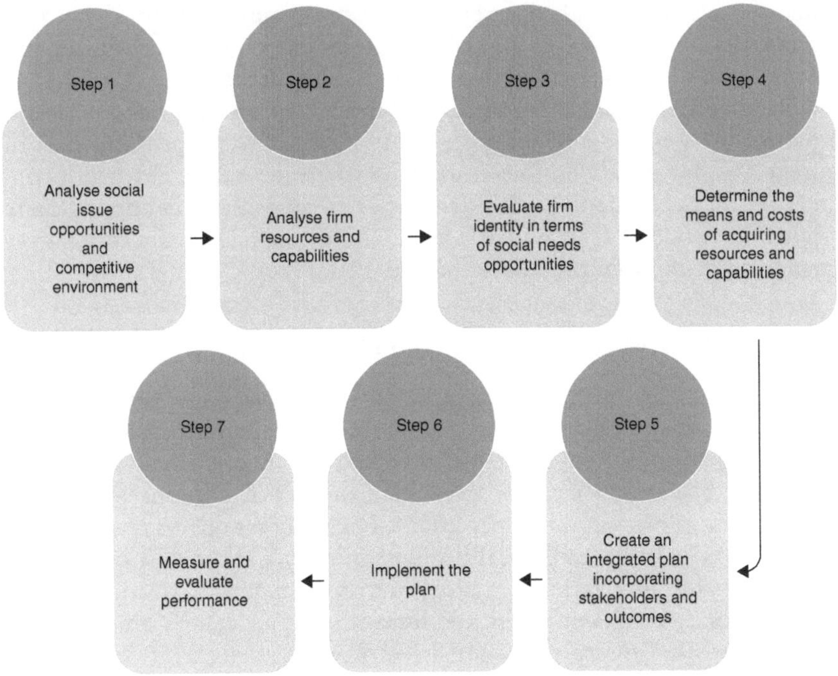

Figure 3.2 Seven-step model of corporate social strategy. (Adapted from Husted and Allen 2010.)

'creates a win-win situation in which both the company and one or more stakeholder groups benefit' (Lantos 2001). Husted and Allen (2010) propose a seven-step model of corporate social strategy (Figure 3.2) based in strategic analysis literature (Husted and Allen 2010).

Some scholars look into CSR from the 'business case' perspective and claim that strategic CSR can improve corporate competitiveness and, subsequently companies' economic and financial performance (Burke and Logsdon 1996; Zadek 2000; Kurucz, Colbert et al. 2008; Carroll and Shabana 2010). Porter and Kramer, the pioneers of connecting CSR and competitiveness (Porter and Kramer 2002), claim that companies can improve their long-term performance by connecting the company's financial and societal goals. They further argue that a strategic approach to corporate philanthropy can align both economic and social objectives (Porter and Kramer 2002). Through strategic CSR a company could make most significant societal and environment impact and harvest the greatest business benefits (Porter and Kramer 2006). Husted and Allen (2010) argue that companies that include social action programmes in the strategic decision-making process achieve better results in economic and social value creation (Husted and Allen

2010). Moreover, strategic CSR could allow companies to achieve a unique business position through differentiation from competitors in a way that lowers costs or better serves a particular set of customer needs (Porter and Kramer 2006). However, because of the increasing stakeholder pressure and benefits that CSR brings, a company's survival in modern society 'seems to require an awareness of social responsibility as an indispensable part of strategy' (Galbreath 2006). Therefore, today there are hardly any major consumer products companies that do not actively innovate to develop socially responsible products (Iyer and Soberman 2016). Being socially responsible is much more important than ever before.

3.3 The link between innovation and CSR

CSR initiatives are meant to be applied to all company activities and business and, therefore, also to innovation processes. Nevertheless, although some companies explore potential benefits of linking innovation and CSR, many businesses perceive investment in CSR as an unnecessary and costly burden (Porter and Kramer 2006). Higher costs due to the trade-off between returns from traditional business models and the cost of changes into responsible innovation practices and investment constraints are an argument for disconnecting CSR and innovation, particularly among small and medium-sized enterprises (SMEs) (Gurzawska et al. 2015; Gurzawska, Mäkinen et al. 2017). As a result, innovation activities are usually excluded from a company's responsibility agenda. There are several reasons for this situation and arguments against them, including the extent to which innovation and CSR overlap and the question of whether and how responsible innovation practices pay off, whether through value creation, competitive advantage, strengthening companies' reputation, enhancing the company's networks and stakeholder relationships, and therefore expanding intangible resources such as knowledge, co-creation and innovation capability. This section discusses these arguments and proposes a counterbalance for these claims.

3.3.1 Conceptual perspective

From a conceptual point of view not every corporate innovation should be perceived through the lens of responsibility, nor should CSR be solely focused on innovation. This is because not every corporate innovation raises societal, ethical, human rights or environmental issues. At the same time, CSR covers all aspects of a company's activity but does not exclusively relate to a company's innovation activities, thus CSR tools or actions are generally not designed specifically for innovation. Nevertheless, the sphere where these two concepts converge allows for the search for responsible corporate innovation. Figure 3.3 demonstrates the relationship between corporate innovation and CSR, showing the responsibility juncture.

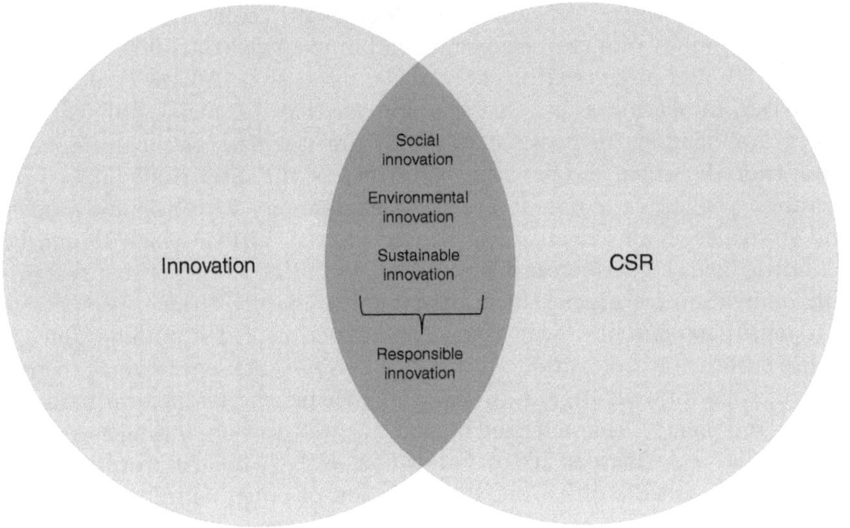

Figure 3.3 Conceptual relationship between corporate social responsibility (CSR) and innovation.

The merging point of corporate innovation and CSR encompasses a variety of well-known concepts that capture responsible innovation in various forms. Over the last 20 years, we have witnessed an expansion of literature, business activities and cross-sectoral exchanges, deliberately engineering societal and environmental responsibilities and objectives. As a result, several responsible innovation concepts exist that tackle various aspects of corporate responsibility. Social innovation, environmental and eco-innovation and sustainable innovation are among the most commonly discussed. Social innovation has been mainly developed by practitioners to 'meet pressing social needs and to improve human and environmental well-being' (Choi and Majumdar 2014). In the business context, it can be defined as a process where companies take 'community needs as opportunities to develop ideas and demonstrate business technologies, to find and serve new markets, and to solve long-standing business problems' (Kanter 1999). Environmental and eco-innovation is an innovation that reflects the concept's explicit emphasis on a reduction of environmental impact (OECD 2009). Regarding sustainable innovation, as emphasised by Adams et al. (2016), a variety of conceptualisations of sustainable innovation exist; however we lack a clear definition of sustainability. This confusion is reinforced by an array of labels applied to sustainable innovation, such as CSR; green, eco- or ecological innovation; social environmental management; and responsible innovation (Adams et al. 2016). While some authors argue for a responsible approach and give equivalence to environmental, social and economic dimensions in sustainability (Hansen et al. 2009; Longoni and Cagliano 2018), the majority of previous work focuses on ecological sustainability such as eco-innovation and environmental innovation (Carrillo-Hermosilla et al. 2010) and often overlooks the social dimension (Adams et al. 2016). Moreover, Lubberink et al. (2017) point out that the social, political and ethical implications of possible solutions are not part of the sustainable innovation discourse and practice. Nevertheless, the sustainable innovation literature provides attempts to connect sustainability with business models and strategies (Boons et al. 2013; Koistinen et al. 2018), which should be acknowledged. For instance, Iñigo and Albareda (2016) propose a sustainable innovation framework from the complex adaptive system perspective that aims to explain how firms engage and experiment. The authors conceptualise sustainable innovation around five components: operational, instrumental, collaborative, organisational and holistic; the interaction between these components is characterised by non-linearity, self-organising and emergence (Iñigo and Albareda 2016). Patala et el. (2016) build a process framework, which consists of: (1) the identification of potential impacts; (2) the identification of customer value creation mechanisms; (3) the choosing of key indicators; (4) life cycle value modelling; and (5) the demonstration of life cycle value.

Recently, academics and policy makers often refer to responsible innovation in the context of RRI. The most well-known definition of RRI by Von Schomberg (2013) defines RRI as:

> a transparent, interactive process by which societal actors and innovators become mutually responsive to each other with a view to the (ethical) acceptability, sustainability and societal desirability of the innovation process and its marketable products (in order to allow a proper embedding of scientific and technological advances in our society).
>
> (Von Schomberg 2013)

Besides Von Schomberg's definition, there is a variety of other definitions and approaches to RRI and RRI is operationalised in different ways (Sutcliffe 2011; Stilgoe et al. 2013; Taebi, Correlje et al. 2014; Foley, Bernstein et al. 2016; Lindner, Kuhlmann et al. 2016; Burget, Bardone et al. 2017; Ribeiro, Smith et al. 2017; Reber 2018; Rip 2018). With regard to further specification of RRI dimensions, there are those that tend to recur in various interpretations of the concept, and those that are more idiosyncratic (Gurzawska, Mäkinen et al. 2017). Responsible innovation in the RRI context is thought to focus on inclusion (also called engagement, or involvement of society), anticipation (assessment at an early stage in research and innovation (R&I) of benefits and risks), reflexivity (reflecting on values and beliefs during R&I) and responsiveness (the ability to change routines, structures and systems to adapt to changing circumstances and new insights: Gurzawska, Mäkinen et al. 2017). The European Commission has provided RRI orientation in the form of six policy keys, i.e. RRI is R&I that: (1) fosters R&I processes that are collaborative and multi-actor; (2) incorporates ethical principles so as to ensure compatibility with fundamental values; (3) promotes science literacy and science education; (4) promotes gender equality; (5) promotes open access to scientific knowledge; and (6) is guided by transparent, accountable and coherent multi-stakeholder governance (European Commission 2012). Therefore, responsible innovation should be societally desirable, sustainable and ethically acceptable (von Schomberg 2013). RRI and CSR share an emphasis on companies' responsibilities towards social goods as well as on stakeholder engagement. Nevertheless, despite some similarities, the concepts are somehow different (Gurzawska, Mäkinen et al. 2017). Firstly, while RRI is largely a top-down approach created in the policy world (e.g. Horizon 2020 funding), CSR can be characterised as a bottom-up approach where CSR policies function as a self-regulating mechanism for business. Secondly, RRI focuses on ethics appraisal and potential and actual social impact; CSR rather concentrates on the impact on community and environment (Gauttier et al. 2017). Lastly, while RRI is about R&I, CSR is generally applicable to all company activities.

Various other forms of a conjunction between CSR and innovation exist, such as social design, socially responsible design (SRD), eco-design, design for values and open innovation. While there are differences between how responsibility is conceptualised and defined in these different concepts, i.e. social innovation, environmental innovation, sustainable innovation and RRI (Lubberink et al. 2017), this chapter construes responsible corporate innovation as an umbrella term for any innovation centred around various forms of corporate responsibility, including sustainability, societal, ethical, human rights and environmental issues. A stronger understanding of scattered and often overlapping terminology can ultimately advance the integration of different disciplines. Therefore, for the purposes of this research, responsible innovation encompasses social innovation, environmental innovation, sustainable innovation, responsible innovation in the context of RRI and other concepts of innovation that centre around various forms of responsibility.

3.3.2 Bi-directional link between CSR and innovation

There is a considerable literature on the way in which, across different sectors and through time, companies have adjusted to the pressures and opportunities presented by the need to become more socially responsible and sustainable, taking into account stakeholder expectations and more responsible products, processes and services that reflect these changes (Maxfield 2008; Gugler and Shi 2009; Hanke and Stark 2009; Nidumolu, Prahalad et al. 2009; Berkhout 2013; Lai, Lin et al. 2015; Sánchez and Benito-Hernández 2015; Herrera 2016; Ueki, Jeenanunta et al. 2016). The European Commission believes that CSR makes companies more sustainable, competitive and innovative (European Commission (a)). It also recognises the importance of CSR in companies' ability to innovate, as well as in risk management, cost savings, access to capital, customer relationships and human resources management (European Commission (a)). Additionally, responsible innovation encourages a move beyond defensive and proactive CSR into a more rewarding synthesis between social and commercial concerns fulfilling public needs and private interests (Midttun 2009). According to Maxfield (2008), CSR activities that are tightly linked to innovation functions might bring more financial benefit than those oriented toward public relations, marketing and human resource management (Maxfield 2008).

A vast majority of CSR and strategic CSR literature focuses on the linkage between CSR and competitiveness, highlighting the need to go beyond mere window-dressing work and investing in strategic CSR. In this view strategic CSR is linked to innovation and therefore depends on an 'innovative way to enhance efficiency in a socially friendly manner' (Gugler and Shi 2009). Grayson and Hodges (2017) explain that CSR creates largely untapped opportunities for product innovation, market development and non-traditional business models (Grayson and Hodges 2017). Moreover, CSR plays a role in differentiation strategies at the product and company levels through innovation, for example, by developing products that are socially responsible, as has been done by companies such as Ben & Jerry's, the Body Shop and Health Valley (McWilliams and Siegel 2000). The relationship between CSR and innovation appears to be bi-directional. On the one hand, companies can use social programmes as a way to foster product and process innovation (Kanter 1999). In addition, social and environmental programmes may help generate competitively valuable resources for the company (Sharma and Vredenburg, 1998). On the other hand, companies that have an ability for continuous innovation are more likely to be able to leverage that same resource in other arenas, such as the development and implementation of social strategy. McWilliams and Segal (2000) argue that CSR and R&D investment are highly correlated, since many aspects of CSR create either a product or process innovation (McWilliams and Siegel 2000). According to the study on the connection between CSR and company innovation by Luo and Du (2015), the positive relationship is stronger for companies with higher R&D investment and where companies operate in more competitive markets (Luo and Du 2015).

According to MacGregor and Fontrodona (2008) there is a bi-directional system between CSR and innovation based on CSR-driven innovation and innovation-driven CSR (MacGregor and Fontrodona 2008). The authors point out that CSR-driven innovation is about 'doing the rights things', and innovation-driven CSR is about 'doing things right' (MacGregor and Fontrodona 2008). Findings of the pan-Nordic project 'CSR-driven innovation' point out that, while some companies identify the societal or environmental need before beginning to think about the business model and profit generation, for other companies social performance is a means for

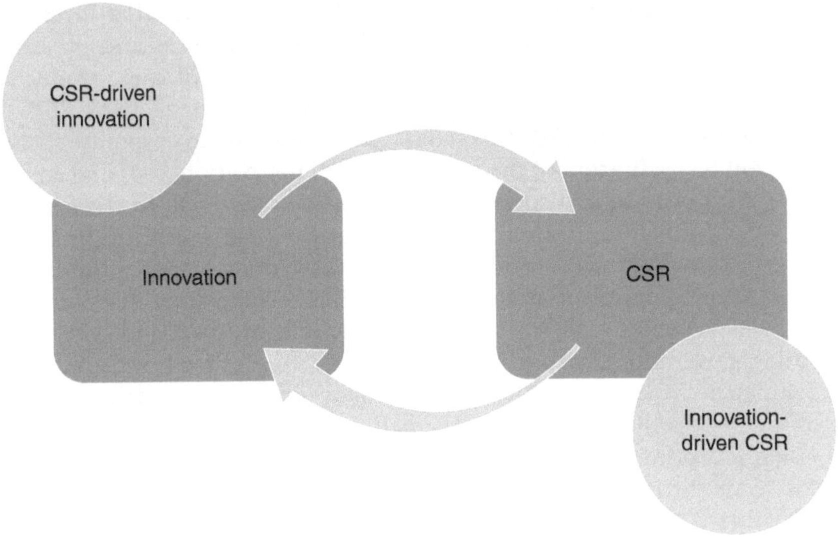

Figure 3.4 Bi-directional relation between corporate social responsibility (CSR) and innovation. (Adapted from MacGregor and Fontrodona 2008.)

achieving the goal of creating private profit (Hockerts, Morsing et al. 2009). Therefore, CSR-driven innovation is a situation when a socially or environmentally important need contributes to the creation of a new innovative product, service or process; and innovation-driven CSR refers to a situation when a technological innovation that already exists allows it to be used to achieve an important societal or environmental goal (Pyszka 2012).

Figure 3.4 presents this two-way spiral model of CSR and innovation based on CSR-driven innovation and innovation-driven CSR, that I discuss in the next subsections.

3.3.2.1 CSR-driven innovation

Globalisation, particularly thanks to the internet, has brought tremendous changes for most companies and society. Internet-connected stakeholders have become more aware of the products, services and processes that companies can offer and their marketing practices. This has caused an increase in stakeholder pressure on companies to take greater responsibility as part of their corporate citizenship. As a consequence, companies more often innovate in order to find solutions for current and emerging societal and environmental challenges. This phenomenon is known as CSR-driven innovation. CSR-driven innovation is 'an innovative process which aims at designing a profitable product or service which in an innovative and user-oriented way can prove beneficial to the surrounding environment and society' (Nordic Innovation Centre 2010). Companies that invest in CSR-driven innovation are characterised by 'the willingness and the capacity to discover, adopt, evaluate and exploit new technologies, products, services or processes for environmental and societal benefit' (Brik 2007). The literature discusses CSR-driven innovation referring to corporate social innovation (CSI), bottom of the pyramid (BOP) innovation, eco- and environmental innovation and social entrepreneurship (Rexhepi, Kurtishi et al. 2013). CSR-driven innovation aims at 'creating a successful business by having sustainability as a focal point when developing a new product or service' (Nordic Innovation Centre 2010). These innovative products or services may tackle such societal and environmental problems as global warming, poverty or diseases. An example is the case of the United Nations Development Project (UNDP) that has introduced the first eco-driving simulator among transport drivers in Azerbaijan, aiming to reduce the carbon emission footprint and make environmentally friendly driving habits more mainstream through training and the application of smart technologies (UNDP 2018).

Helping to address global problems may also be profitable for business. Many businesses, therefore, have already implemented a new approach to innovation, as in the case of Digital Green, addressing poverty by connecting smallholder farmers with their peers via video training conducted in local languages (Digital Green). In the automotive industry, CSR-driven innovation is about investment in long-term innovations that will reduce the threat of climate change, including electric, hybrid, fuel cell and solar vehicles (Khaledabadi and Magnusson 2008). The energy sector is exploring alternative sources of energy, including solar panels, wave power, wind turbines and bioenergy. One of the energy sector CSR-driven innovations is the US company SolarCity, which from 2008 was providing comprehensive solar power solutions for energy-efficient lighting as part of a sustainable cities initiative. Thanks to integrated sales, financing, design, installation, monitoring and efficient services, customers could get cleaner and more affordable energy. In 2016 the company merged with Tesla Inc. and introduced the Tesla Solar Roof based on solar roof tiles, made of glass (Tesla).

Moreover, companies develop innovation and business models based on consumers' preferences regarding social responsibility (Iyer and Soberman 2016). CSR-driven innovation may create an environment in which consumers identify with a product or service or brand because of its socially responsible approach, ultimately creating a strategic business model based on responsible innovation. One example of a company placing responsibility at the core of its business is Fairphone. This Dutch social enterprise engages in fairer electronics by developing a business model that puts ethical values first (Fairphone (a)). Fairphone produces repairable and recyclable phones made of conflict-free minerals (natural resources not extracted in a conflict zone and sold to perpetuate the fighting), manufactured with respect for workers' rights and wellbeing. The Fairphone community consists of 100,000 owners of fair smartphones, over 136,000 fans on Facebook and thousands of Twitter and Instagram followers (Fairphone (b)). Nearly every aspect of the company's value chain reinforces the societal and environmental dimensions of its value proposition, distinguishing Fairphone from its competitors.

Another example is Lush Fresh Handmade Cosmetics (Lush Cosmetics), a British cosmetics business with a global brand. Lush offers 100% vegetarian cosmetics no-packaging-required and when packaging is unavoidable, they use recyclable or compostable packaging. The company supports human rights, environmental conservation and animal welfare through (for instance) developing a palm oil-free soap base.[1]

The last example of a strategic business model based on responsible innovation is a Dutch company called Rural Spark. Rural Spark provides a smart energy grid based on the concept of smartly distributed energy networks for rural villagers. For a monthly subscription, villagers rent a Rural Spark energy kit and become local energy suppliers who generate, use and sell energy (Rural Spark). Rural Spark provides access to clean, safe and sustainable energy; it also leapfrogs the outdated, top-down centralised energy grids and empowers users and encourages entrepreneurship (Rural Spark).

3.3.2.2 Innovation-driven CSR

Empirical studies confirm that R&D-intensive companies are more competitive on the market (Kinkel, Lay et al. 2005). R&D is a part of innovation situated at the front end of the innovation life cycle. Studies by McWilliams and Siegel (2000) and Bansal (2005) show a positive correlation between CSR and R&D intensity 'because both are associated with product and process innovation' (McWilliams and Siegel 2000). Other studies, such as those by Bouquet and Deutsch (2008) and Hull and Rothenberg (2008), seem to confirm this correlation. Padgett and Galan (2010) extend these findings and show that R&D intensity positively affects CSR in a way that R&D is perceived as a form of investment resulting in increased knowledge that leads to product and process innovation (Padgett and Galan 2010). Furthermore, they claim that these innovations can lead to CSR-related processes and products (Padgett and Galan 2010). This model is known as innovation-driven CSR. It is a situation in which a technological innovation already exists and can be used for the realisation of societal and environmental objectives. Therefore, it is a technological drive or an entrepreneur's desire to develop new products that drives the innovation (Hockerts, Morsing et al. 2009). However, through the innovation processes, a company may improve its effectiveness and efficiency, e.g. through minimising its CO_2 footprint and water footprint and/or reducing the number of casualties or fatal accident rates. These actions, as Padgett and Galan (2010) claim, should be taken into consideration when developing companies' CSR strategies, because process and product innovations may already be involved in CSR activities (Padgett and Galan 2010). Therefore, the authors suggest that an

> innovative firm should focus their efforts on identifying opportunities in their R&D processes to initiate related CSR activities. This will allow the company to manage costs more effectively and determine whether other CSR activities might be necessary to meet stakeholder expectations.
>
> (Padgett and Galan 2010)

Innovation-driven CSR is a new phenomenon that requires further empirical study on its functioning and consequences. Nevertheless, it shows that the interconnectedness of CSR and innovation is not linear, but this process is more complex as there are numerous feedback loops. CSR has an indirect influence on innovation. It may give rise to new products and processes. Innovation, however, can also influence CSR instruments.

3.3.3 CSR, innovation and value creation

Companies engage in social responsibility and go 'beyond compliance' for various reasons, including market demand, cost advantage, differentiation strategy, upgrading strategy, to build intangible brand value around social responsibility or sustainability reputation, to reduce regulatory and other risks (Berkhout 2013). According to Burke and Logsdon (1996) 'the ultimate measure of strategic benefits from CSR activities is the value they create for the firm', where value creation is understood as having identifiable, measurable economic benefits that the company expects to receive (Burke and Logsdon 1996). Value creation is crucial for strategic success (Tantalo and Priem 2016).

Burke and Logsdon (1996) identify five CSR behaviours that bring strategic benefits: (1) philanthropy (e.g. engineering research fellowship and community support), ensuring customer loyalty and future purchasers; (2) direct or indirect employee benefits (e.g. flexible working hours, health and wellness) may improve productivity through improved employee loyalty and morale; (3) environmental management may lead to product and process innovation (e.g. new 'green' products) and improved public relations and/or marketing advantage, ultimately opening up new markets; (4) through a political activity companies may achieve favourable changes in economic or social regulations and create new market opportunities or geographical market opportunities; and lastly (5) product- or service-related characteristics, innovations or processes such as product reformulation (e.g. improved 'green' design) may lead to new markets, first-to-market or leadership benefits and provide edge in meeting emergency needs (Burke and Logsdon 1996).

Lubin and Etsy (2010) claim that CSR-related issues are one of the so-called 'megatrends' that create incipient societal and economic shifts, such as globalisation, the rise of the information society and the move from hierarchical organisations to networks. These transformations arise from technological innovation or from new ways of doing business (Lubin and Esty 2010). Nowadays, thousands of companies are placing strategic bets on innovation in CSR-related issues such as renewable power and pollution control (Lubin and Esty 2010). This would not be possible without innovative technologies and business approaches. In the sustainable innovation literature stream, strongly sustainable business models do no harm but create positive environmental, social and economic value (Koistinen et al. 2018). Accordingly, strongly sustainable companies take financial, societal and environmental costs into account and measure financial rewards, social benefits and environmental regeneration (Koistinen et al. 2018). Nevertheless, value should be created to the whole range of stakeholders and the natural environment (Koistinen et al. 2018).

In line with this assertion, Patala et al. (2016) introduce a concept of sustainable value propositions as 'a promise on the economic, environmental and social benefits that a firm's offering delivers to customers and society at large, considering both short-term profits and long-term sustainability' (Patala et al. 2016). From the system's point of view, we should consider value creation as not only economic, but as a psychological, sociological and ecological concept, which includes value for the organisation, customers, ecosystem and society (Den Ouden 2011).

Value creation through responsible corporate innovation can be placed in a broader discussion on measurable benefits of CSR. Financial benefits of responsible corporate behaviour can be found particularly in the areas of human resources, reputation and branding, reduction of risk and operational cost (McElhaney 2009). Moreover, Gugler and Shi (2009) claim that the economic interests offered by CSR, such as better access to market, finance and business; enhanced intangible assets, reputation, community relations; and reduced risk from regulatory sanction could encourage companies to structural changes, including innovative processes and technological upgrading. As a result, these transformations enhance productivity and efficiency, and ultimately compensate for the initial costs and enable competitiveness (Gugler and Shi 2009). Nevertheless, the discussion on the relationship between CSR and a company's performance seems to be unresolved. While some studies disagree on the link between corporate social performance (CSP) and corporate financial performance (CFP), some seem to indicate the existence of a positive relationship between CSP and CFP (Margolis and Walsh 2003; Orlitzky, Schmidt et al. 2003, Carroll and Shabana 2010; ING 2018), while some other studies point out inconsistencies (Griffin and Mahon 1997; Roman, Hayibor et al. 1999).

According to the study by Bonini et al. (2009), CSR-related programmes create measurable value through return on capital, risk management and quality of management and growth. This study is particularly interesting in the context of CSR and innovation, showing that companies' growth is ensured thanks to opportunities to access new markets and new customers, cutting-edge technologies and innovative products/services for unmet societal or environmental needs and the possibility to use these products/services for business purposes, such as patents and proprietary knowledge that ultimately lead to higher brand loyalty, reputation and goodwill with stakeholders (Bonini, Koller et al. 2009). Therefore, innovation is one of the key pathways through which CSR creates measurable business value (Bonini, Koller et al. 2009).

Similar findings can be found in the study by Husted and Allen (2010), who emphasise that 'the firm's social projects allow it to achieve measurable social objectives as well as improved corporate financial performance when the social action is linked to product and service innovation, process innovation, or corporate reputation' (Husted and Allen 2010). Martinez-Conesa et al. (2017), in their recent study, argue that 'innovation may help to ensure the sustainability of a more responsible approach to business, resulting in system level solutions that are at the same time, responsible and profitable' (Martinez-Conesa, Soto-Acosta et al. 2017). Therefore, it is crucial for companies to integrate different business activities and tie them in with the firm strategy as this is the only way to generate value (Martinez-Conesa, Soto-Acosta et al. 2017).

3.3.4 Responsible innovation and stakeholder management

However, to generate value companies need to transform their approach to CSR from a bolt-on activity to built-in to business strategy activity supporting business purpose and objectives (Grayson and Hodges 2017). Studies by Husted et al. (2007) show that a company's resources for continuous innovation are significantly related to the use of strategic social positioning. According to a study by Luo and Du (2012), in companies where CSR is not a peripheral activity, it can become a pivotal component of competitiveness and growth. As a result CSR programmes can make a company more innovative (Luo and Du 2012). In their study of 128 companies in all major industry sectors, they found that companies that are in the top third in terms of CSR activities brought out, on average, 47 new products a year, while companies in the bottom third brought out only 12 (Luo and Du 2012). The authors conceptualise that the reason for this is the fact that CSR builds broader and deeper relationship networks with external stakeholders, such as customers, suppliers, non-governmental organisations (NGOs) and governments, facilitating the sharing and exchange of external knowledge of its stakeholders (Luo and Du 2012, 2015). From the strategy point of view, stakeholders' external knowledge complements the company's internal knowledge and promotes a company's innovation (Luo and Du 2015).

According to Hanke and Stark (2009) the internal culture of a company (e.g. the way human resource development and organisational learning processes are performed) affects the level of innovation and the organisation's development with respect to its environment. For instance, in the context of sustainable vehicles, governmental regulations can foster innovations in sustainable vehicles; at the same time, sustainable vehicle development can foster environmental regulations (Khaledabadi and Magnusson 2008). Therefore, the relationship between the company and its key external stakeholders (e.g. customers, governments, society) enables each sector to gain profit by mutual learning and exchange, and through developing CSI (Hanke and Stark 2009).

Amos and Awuah (2017) argue that a company could achieve its strategic objectives and improve competitiveness when blending in the potential of stakeholders. A company may achieve a unique competitive position in the local market through a collaboration with stakeholders (e.g. the local community and NGOs), at the same time creating and delivering societal and economic benefits (Amos and Baffour Awuah 2017).

The integration of CSR and multiple stakeholders into innovation activities leads to new innovations (Hansen et al. 2009) and bundles of products and services that suit local market conditions. As discussed in the previous subsection, responsible innovation is about creating value for the organisation, customers, ecosystem and society (Den Ouden 2011). Responsible innovation calls for integration of stakeholders, especially the people who might be affected by the innovation (Adams et al. 2016; Lubberink et al. 2017). From a business strategy perspective, the success of innovations depends ultimately on consumers' acceptance and, therefore, to succeed a company firstly needs to understand customer needs and then develop products that meet those needs (Hauser, Tellis et al. 2006). Innovation is about identifying opportunities and creating strategies to fulfil customer needs and expectations (Husted and Allen 2010). Responding to customers' social needs may stimulate innovation (Husted and Allen 2010). These findings strongly reflect one of the approaches to CSR, namely stakeholder theory that focuses on 'managing potential conflict stemming from divergent interests' (Frooman 1999).

Furthermore, companies with a strategic priority to innovation can use CSR as an effective means to reduce information asymmetry between themselves and stakeholders (Shen, Tang et al. 2016). Moreover, the development of regional and supra-regional networks and efficient network governance of different actors may be vital for an innovative and sustainable CSR strategy (Hanke and Stark 2009). Unsurprisingly, Husted and Allen (2007) found that there is a high correlation between stakeholder integration and continuous innovation (Husted and Allen 2007). Nevertheless, connecting innovation and CSR raises some challenges and uncertainties, which are discussed in the next subsection.

3.3.5 Challenges and limitations in connecting CSR and innovation

The link between CSR and innovation is a relatively new topic. Midttun points out that innovation and CSR are complex, multidimensional phenomena/

concepts/fields (Midttun 2006). He identifies an incompatibility between the dynamic nature of innovation and the static character of CSR. Therefore, he calls for a dynamic reinterpretation of CSR which can be better aligned with the 'disruptive' innovation literature. This approach may also 'provide important insights into the socio-economic realignment necessary to accommodate new technology and business models' (Midttun 2006). Furthermore, despite some theoretical considerations, the empirical evidence is scarce and inconclusive (Gallego-Alvarez, Manuel Prado-Lorenzo et al. 2011; Luo and Du 2015; Shen, Tang et al. 2016; Halkos and Skouloudis 2018). On the one hand, empirical studies on the relationship between CSR and innovation by Gallego-Alvarez et al. (Gallego-Alvarez, Manuel Prado-Lorenzo et al. 2011) show that the bi-directional relationship between CSR and innovation is negative. These findings seem to be endorsed by Halkos and Skouloudis (2018). On the other hand, findings by Luo and Du (2015) and by Shen et al. (2016) demonstrate that CSR activities boost innovation. Moreover, the pan-Nordic project 'CSR-driven innovation' presents several success stories of businesses engaged in CSR-driven innovation. We lack synergies in a system approach, between the external (a system-level) and internal business (a company-level) environments to achieve responsible corporate innovation. There should be an interplay between policy-oriented external environments (e.g. legal and governance frameworks) and business strategy change, for instance through cooperation between the public and private sectors in the form of private–public partnerships (Koistinen et al. 2018).

Taking into consideration that responsible corporate innovation management raises both opportunities and challenges, further studies are required, particularly empirical studies, to investigate the nature, benefits and challenges, measurement metrics and business models for responsible innovation. While difficulties remain, undoubtedly some of the most successful corporations are also among the most socially responsible.

3.4 Strategic responsible innovation management (StRIM)

When considering the opportunities and benefits of linking innovation and CSR, this study proposes a new approach, called strategic responsible innovation management (StRIM). I argue that a responsible innovation strategy model based on the commitment of the company, in conjunction with the strategic use of responsible innovation, can lead to a competitive advantage and value creation, while incorporating key market and non-market stakeholders. In developing the model, I build on the concepts discussed in Section 3.3: innovation management; strategic CSR(including Husted and Allen's seven-step model of corporate social strategy); recent developments in business models in the sustainable innovation literature stream (including models by Hansen 2009; Iñigo and Albareda 2016; Patala et al. 2016); value creation; CSR-driven innovation and innovation-driven CSR; and the multi-stakeholder approach.

Deriving from RRI, I argue that responsible corporate innovation is a strategic concept (Von Schomberg 2013; Lindner, Kuhlmann et al. 2016) related to strategic CSR that imposes several demands on the way in which innovation is organised. Firstly, innovation should be guided by the principles of good governance, which includes anticipation, openness and transparency. Secondly, responsible innovation requires the participation of a variety of stakeholders in the innovation process. Thirdly, societal and environmental issues should be carefully considered, evaluated and controlled throughout the innovation process. Responsibility should be embedded in both the process of innovation (e.g. fair labour conditions, ethical sourcing, avoiding animal cruelty) as well as its outcomes (e.g. run-tracking device ensuring users' privacy, non-discriminatory software systems for predictive policing). If innovation could lead to negative societal or environmental consequences or conflicts with ethical criteria, including the fundamental values that societies uphold in their constitutions and legal frameworks, mitigation actions should be undertaken (Gurzawska, Mäkinen et al. 2017).

The StRIM approach is built on the traditional strategic management process. The strategic process of a company has three components: (1) planning; (2) implementation; and (3) evaluation and control (White and Bruton 2010). These activities should be performed simultaneously and continuously. A successful company should manage its strategies in three main areas: (1) the company's internal environment, including resources and capabilities; (2) the external environment within which the organisation operates; and (3) the company's ability to add value to what it does (Lynch 2015). The internal environment involves departments, management teams and individual employees, but also resources and capabilities that form the company. The external environment refers to external forces that impact the company. Figure 3.5 presents the strategic management scheme based on a system approach presenting the company as an association of interrelated and interdependent parts (White and Bruton 2010).

The systems approach to StRIM involves a framework of strategic process components and three main areas that the company needs to manage and feed back during the entire process. Decisions made on each stage of the strategy (planning, implementation and evaluation and control) feed back into the internal environment, resources and capabilities; the external environment; and value creation. Changes in the company's internal environment (e.g. reorganisation in the R&D department, new more sustainable production methods, changes in a company's identity and a culture embodied in a new code of conduct); the external environment (e.g. new customer group, new legal regulations on CO_2 emission threshold); and in the value the company creates will all have an impact on subsequent strategic decisions which are represented in the scheme. This approach reflects the non-linearity of the process with a bi-directional link between innovation and CSR with feedback loops. The next subsections discuss individual stages of the StRIM approach.

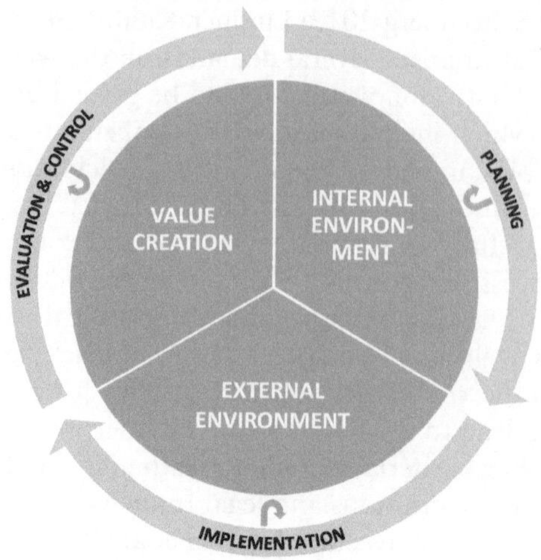

Figure 3.5 Strategic responsible innovation management (StRIM) scheme. (Adapted from White and Bruton, 2010, and Lynch, 2015.)

3.4.1 Planning

Starting from the outer circle, planning of the strategy includes strategic analysis and strategic development.

In the strategic analysis phase, a company should examine the internal and external environment and the links between them and identify its vision, mission and objectives (Lynch 2015). In the internal environment analysis, a company should explore the internal resources (tangible and intangible) and capabilities, what their role is, added value and competitive advantage and how they can be improved over time (Lynch 2015).

The internal environment consists of the company's culture, structures and processes. Adapting Husted and colleagues' approach to social strategy, StRIM requires the integration of strategic business and strategic responsibility actions (Husted, Allen et al. 2015). While business strategies employ a company's resources and capabilities to achieve purely market-based competitive objectives (Husted, Allen et al. 2015), StRIM would use a company's resources and capabilities to meet both financial as well as societal and environmental objectives. Therefore, the purpose of StRIM is to create value for the company, especially economic value, by embracing societal and environmental objectives. Responsible innovation and corporate responsibility can be perceived as specific intangible resources that provide benefits to companies (McWilliams and Siegel 2011). Components of corporate culture, such as corporate values and philosophy, are specific resources essential to a company's identity (Albert and Whetten 1985). They can provide a competitive advantage and help in inclusion of non-economic objectives within a company's strategy (Husted and Allen 2007). The company's culture and identity should be evaluated in terms of societal and environmental needs and opportunities (Husted and Allen 2010). The successful integration of innovation with strategic concerns should begin with a company's capabilities, because the business ultimately develops its competitive advantage through capabilities (Barney 1991). Capabilities are skills that a company develops (White and Bruton 2010). The capabilities of a company can be classified as either technical or market capabilities. Technical capabilities address how the company approaches technology it already has or wishes to have in the future, and market capabilities are market-relevant skills that indirectly impact the technology of the company (White and Bruton 2010).

The external environment analysis determines what is happening or is likely to happen outside the company (external environment), including societal and environmental risks and opportunities, competitive environment and relationships with non-market stakeholders. In this stage, a company should identify external factors that could affect or be affected by the innovation process, for instance, economic, political and technological developments, potential alliances, networks and partnerships leading to sustainable co-operation, competitors and customers' preferences. The external environment analysis is crucial, because a company can impact or be impacted by its broader environment (White and Bruton 2010). Based on the internal and external environment insights, a company should develop and review its vision, mission and strategic objectives.

The second part of planning is the strategic development that involves the identification and rational selection of options available to achieve the agreed objectives and determination of the strategy, its structure and style (including organisational structure and people) (Lynch 2015), taking into consideration value creation. These objectives should be intrinsic with a company's resources and capabilities, mission and vision and the external environment.

As discussed above, responsible innovation requires embedding the principles of good governance that include anticipation, openness, transparency, and accountability in the corporate vision, mission, objectives and corporate culture rooted in beliefs and value systems shared by employees. Moreover, relevant internal and external stakeholders should be engaged in responsible innovation planning. They can improve the quality of the environmental analysis and strategic development, providing specific knowledge and experience, and through communicating their preferences, needs and concerns. Nevertheless, the company should manage and prioritise these various interests.

3.4.2 Implementation

The implementation stage is a process of applying the chosen strategy in practice. The success of the implementation stage depends on three specific policies and practices: a company's formal organisational structure, its formal and informal management control systems and its employee compensation policies (Hesterly and Barney 2008). Responsible innovation should be integrated along the whole value chain and, therefore, into the governance of the company and into existing management systems (Gurzawska, Mäkinen et al. 2017). Efforts to create innovations that are socially and environmentally responsible should be treated and managed as core business strategy, just as are the strategies of capital expenditure, talent management and marketing. Responsible innovation principles that are embedded in the governance of a company might improve integration of the vision, mission and objectives of the company's personnel with those of the corporate policy (Chatfield, Borsella et al. 2017).

Senior leadership and management of the company should organise innovation internally to pursue responsible practices and behaviours when developing new products, processes and services (Responsible-Industry 2017). Responsible corporate innovation has little effect on strategy if individuals and leaders within the organisation are not committed to responsible behaviour. Therefore, senior leadership and management of the company, including the board of directors, must make an authentic, firm and public commitment to responsible innovation efforts, and engage with them (McElhaney 2009). A clear commitment towards responsible innovation principles can build a consistent picture of corporate values as an ideological system that aligns employees to strategic objectives and binds them to these corporate goals (Berkhoud 2013). Aligning employees' values with organisational values can support and nurture responsible innovations (Chatfield, Iatridis et al. 2017). According to Grant (2007) a company that cares about user needs and societal welfare can spark motivation, positively affect employees' actions and behaviour (Grant 2007), enhance their sense of having 'meaningful work' and enhance employee engagement (Gurzawska, Mäkinen et al. 2017). It is not only beneficial for an employee, but also for a company. This is because employee engagement is correlated with higher productivity, costs and sales, which are the main performance indicators (Gurzawska, Mäkinen et al. 2017). A company can incentivise employees to pursue responsible corporate innovation, because a properly designed reward and incentive system is crucial for creating motivation and commitment (De Kluyver and Pearce 2006). A company can do this through awareness raising, an integration of ethical thinking into the design/production process, advocating and encouraging employees to maintain a responsible attitude and discouraging/stigmatising unethical behaviour (Responsible-Industry Project Consortium, 2017[BIB114]; Gurzawska, Mäkinen et al. 2017). Management should also adopt social responsibility governance tools to support the strategy implementation. These tools should be based on co-creation and shared responsibility of all stakeholders (Gurzawska, Mäkinen et al. 2017).

3.4.3 Evaluation and control

A strategy is effective when it creates value for shareholders, partners, suppliers, employees and the community, yet it delivers customer value by satisfying their needs, including societal and environmental needs, better than rivals. The evaluation and control stage focuses on monitoring innovation to ensure that it meets the desired outcomes and creates value, both financial and social. It is necessary that after an innovation is implemented, the company monitors changes that may affect innovation, making it irresponsible and unethical, technologically obsolete, replaceable or competitively weak (White and Bruton 2010).

Every company should measure its performance and created value. Performance metrics serve as a powerful management tool in ensuring that the company focuses on accomplishing its mission and objectives and creating incentives for staff and managers (Sawhill and Williamson 2001). As discussed in Subsection 3.3.3, innovation linked to CSR is one of the key ways through which a company can achieve measurable social objectives and create measurable business value. Nevertheless, the responsible innovation metrics should be selected carefully to support the strategy, otherwise the strategy cannot be delivered. The key here is to develop metrics based on the company's context which is not only internal but also relevant to its key external partners and ecosystems. Measuring societal and environmental objectives may be challenging; nevertheless non-profit performance metrics could provide some lessons learnt about approaches to quantifying success, even for highly ambitious and abstract goals. Sawhill and Williamson (2001) determine three kinds of performance metrics: (1) metrics measuring companies' successes in mobilising resources; (2) metrics measuring staff effectiveness on the job; and (3) metrics measuring progress in fulfilling companies' missions.

StRIM offers a more conscious and integrated approach connecting companies' innovation and CSR strategies. It could help to arrange patterns of organisational behaviour in terms of strategising responsible corporate innovation. Following Mintzberg et al. (1998), this strategy should be developed as 'a transformational process based on learning and growth, both of the informal (culture, vision, position, people) and formal (programs, products, structure, system) parts of an organisation' (Mintzberg et el. 1998). This business strategy should be integrated with core business objectives and embedded in day-to-day business culture and operations. Furthermore, it should encourage stakeholder dialogue and 'social learning' (multi-stakeholder approach). Such an approach would foster the responsible development of innovations that are profitable for companies, accepted by society and relevant to societal and environmental problems. Nevertheless, a strategic approach to responsible innovation raises some challenges and uncertainties, which are discussed in the next subsection.

3.5 Conclusions

A specific contribution of this chapter was to develop a better account of how companies may create economic and social value through integrating responsible innovation in their strategies. This study explains how the link between innovation and CSR may assist in improving a company's competitiveness, value creation and stakeholder management. In this chapter, I connect innovation and CSR in order to foster the responsible development of product, process, organisation and marketing innovation. Firstly, I delineate the field by defining the concepts underpinning responsible innovation, including innovation, CSR and their relation to business strategy. This analysis allows me to provide recommendations for ways in which companies can develop strategies for responsible corporate innovation management. By identifying the key analytical factors (innovation management, responsible innovation, strategic CSR, CSR-driven innovation and innovation-driven CSR, multi-stakeholder approach) it is possible to recognise several strands or connections that help frame understanding of the relationship between innovation and CSR. Both innovation and CSR should be perceived as a strategic tool and a goal. Therefore, companies should explore innovations that are accepted by society and address societal and environmental problems. I argue that the concept of CSR enriches the innovation process by emphasising the interdependence of business and society. At the same time, CSR activities are tightly linked to innovation functions that might ensure a competitive advantage and therefore might be more profitable than those oriented toward public relations, marketing and human resource management. I propose a new approach, called StRIM, that is intertwined with companies' social responsibility. This approach is intended to redefine companies' perceptions of a 'successful innovation' by shifting the focus from a company's financial success to sustainable outcomes, for both business and society. A strategy is unique for an organisation, therefore StRIM can help to develop strategies best suited to the company's continuous success. In this way, responsible innovation will create and generate revenue, not just minimise costs and risks. The conceptual framework developed in this chapter may support companies to reflect on their relations with other parts of society. The framework may also be helpful to answer questions on CSR strategising. Nevertheless, there will be further questions on the deep-rooted values and beliefs in companies, which are responsible for the acceptance (and non-acceptance) of an organisational engagement. Hence, the conceptual framework serves as a first attempt to arrange patterns of organisational behaviour in responsible innovation strategising.

Funding

The research leading to these results received funding from the European Community's Seventh Framework Programme (FP7/2007–2013) under grant agreement no. 612231 (SATORI).

Acknowledgements

The author would like to acknowledge the contribution of all the project's participants and all the project's activities to the ideas that underpin this paper.

Copyright

Note

1 The palm oil industry has been linked to deforestation, habitat degradation, climate change, animal cruelty and indigenous rights abuses in the countries where it is produced. See e.g. *Say No to Palm Oil*: www.saynotopalmoil.com/

References

Adams, R., S. Jeanrenaud, J. Bessant, D. Denyer, and P. Overy (2016). Sustainability-oriented innovation: A systematic review. *International Journal of Management* Reviews, *18*(2), 180–205. doi:10.1111/ijmr.12068.

Albert, S. and D. A. Whetten (1985). Organizational identity. In Hatch, M.J. & Schultz, M. (Eds.), *Organizational identity: A reader* (Oxford Management Readers). Oxford, UK: Oxford University Press.

Amos, G. J. and G. Baffour Awuah (2017). In search of competitiveness through innovation-driven CSR initiatives in Multinational Enterprise subsidiaries in developing countries. *Journal of Developing Country Studies*, *7*(2), 161–173.

Baker, M. J. (2014). *Marketing strategy and management*. Macmillan International Higher Education.

Bansal, P. (2005). Evolving sustainably: A longitudinal study of corporate sustainable development. *Strategic Management Journal*, *26*(3), 197–218.

Barney, J. (1991). Firm resources and sustained competitive advantage. *Journal of Management*, *17*(1), 99–120. https://doi.org/10.1177/014920639101700108

Baron, D. P. (2001). Private politics, corporate social responsibility, and integrated strategy. *Journal of Economics & Management Strategy*, *10*(1), 7–45.

Basu, K. and G. Palazzo (2008). Corporate social responsibility: A process model of sensemaking. *Academy of Management Review*, *33*(1), 122–136.

Bennett, R. C. and R. G. Cooper (1981). The misuse of marketing: An American tragedy. *Business Horizons*, *24*(6), 51–61.

Berkhout, F. (2013). Sustainable Innovation Management. In Dodgson, M., Gann, D. M., & Phillips, N. (Eds.), *The Oxford handbook of innovation management*. Oxford, UK: Oxford University Press.

Bonini, S. T. M. Koller, and P. H. Mirvis (2009). Valuing social responsibility programs. *McKinsey on Finance*, *32*(Summer), 11–18. Retrieved from www.mckinsey.com/business-functions/strategy-and-corporate-finance/our-insights/valuing-social-responsibility-programs

Boons, F., C. Montalvo, J. Quist, and M. Wagner (2013). Sustainable innovation, business models and economic performance: An overview. *Journal of Cleaner Production*, *45*, 1–8.

Bouquet, C. and Y. Deutsch (2008). The impact of corporate social performance on a firm's multinationality. *Journal of Business Ethics*, *80*(4), 755–769.

Brik, A. B. (2007). CSR Innovation: Its antecedents and impact on market orientation. Paper presented at the AIB-UK Conference 2007, London. As cited in Gugler, P. and J. Y. Shi (2009). Corporate social responsibility for developing country multinational corporations: Lost war in pertaining global competitiveness? *Journal of Business Ethics*, *87*(1), 3–24.

Burget, M., E. Bardone, and M. Pedaste (2017). Definitions and conceptual dimensions of responsible research and innovation: A literature review. *Science and Engineering Ethics*, *23*(1), 1–19.

Burke, L., and J. M. Logsdon (1996). How corporate social responsibility pays off. *Long Range Planning*, *29*(4), 495–502.

Carrillo-Hermosilla, J., P. Del Río, and T. Könnölä (2010). Diversity of eco-innovations: Reflections from selected case studies. *Journal of Cleaner Production*, *18*(10–11), 1073–1083.

Carroll, A. B. (1979). A three-dimensional conceptual model of corporate performance. *Academy of Management Review*, *4*(4), 497–505.

Carroll, A. B. (1991). The pyramid of corporate social responsibility: Toward the moral management of organizational stakeholders. *Business Horizons*, *34*(4), 39–48.

Carroll, A. B., and K. M. Shabana (2010). The business case for corporate social responsibility: A review of concepts, research and practice. *International Journal of Management Reviews*, *12*(1), 85–105.

Chatfield, K., E. Borsella, E. Mantovani, A. Porcari, and B. Stahl (2017). An investigation into risk perception in the ICT industry as a core component of responsible research and innovation. *Sustainability*, *9*(8), 1424.

Chatfield, K., K. Iatridis, B. Stahl, and N. Paspallis (2017). Innovating responsibly in ICT for ageing: Drivers, obstacles and implementation. *Sustainability*, *9*(6), 971.

Choi, N., and S. Majumdar (2014). Social innovation: Towards a conceptualisation. In Majumdar, S., Guha, S., and Marakkath, N. (Eds.), *Technology and innovation for social change* (pp. 7–34). India: Springer. doi:10.1007/978-81-322-2071-8. Retrieved from https://link.springer.com/book/10.1007%2F978-81-322-2071-8 (accessed 15 May 2019).

Conway, S. and F. Steward (2009). *Managing and shaping innovation*. Oxford, UK: Oxford University Press.

Crane, A. and D. Matten (2016). *Business ethics: Managing corporate citizenship and sustainability in the age of globalization*. Oxford, UK: Oxford University Press.

Dahlsrud, A. (2008). How corporate social responsibility is defined: An analysis of 37 definitions. *Corporate Social Responsibility and Environmental Management*, *15*(1), 1–13.

Davila, T., M. Epstein, and R. Shelton (2012). *Making innovation work: How to manage it, measure it, and profit from it*. Upper Saddle River, New Jersey: FT Press.

De Kluyver, C. A. and J. A. Pearce (2006). *Strategy: A view from the top. An executive perspective*. Upper Saddle River, New Jersey: Pearson.

De Weerd-Nederhof, P. (2007). Innovation management, strategy and implementation using the pentathlon framework. Edited by Keith Goffin and Rick Mitchell. *R&d Management*, *37*(1), 90–92.

Den Ouden, E. (2011). *Innovation design: Creating value for people, organizations and society*. Springer Science & Business Media.

Digital Green. Retrieved from www.digitalgreen.org/ (accessed: 15 May 2018).

Dodgson, M., D. M. Gann, and N. Phillips, (Eds.). (2013). *The Oxford handbook of innovation management*. Oxford: Oxford University Press.

European Commission (2012). *Responsible Research and Innovation: Europe's ability to respond to societal challenges*. Publications Office of the European Union, Luxembourg. Retrieved from https://ec.europa.eu/research/swafs/pdf/pub_rri/KI0214595ENC.pdf

European Commission (a). *Corporate social responsibility (CSR)*. Retrieved from http://ec.europa.eu/growth/industry/corporate-social-responsibility_en (accessed: 15 April 2018).

Fairphone (a). Retrieved from www.fairphone.com/about/ (accessed: 23 December 2017).

Fairphone (b). Retrieved from www.facebook.com/Fairphone/?fref=ts; www.fairphone.com/we-are-fairphone/ (accessed: 23 December 2017).

Foley, R. W., M. J. Bernstein, and A. Wiek (2016). Towards an alignment of activities, aspirations and stakeholders for responsible innovation. *Journal of Responsible Innovation*, *3*(3), 209–232.

Frooman, J. (1999). Stakeholder influence strategies. *Academy of Management Review*, *24*(2), 191–205.

Galbreath, J. (2006). Corporate social responsibility strategy: Strategic options, global considerations. *Corporate Governance: The International Journal of Business in Society*, *6*(2), 175–187.

Gallego-Alvarez, I., J. Manuel Prado-Lorenzo, and I. M. García-Sánchez (2011). Corporate social responsibility and innovation: A resource-based theory. *Management Decision*, *49*(10), 1709–1727.

Garriga, E., and D. Melé (2004). Corporate social responsibility theories: Mapping the territory. *Journal of Business Ethics*, *53*(1–2), 51–71.

Gauttier, S., J. H. Søraker, C. Arora, P. A. E., Brey, and M. Mäkinen (2017). *Models of RRI in Industry*. Deliverable 3.3, Responsible Industry Project. Retrieved from: www.responsible-industry.eu/ (accessed: 1 May 2019).

Grant, A. M. (2007). Relational job design and the motivation to make a prosocial difference. *Academy of Management Review*, *32*(2), 393–417.

Grayson, D., and Hodges, A. (2017). *Corporate social opportunity!: Seven steps to make corporate social responsibility work for your business*. London: Routledge. Retrieved from: www.taylorfrancis.com/books/9781351280884

Griffin, J. J., and J. F. Mahon (1997). The corporate social performance and corporate financial performance debate: Twenty-five years of incomparable research. *Business & Society*, *36*(1), 5–31.

Gugler, P., and J. Y. Shi (2009). Corporate social responsibility for developing country multinational corporations: Lost war in pertaining global competitiveness? *Journal of Business Ethics*, *87*(1), 3–24.

Gurzawska, A., R. Cardone, A. Porcari, E. Mantovani, and P. Brey (2015). *Ethics assessment in different types of organizations: Industry*, SATORI Deliverable 1.1: Ethical Assessment of R&I: A Comparative Analysis; Annex 3h:, SATORI Project. Retrieved from: http://satoriproject.eu/media/3.h-Industry.pdf (accessed: 23 December 2017).

Gurzawska, A., M. Mäkinen, and P. Brey (2017). Implementation of Responsible Research and Innovation (RRI) practices in industry: Providing the right incentives. *Sustainability*, *9*(10), 1759.

Halkos, G., and A. Skouloudis (2018). Corporate social responsibility and innovative capacity: Intersection in a macro-level perspective. *Journal of Cleaner Production*, *182*, 291–300.

Hanekamp, G. (2007). Business ethics of innovation. An introduction. In Hanekamp, G. & Wütscher, F. (Eds.), *Business ethics of innovation* (pp. 1–6). Springer-Verlag Berlin Heidelberg. Retrieved from: https://link.springer.com/book/10.1007/978-3-540-72310-3#about

Hanke, T., and W. Stark (2009). Strategy development: Conceptual framework on corporate social responsibility. *Journal of Business Ethics*, *85*(3), 507.

Hansen, E. G., F. Grosse-Dunker, and R. Reichwald(2009). Sustainability innovation cube – a framework to evaluate sustainability-oriented innovations. *International Journal of Innovation Management*, *13*(04), 683–713.

Hauser, J., G. J. Tellis, and A. Griffin (2006). Research on innovation: A review and agenda for marketing science. *Marketing Science*, *25*(6), 687–717.

Herrera, M. E. B. (2016). Innovation for impact: Business innovation for inclusive growth. *Journal of Business Research*, *69*(5), 1725–1730.

Hesterly, B. and J. Barney (2008). *Strategic management and competitive advantage.* Essex: Pearson Education Limited. Retrieved from: www.pearson.com/us/higher-education/product/Barney-Strategic-Management-and-Competitive-Advantage-Concepts-2nd-Edition/9780136135203.html

Hockerts, K., M. Morsing, J. Eder-Hansen, P. Krull, A. Midttun, M. Halme, ... and P. Nurmi (2009). *CSR-driven innovation: Towards the social purpose business.* Center for Corporate Social Responsibility, CBS.

Hull, C. E., and S. Rothenberg (2008). Firm performance: The interactions of corporate social performance with innovation and industry differentiation. *Strategic Management Journal*, *29*(7), 781–789.

Husted, B. W. and D. B. Allen (2000). Is it ethical to use ethics as strategy? Business challenging business ethics: New instruments for coping with diversity. In Sójka, J., & Wempe, J. (Eds.), *International business* (pp. 21–31). Netherlands: Springer. Retrieved from: www.springer.com/gp/book/9780792365860

Husted, B. W., and D. B. Allen (2007). Corporate social strategy in multinational enterprises: Antecedents and value creation. *Journal of Business Ethics*, *74*(4), 345–361.

Husted, B. W., and D. B. Allen (2010). *Corporate social strategy: Stakeholder engagement and competitive advantage.* Cambridge, UK; New York: Cambridge University Press.

Husted, B. W., D. B. Allen, and N. Kock (2015). Value creation through social strategy. *Business & Society*, *54*(2), 147–186.

ING. (2008). *From sustainability to business value finance as a catalyst.* Retrieved from: www.ingwb.com/media/2266556/ing-sustainability-study-2018.pdf (accessed: 15 April 2018).

Iñigo, E. A., and L. Albareda (2016). Understanding sustainable innovation as a complex adaptive system: A systemic approach to the firm. *Journal of Cleaner Production*, *126*, 1–20.

Iyer, G., and D. A. Soberman (2016). Social responsibility and product innovation. *Marketing Science*, *35*(5), 727–742.

Kanter, R. M. (1999). From spare change to real change: The social sector as beta site for business innovation. *Harvard Business Review*, *77*(3), 122–123.

Khaledabadi, H. J., and T. Magnusson (2008). Corporate social responsibility and knowledge management implications in sustainable vehicle innovation and development. *Communications of the IBIMA*, 6, 8–14.

Kinkel, S., G. Lay, and J. Wengel (2005). *Innovation: more than research and development. Growth opportunities on different innovation paths* (No. 33e). Bulletins manufacturing innovation survey.

Koistinen K., M. Laukkanen, M. Mikkilä, J, Huiskonen, and L. Linnanen (2018). Sustainable system value creation: Development of preliminary frameworks for a business model change within a systemic transition process. In: Moratis L., Melissen F., & Idowu S. (Eds.), *Sustainable business models. CSR, sustainability, ethics & governance*. Cham: Springer. Retrieved from: http://link-springer-com-443.webvpn.fjmu.edu.cn/chapter/10.1007%2F978-3-319-73503-0_6#citeas.

Kranzberg, M. (1986). Technology and history: "Kranzberg's laws". *Technology and Culture*, *27*(3), 544–560.

Kurucz, E. C., A., Crane, A. McWilliams, D. Matten, J. Moon, & D. S. Siegel (2008). The business case for corporate social responsibility. In Crane, A., McWilliams, A., Matten, D., Moon, J., and Siegel, D. S. (Eds.), *The Oxford handbook of corporate social responsibility* (pp. 83–112). Oxford, UK: Oxford University Press.

Lai, W. H., C. C. Lin, and T. C. Wang (2015). Exploring the interoperability of innovation capability and corporate sustainability. *Journal of Business Research*, *68*(4), 867–871.

Lantos, G. P. (2001). The boundaries of strategic corporate social responsibility. *Journal of Consumer Marketing*, *18*(7), 595–632.

Liedtka, J. (2000). In defense of strategy as design. *California Management Review*, *42*(3), 8–30.

Lindner, R., S. Kuhlmann, S. Randles, B. Bedsted, G. Gorgoni, E. Griessler, ... and N. Mejlgaard (2016). Navigating towards shared responsibility. In Research and Innovation: Approach, Process and Results of the Res-AGorA Project. Retrieved from: http://pure.au.dk/portal/files/98634660/RES_AGorA_ebook.pdf.

Longoni, A., and R. Cagliano (2018). Sustainable innovativeness and the triple bottom line: The role of organizational time perspective. *Journal of Business Ethics*, *151*(4), 1097–1120.

Lubberink, R. J. B., V. Blok, J. A. C., van Ophem, and S. W. F. Omta (2017). A framework for responsible innovation in the business context: Lessons from responsible-, social- and sustainable innovation. In L. Asveld, R. van Dam-Mieras, T. Swierstra, S. Lavrijssen, K. Linse, & J. van den Hoven (Eds.), *Responsible innovation 3: A European agenda?* (pp. 181–207). Springer, Cham: Springer. Retrieved from: https://research.wur.nl/en/publications/a-framework-for-responsible-innovation-in-the-business-context-le

Lubin, D. A., and D. C. Esty (2010). The sustainability imperative. *Harvard Business Review*, *88*(5), 42–50.

Luo, X., and S. Du (2012). Good companies launch more new products. *Harvard Business Review*, *90*(4), 28.

Luo, X., and S. Du (2015). Exploring the relationship between corporate social responsibility and firm innovation. *Marketing Letters*, *26*(4), 703–714.

Lush Cosmetics. *Our environmental policy*. Retrieved from: https://uk.lush.com/article/our-environmental-policy (accessed: 15 June 2015).

Lynch, R. (2015). *Strategic management*. Seventh edition. Harlow, UK: Pearson Education Limited. Retrieved from: www.pearson.com/uk/educators/higher-education-educators/program/Lynch-Strategic-Management-7th-Edition/PGM1069920.html

MacGregor, S. P., and J. Fontrodona (2008). *Exploring the fit between CSR and innovation* (no. 759). IESE Business School Working Paper.

Maignan, I., and D. A. Ralston (2002). Corporate social responsibility in Europe and the US: Insights from businesses' self-presentations. *Journal of International Business Studies*, *33*(3), 497–514.

Margolis, J. D., and J. P. Walsh (2003). Misery loves companies: Rethinking social initiatives by business. *Administrative Science Quarterly*, *48*(2), 268–305.

Martinez-Conesa, I., P. Soto-Acosta, and M. Palacios-Manzano (2017). Corporate social responsibility and its effect on innovation and firm performance: An empirical research in SMEs. *Journal of Cleaner Production*, *142*, 2374–2383.

Maxfield, S. (2008). Reconciling corporate citizenship and competitive strategy: Insights from economic theory. *Journal of Business Ethics*, *80*(2), 367–377.

McElhaney, K. (2009). A strategic approach to corporate social responsibility. *Leader to Leader*, *52*(1), 30–36.

McElhaney, K. (2009). *Just good business: The strategic guide to aligning corporate responsibility and brand*. San Francisco, CA: Berrett-Koehler Publishers. Retrieved from: https://books.google.ht/books?id=uYR0zHbSTbsC&pg=PR3&source=gbs_selected_pages&cad=2#v=onepage&q&f=false

McWilliams, A., and D. Siegel (2000). Corporate social responsibility and financial performance: Correlation or misspecification? *Strategic Management Journal*, *21*(5), 603–609.

McWilliams, A., and D. S. Siegel (2011). Creating and capturing value: Strategic corporate social responsibility, resource-based theory, and sustainable competitive advantage. *Journal of Management*, *37*(5), 1480–1495.

McWilliams, A., D. S. Siegel, and P. M. Wright (2006). Corporate social responsibility: Strategic implications. *Journal of Management Studies*, *43*(1), 1–18.

Midttun, A. (2006). CSR and innovation, compatibility or contradiction? Towards a dynamic reinterpretation of CSR. *The Norwegian School of Management, CSR paper*, 3.

Midttun, A. (2009). Strategic CSR innovation – serving societal and individual needs. Retrieved from: https://biopen.bi.no/bi-xmlui/handle/11250/940 93 (accessed 15 May 2018).

Milling, P. M., and J. Stumpfe (2000, August). Product and process innovation: A system dynamics-based analysis of the interdependencies. In *18th International Conference of the System Dynamics Society Sustainability in the Third Millennium, Bergen, Norway*.

Mintzberg, H., and A. McHugh (1985). Strategy formation in an adhocracy. *Administrative Science Quarterly*, 30(2), 160–197. Retrieved from: www.jstor.org/stable/2393104

Mintzberg, H., B. Ahlstrand and J. Lampel (1998). *Strategy safari*. Prentice Hall, London. Cited in Hanke, T., and Stark, W. (2009). Strategy development: Conceptual framework on corporate social responsibility. *Journal of Business Ethics*, *85*, 507–516.

Muller, A., L. Välikangas, and P. Merlyn (2005). Metrics for innovation: Guidelines for developing a customized suite of innovation metrics. *Strategy & Leadership*, *33*(1), 37–45.

Nag, R., D. C. Hambrick, and M. J. Chen (2007). What is strategic management, really? Inductive derivation of a consensus definition of the field. *Strategic Management Journal*, *28*(9), 935–955.

Nidumolu, R., C. K. Prahalad, and M. R. Rangaswami (2009). Why sustainability is now the key driver of innovation. *Harvard Business Review*, 87(9), 56–64. Retrieved from: https://hbr.org/2009/09/why-sustainability-is-now-the-key-driver-of-innovation (accessed 15 May 2018).

Nordic Innovation Centre (2010). *CSR-driven innovation – combining design and business in a profitable and sustainable way*. August 2010. Retrieved from: http://nordicinnovation.org/Global/_Publications/Reports/2010/CSR-DRIVEN%20INNOVATION%20-%20Combining%20design%20and%20business%20in%20a%20profitable%20and%20sustainable%20way.pdf (accessed: 15 April 2018).

OECD. (2005). Oslo manual. Guidelines for collecting and interpreting innovation data. Retrieved from: www.oecd.org/science/inno/2367614.pdf (Accessed: 15 April 2018).

OECD. (2009). Sustainable manufacturing and eco-innovation. OECD. Retrieved from: www.oecd.org/env/consumption-innovation/42957785.pdf (accessed: 15 April 2018).

Orlitzky, M., F. L. Schmidt, and S. L. Rynes (2003). Corporate social and financial performance: A meta-analysis. *Organization Studies*, *24*(3), 403–441.

Padgett, R. C., and J. I. Galan (2010). The effect of R&D intensity on corporate social responsibility. *Journal of Business Ethics*, *93*(3), 407–418.

Patala, S., A. Jalkala, J. Keränen, S. Väisänen, V. Tuominen, and R. Soukka (2016). Sustainable value propositions: Framework and implications for technology suppliers. *Industrial Marketing Management*, *59*, 144–156.

Porter, M. E. (1985). *The Competitive advantage: Creating and sustaining superior performance*. New York: Free Press. Retrieved from: www.hbs.edu/faculty/Pages/item.aspx?num=193

Porter, M. E. and M. R. Kramer (2002). The competitive advantage of corporate philanthropy. *Harvard Business Review*, 80(12), 56–68.

Porter, M. and M. Kramer (2006). Strategy and society: The link between competitive advantage and corporate social responsibility. *Harvard Business Review*, 84(12), 78–92.

Powell, T. C. (2001). Competitive advantage: Logical and philosophical considerations. *Strategic Management Journal*, 22(9), 875–888.

Pyszka, A. (2012). Społecznie odpowiedzialne innowacje-konieczność czy moda? *Innowacje w zarządzaniu i inżynierii produkcji. Edited by R. Knosal*, 108–18. Retrieved from: www.ptzp.org.pl/files/konferencje/kzz/artyk_pdf_2012/p011.pdf

Reber, B. (2018). RRI as the inheritor of deliberative democracy and the precautionary principle. *Journal of Responsible Innovation*, *5*(1), 38–64.

Responsible-Industry Project Consortium (2017). Responsible-industry guide for the implementation of Responsible Research and Innovation (RRI) in the industrial context. Retrieved from: www.responsible-industry.eu/dissemination/deliverables (accessed: 15 May 2018).

Reuters. (2018). China to bar people with bad 'social credit' from planes, trains, March 16, 2018. Retrieved from: www.reuters.com/article/us-china-credit/china-to-bar-people-with-bad-social-credit-from-planes-trains-idUSKCN1GS10S (accessed: 16 May 2018).

Rexhepi, G., S. Kurtishi, and G. Bexheti (2013). Corporate social responsibility (CSR) and innovation – the drivers of business growth?. *Procedia-Social and Behavioral Sciences*, *75*, 532–541.

Ribeiro, B. E., R. D. Smith, and K. Millar (2017). A mobilising concept? Unpacking academic representations of responsible research and innovation. *Science and Engineering Ethics*, *23*(1), 81–103.

Rip, A. (2018). The past and future of RRI. In *Futures of science and technology in society*. Technikzukünfte, Wissenschaft und Gesellschaft / Futures of Technology, Science and Society. (pp. 115–133). Wiesbaden: Springer VS. Retrieved from: https://link.springer.com/chapter/10.1007/978-3-658-21754-9_7#citeas

Roman, R. M., S. Hayibor, and B. R. Agle (1999). The relationship between social and financial performance: Repainting a portrait. *Business & Society*, *38*(1), 109–125.

Rural Spark. Retrieved from: www.ruralspark.com/ (accessed: 15 May 2018).

Sánchez, P. E., and S. Benito-Hernández (2015). CSR policies: Effects on labour productivity in Spanish micro and small manufacturing companies. *Journal of Business Ethics*, *128*(4), 705–724.

Sawhill, J., and D. Williamson (2001). Measuring what matters in nonprofits. *McKinsey Quarterly*, (2), 98–107. Retrieved from: www.mckinsey.com/industries/public-and-social-sector/our-insights/measuring-what-matters-in-nonprofits

Sharma, S., and H. Vredenburg. (1998). Proactive corporate environmental strategy and the development of competitively valuable organizational capabilities. Strategic Management Journal, 19(8), 729–753.

Shen, R., Y. Tang, and Y. Zhang (2016). Does firm innovation affect corporate social responsibility? (No. 16-096). *Harvard Business School Working Papers*. Retrieved from http://hdl.handle.net/1765/99614

Stilgoe, J., R. Owen, and P. Macnaghten (2013). Developing a framework for responsible innovation. *Research Policy*, *42*(9), 1568–1580.

Sutcliffe, H. (2011). A report on responsible research and innovation. *MATTER and the European Commission*. Retrieved from https://ec.europa.eu/research/science-society/document_library/pdf_06/rri-report-hilary-sutcliffe_en.pdf

Taebi, B., A. Correlje, E. Cuppen, M. Dignum, and U. Pesch (2014). Responsible innovation as an endorsement of public values: The need for interdisciplinary research. *Journal of Responsible Innovation*, *1*(1), 118–124.

Tantalo, C., and R. L. Priem (2016). Value creation through stakeholder synergy. *Strategic Management Journal*, *37*(2), 314–329.

Tesla. Solar Roof. Retrieved from: www.tesla.com/en_EU/solarroof (accessed: 15 May 2018).

Trott, P. (2008). *Innovation management and new product development*. Pearson Education.

Ueki, Y., C. Jeenanunta, T. Machikita, and M. Tsuji (2016). Does safety-oriented corporate social responsibility promote innovation in the Thai trucking industry? *Journal of Business Research*, *69*(11), 5371–5376.

UNDP (2018). A new approach to an old problem. Clean technology innovation in response to climate change, 2 March 2018. Retrieved from: www.az.undp.org/content/azerbaijan/en/home/presscenter/pressreleases/2018/03/02/a-new-approach-to-an-old-problem-clean-technology-innovation-in-response-to-climate-change.html (accessed: 15 May 2018).

Von Schomberg, R. (2013). A vision of responsible research and innovation. In Owen, R., Bessant, J., & Heintz, M. (Eds.), *Responsible Innovation: Managing the Responsible Emergence of Science and Innovation in Society*, 51–74. https://doi.org/10.1002/9781118551424.ch3

von Weltzien Høivik, H., and D. Shankar (2011). How can SMEs in a cluster respond to global demands for corporate responsibility? *Journal of Business Ethics*, *101*(2), 175–195.

White, M. A., and G. D. Bruton (2010). *The management of technology and innovation: A strategic approach*. Cengage Learning.

Whittington, R. (2001). *What is strategy and does it matter?* Cengage Learning EMEA.

Yin, J., and D. Jamali (2016). Strategic corporate social responsibility of multinational companies subsidiaries in emerging markets: Evidence from China. *Long Range Planning*, 49(5), 541–558.

Zadek, S. (2000). *Doing Good and Doing Well: Making the Business Case for Corporate Citizenship*. Conference Board.

4 On the challenges and drivers of implementing responsible innovation in foodpreneurial SMEs

Cristina Covello and Konstantinos Iatridis

4.1 Introduction

Responsible research and innovation or responsible innovation (RI) has recently attracted a lot of scholarly interest, leading to fruitful streams of research ranging from discussions about the responsibility of researchers in research and development (R&D) teams (Pandza & Ellwood, 2013) to theoretical frameworks of RI (Genus & Stirling, 2018; Stilgoe et al., 2013; Voegtlin & Scherer, 2017) and risk management approaches to lessen the negative externalities of innovations developed in publicly funded academic research (Owen et al., 2013). At the same time, driven by the fact that the vast majority of research and innovation takes place in the private sector, a burgeoning stream of research focuses on the implementation of RI in industry (Auer & Jarmai, 2018; Dreyer et al., 2017; Hemphill, 2016; Iatridis & Schroeder, 2016; Martinuzzi et al., 2018; Stahl et al., 2017; van de Poel et al., 2017). The insights generated from these studies have diffused the concept at industry level but there is a dearth of research focusing on RI and small and medium-sized enterprises (SMEs). This is an important gap in the literature as the vast majority of firms fall into this category. In the words of the European Commission, SMEs are the "backbone of the European economy", making up 99% of the businesses in the EU (European_Commission, 2018). Therefore, learning more about the challenges and drivers SMEs might face when implementing RI presents an opportunity for RI to ensure that SME strategy is aligned with the European Commission's vision for growth.

In this chapter, we aim to enhance knowledge of this topic by focusing on a specific category of SMEs active in the food industry. These firms, also known as "foodpreneurial", are SMEs with innovative food ideas, from artisan products to cutting-edge food technology. Foodpreneurial SMEs represent an interesting case for research because businesses of this type tend to have a start-up mentality and use socially responsible business models to tackle issues as disparate as social inclusion and fighting disease. At the same time, foodpreneurial SMEs are disrupting big food and beverage producers at an alarming rate. In the USA, the top 25 food and beverage producers lost $18B in market share between 2009 and 2015 (Kowitt, 2015). Equally in the UK, such SMEs seem to play a gradually increasing role. For instance, Pip and Nut, a UK nut butter brand that claims to be a healthy alternative to sugar and palm oil-filled conventional nut butters, is a prime example of this type of challenger brand. In just four years, the company has captured approximately 13% of the UK nut butter market (Newsdesk, 2018). Perhaps foodpreneurial SMEs have found a way to see the "grand challenges" of our time as an opportunity rather than a constraint. They are establishing innovative business models that put responsibility higher on the agenda than large firms. For this reason, there may be a natural tendency for such SMEs to innovate responsibly.

This chapter contributes to the literature both empirically and theoretically. Empirically, we introduce new interview data with 19 foodpreneurial SMEs operating in London, UK. Theoretically, we contribute to the literature by providing insights on the challenges and drivers SMEs face when implementing RI. Our results suggest that lack of awareness, time, labour and money imposes obstacles to RI. Inclusion proves to be equally tricky, as most respondents had difficulties identifying stakeholders and articulating how they engaged with them. Respectively, and opposite to the enlightened self-interest perspective found in many orthodox economic arguments, our findings highlight personal values and a concern for societal and environmental wellbeing as business drivers. This illustrates that, despite contemporary pressures for quarterly-based profit maximisation, there are SMEs that resist such demands and attribute the same, if not higher, significance to societal and environmental issues as they do to profit.

The chapter is organised as follows. First, we discuss the literature of RI in industry to act as a frame of reference with which to compare RI to already embedded RI practices in the SMEs studied. We then discuss the research methodology and findings of our analysis, with a focus on the challenges and drivers of implementing RI in the aforementioned SMEs. We conclude by discussing the implications of the study, its limitations and directions for future research.

4.2 Literature review

RI prioritises the integration of responsibility in R&D processes. The literature suggests that innovation outcomes can be considered responsible when they are environmentally sustainable, ethically acceptable and socially desirable (von Schomberg, 2013). Environmental sustainability prioritises a responsible use of natural resources during the innovation processes while ethical acceptability proposes that innovation outcomes need to be in line with widely accepted norms and values in society. In turn, societal desirability suggests that innovation outcomes should contribute to the solution of key challenges facing humanity.

To achieve these ends, the literature suggests that innovators should anticipate the outcomes of innovation and possible consequences (both intended and unintended) for broader groups. Additionally, RI highlights the need for auditing and reviewing mechanisms during innovation processes to constantly reflect on each stage of an innovation, ensuring potential problems are identified and managed appropriately and timely, or prevented altogether. The inclusion of stakeholder groups at the early stages of innovation is perceived as one of the most important aspects of RI, as early engagement is believed to improve the chances for the innovation outcome to be in line with stakeholders' social, environmental and ethical demands. It is important to notice that in the private sector inclusion is based on the assumption that the most effective approach for companies is to prioritise the various stakeholder demands based on stakeholder salience (Mitchel et al., 1997). Furthermore, RI demands that researchers and innovators respond openly and transparently to all those, directly or indirectly, affected by the outcome of innovation (Stilgoe et al., 2013).

Recently, scholars have started addressing how some of the abovementioned points can be undertaken by industry. For instance, Lubberink et al. (2017) in their literature review of RI in industry found that existing studies focusing on inclusion prioritise clients and end-users, at the expense of including wider social groups in innovation processes. In turn, Van de Poel et al. (2017) discuss ways of developing key performance indicators, which can facilitate auditing and monitoring RI implementation. Their insights provide a useful first step for assessing and benchmarking RI implementation in industry. Equally, Gurzawska et al. (2017) argue that RI integration into corporate processes can be achieved by broadening the inclusion approach and taking into account the views of not only consumers but also employees and institutional actors. Respectively, Garst et al.'s (2017) study focuses on firms operating in the food industry, and analyses corporate motives for adopting RI practices. The authors conclude that both internal motives, such as profit maximisation and moral duty, as well as external drivers, such as regulation and legitimacy, play an important role in firms' decision to engage with RI. Similarly, Chatfield et al. (2017) focus on corporate motives but also take into account obstacles in the implementation of RI. Although they argue that economic motives play an important role in the adoption of RI, these authors found that there are cases of companies that have been able to balance financial and altruistic goals.

At the same time, a handful of studies have focused on the implementation of RI in SMEs. Pavie et al.'s (2014) study argues that SMEs' peculiarities, such as the fact that they are more flexible and less restricted by formal policies and procedures, make them ideal entities for experimenting with emerging technologies that tackle societal challenges. The authors envision SMEs as critical to RI dissemination and suggest that, by working in niche markets, SMEs can experiment until the proposition becomes attractive enough for large incumbents to acquire the SME and bring the innovation to a larger market. Policy instruments that broker relationships between academia, incumbents and start-ups could facilitate experimentation by helping a range of stakeholders interact – from suppliers, to users, to competitors. The authors state that a pre-requisite for SMEs engaging in experimentation is recognising the added value generated by such an approach. However, the authors discuss neither the resource constraints SMEs face nor how SMEs should find the resources to experiment and interact with stakeholders.

Some authors claim that resource constraints endured by SMEs might not necessarily be a bad thing but can rather be seen as a source of innovative thought. For instance, Gibbert et al. (2007) argue that resource constraints actually lead to better innovation as "would-be innovators facing constraints are more likely to find creative analogies and combinations that would otherwise be hidden under a glut of resources" (p. 16). Halme and Korpela's (2014) study of 13 SMEs developing RIs in Scandinavia reached similar findings. They looked at the quality, quantity and combinations of resources in SMEs that have successfully commercialised their innovations to uncover the resource conditions necessary for RI. What they found was that all SMEs needed at least some financial capital to innovate but that SMEs with financial resource constraints could compensate with social networks.

4.3 Method

We adopted a qualitative research framework, whereby we collected and analysed interview data to enrich theory. The sample of informants was purposively selected by the researchers, who identified the respondents as being theoretically relevant to the study's purpose (Silverman, 2006). We chose 19 foodpreneurial SMEs, of which 12 did not have a specific social purpose, four had a clear social purpose mission but they could not be classified as social enterprises and three could be classified as social enterprises (see Table 4.1 for participant and interview details). Seven of the participant SMEs belonged in the primary researcher's professional network, and they were chosen because the researcher was familiar with their innovative and ethical business practices. These seven were studied first, and the findings that emerged informed the selection of subsequent participants. The other 12 participants were found through online research. Their suitability for the project was determined from their websites and they were approached through cold calling. All interviewees held the title of either owner/manager or co-founder, which means that each had in-depth knowledge of the business' purpose, strategy and approach to responsibility. The interviews were conducted in both private and public spaces, all of which offered a quiet and private setting. Each interview was recorded, and the parts of interviews that were deemed interesting and useful were transcribed (Bogdan & Biklin, 1998; Creswell, 2009; Seale & Silverman, 1997; Silverman, 2006). Respondents were guaranteed compliance with ethical restrictions, such as those relating to informed consent, anonymity via the use of pseudonyms and avoidance of harm and deception (Fontana & Frey, 2005), and were informed that they could opt out of the research at any time.

Table 4.1 Profile of respondents

Participant	*Business type*	*Number of employees*	*Interview length (minutes)*
1	Tea blender and wholesaler	3	75
2	Online retailer of craft food	13	60
3	Chocolatier	10	45
4	Cold brewed coffee	7	71
5	Sugar- and sweetener-free fizzy drinks	10	45
6	Tea blender and wholesaler	2	38
7	Hydroponic vegetable-growing appliance	7	38
8	Bean to bar chocolate maker	10	65
9	Beer brand	7	40
10	Bakery	3	46
11	British pulses wholesaler	4	50
12	Honey production and hive rental	2	50
13	Cheesemonger	3	50
14	Hydroponic vegetable grower	7	39
15	Community-supported fishing and online retailer	2	56
16	Food-ordering app	1	46
17	Sauce producer and street food	2	35
18	Honey producer and bee farmer	2	80
19	Craft brewer	3	55

The interview questions (Table 4.2) were framed according to the purpose of the research and focused on the SMEs' motivations to adopt RI and the obstacles they faced during implementation. Interview lengths ranged from 35 minutes to 1 hour and 15 minutes, with most hovering near the 47-minute mark. The interviews were all run by a key member of the research team, who is working in the food industry and could "talk shop" with the respondents. Interviews took place in person and via telephone, on a one-to-one basis. The researcher would recapitulate the interviewees' answers to let the respondent assess the adequacy of the interviewer's interpretation (Healey & Rawlinson, 1994). In line with the relevant literature discussing bias control in interviews and ensuring data reliability (Healey & Rawlinson, 1994; Saunders et al., 2012), following each interview the researcher recorded the information about the location, date and time, setting, and the researcher's immediate impression of how the interview went. To gain further insights into the accuracy of the respondents' answers, the interviewer took reflective notes that included observations of the participants and key ideas pertaining to the research themes. Additionally, the research relied on data from the company's website and relevant secondary data from journals, governmental sources and project deliverables to inform the interview questions and help build rapport between the interviewer and interviewee.

Table 4.2 Interview questions

1. Can you tell me about what you do, about your business and how you got started?
2. What does the term Responsible Research and Innovation mean to you?
3. Can you tell me about your company's innovation strategy? Do you have one? How does it work?
4. Why do you think your business strategy is innovative?
5. Can you provide examples of how your company acts responsibly?
6. Can you think of an ethical dilemma or tricky situation your business has had to resolve? How did you resolve it?
7. What incentives are there for running your business responsibly?
8. Can you talk about how societal challenges affect your business?
9. How do you engage with internal and external stakeholders on these issues?
10. Are you responsive to what you learn from stakeholders?
11. How do you integrate their ideas into your business?
12. If you do not engage with stakeholders what is the reason? What obstacles do you face?
13. Besides engagement what does your company do to make sure research and innovation is done responsibly?
14. In which areas of your business do you foresee societal or ethical risks arising? Do you do anything to overcome these? Who evaluates this?
15. Can you describe how you evaluate what future impacts your innovation decisions might have beyond your firm?
16. How do you anticipate your innovation strategy will develop in the future? What obstacles do you foresee?

Our study's interpretivist approach is informed by previous research (Holstein & Gubrium, 1995) and perceives an interview as a socially constructed event in which participants "articulate on-going interpretive structures" (p. 16). Accordingly, the interviewer acted as an expert facilitator who gave space to respondents to talk about their views through open-ended, topic-based questions (Silverman, 2006). This allowed the respondents to talk about topics that might even contradict each other (e.g. expressing business incentives beyond profit maximisation while at the same time lamenting a lack of resources and the need to balance responsibility with business survival). Such an approach allowed the interviewees to identify with multiple narrative positions (Arsel & Thompson, 2011), and the researcher to explore how the thematically significant narrative constructions link with each other, and how they emerge. This enabled the formulation of a narrative pluralism, with respondents producing various accounts of their perception of what it means to innovate responsibly. As stated earlier, all the participants in this research were either owner/managers or co-founders with a vested interest in how the business is perceived. Therefore, it is possible that the interviews are subject to social desirability bias (Fisher, 1993).

Transcripts were systematically studied through an iterative process of coding and analysis. The first reading of the transcripts was quick and codes were generated indiscriminately. At one point during the initial coding stage this research had 51 codes that pointed to possible themes for further exploration. As research progressed, some codes were renamed or combined with other codes that revealed the same sentiment. After an initial coding, the process became more focused. Here, the researcher looked for data that fit with established codes while remaining open to new codes. Codes that ceased to accumulate data were set aside but not deleted. Deleting the codes would have prohibited their future use. Coding is not a linear process and a further reading of the transcripts may have prompted a new idea, strengthening the deleted code. The process of coding remained interactive and iterative until all the interviews had been coded. Once all the interviews had been coded they were reread. A final edit of the codes revealed the following three emergent themes: (1) unfamiliarity with the RI concept; (2) the challenge of stakeholder inclusion; and (3) beyond profit maximisation, personal incentives and visions of change.

The next section of this chapter is organised around these themes, which are described together with illustrative quotes from the interviewees that encapsulate their meaning. These themes capture the challenges and drivers that foodpreneurial SMEs may encounter when trying to use the RI framework.

4.4 Findings

4.4.1 Unfamiliarity with the RI concept

None of the participants had heard of the terms responsible innovation or Responsible Research and Innovation before the interview. This is not surprising as the concept is relatively new and there has been low diffusion of the concept at an SME level. Most participants defined the term by accounting for the words separately and then combining them to complete a conceptual definition. For example:

> I haven't heard of responsible innovation before but I can imagine what it means as a set of words. Responsible: to consider ethics, morals and social impact. Innovation: the development of new products and services. So I guess responsible innovation would be the development of ethical products and services.
>
> (Participant 9)

Another example illustrates this approach but also reveals a struggle to define the term:

> Responsible innovation – I'm not sure what it means but I guess it would be the two words together. Hmmm – I'm having trouble articulating this. I think sometimes businesses are too focused on profits and

> they put ethics to one side. But actually if they were to focus on the ethics the rest would take care of itself.
>
> (Participant 13)

Difficulty articulating the term was not uncommon. Most participants needed time to pause and think about their answers. This led to a distinction between those who defined RI in terms of desired outcomes. For example:

> I think innovation is inherently good. All innovation has its benefits. In other words there is no bad innovation. I guess when you add the responsible part in you're talking about customer welfare and how innovation benefits society.
>
> (Participant 16)

And those who included a process element as articulated here:

> My understanding of responsible innovation is that it's about developing products and processes in a way that is ethically and environmentally sustainable.
>
> (Participant 11)

Most participants understood that RI is about both processes and outcomes. They recognised RI as not only innovating for society but also how that innovation should be achieved. However, only one participant provided an indication as to what that process might look like.

> I guess I see responsible innovation as building in the responsible element right at the beginning of the innovation process. So it's an added thing to consider. So for instance when creating a new product you would work from the ground up by making sure the ingredients are coming from an ethical source, that the processes to make it are energy efficient, that we use the least amount of packaging as possible. It's about compromise too and doing the best with the resources available.
>
> (Participant 3)

One interviewee had done background research on RI prior to the interview and therefore his definition of the term was most closely related to definitions found in the literature.

> I guess it's about creating a framework for decision-making that tries to encompass all the facets and impacts of innovation – other factors than financial.
>
> (Participant 7)

The distinguishing feature in his definition is the word "framework" which connotes a structured process for making ethical decisions. None of the interviewees mentioned stakeholders or stakeholder engagement in their definitions. As such, the interactive and mutually responsive aspect of RI was absent from the definitions provided.

4.4.2 The challenge of stakeholder inclusion

Deliberation with stakeholders on responsible decision-making proved inconsistent. There seemed to be tension between not having a complete understanding of who a stakeholder might be or how to engage with them, yet acknowledging stakeholder influence and the stakeholder as a source of innovation. Stakeholder engagement was generally informal and framed as networking rather than stakeholder engagement. For example:

> We talk to the Greater London Authority and the Food Team to affect policy, we do a lot of networking and drinks…We do a lot of networking on urban ag [agriculture] and food tech.
>
> (Participant 14)

> I think doing talks and informal networking is the best way to engage with stakeholders.
>
> (Participant 5)

Some participants related to stakeholder engagement in terms of building a community and fostering trust within that community.

> We could definitely improve our engagement with stakeholders. We're in constant contact with the fishermen and with our members about the business and about how it all works and the logistics but we don't engage with them on social issues. Most of what we do is informal and anecdotal. But we have built up trust – not just with our members like I spoke about before but also with the fishermen…We've had to do some cultural navigating for sure – but mostly we've just taken our time to build up a certain level of trus.
>
> (Participant 15)

Perhaps less explicit but still evident in dialogue surrounding stakeholder engagement was the idea of reciprocity.

> We engage with the farmers. We have direct relationships with them and we get to learn about their issues and try and take those on board. So for example one of the guys we deal with in Ecuador has decided to take this project on with monkeys because there's an area that's being deforested that's both a monkey habitat and a cocoa growing area. He's

married these issues and said if we can sell his cocoa then it's a double win. So we've said right – okay – and we've gone to Cocoa Runners who distribute chocolate from lots of different bean to bar chocolate makers to set up a promotion where they get a bunch of different makers on board to try making bars with these beans and see what everyone comes up with – sort of like a competition.

(Participant 8)

This story illustrates an innovative way to engage with stakeholders and suggests reciprocity between the company and the farmers that builds real partnership. These three attributes of stakeholder engagement in SMEs – informality, trust and reciprocity – align with the idea of social capital as defined by Putnam (2001).

Russo and Perrini (2010) also believe that for SMEs stakeholder engagement is linked to the building of social capital and that social capital is a driver of sustainability from the SME perspective. These authors studied how social networks relate to corporate social responsibility (CSR) strategy and assert that SMEs form natural relationships with a range of stakeholders but that they do not exploit these relationships for competitive advantage. However, they argue that SMEs should use their social capital to enhance their CSR credentials through a more focused stakeholder approach (ibid.). Our findings corroborate those of Russo and Perrini (2010) as they suggest that corporate responsibility in SMEs is more closely related to social capital than to stakeholder salience theory (Mitchell et al., 1997).

Although all participants engaged with stakeholders from within their own firms and external stakeholders from the wider community, many participants had trouble identifying their stakeholders. In fact, some were not entirely sure what the term meant. One participant thought that stakeholder was synonymous with shareholder. He asked:

What do you mean by stakeholders? Do you mean my shareholders?

(Participant 13)

Another participant needed clarification:

Stakeholder – who do you define as stakeholders?

(Participant 9)

Still others did not classify their relationship building as stakeholder engagement.

I would totally be willing to engage with internal and external stakeholders on these [social] issues. I'll be honest with you that until now it hadn't occurred to me.

(Participant 1)

This comment came despite the fact that upon further probing the participant revealed he indeed engaged on the social issues that affect his business.

> We do speak about these [social] issues internally, in an informal way.
>
> (Participant 1)

These examples point to a lack of familiarity with the stakeholder engagement concept. Importantly, this lack of knowledge does not create resistance to engagement. The theme of willingness was common amongst participants.

> We're definitely willing to engage with internal and external stakeholders on these issues – even excited about it – but the question is when and how. I don't really know how I would go about engaging on these issues or what I would do with the information. I wouldn't know where to start in terms of knowing what to say or how best to approach these topics.
>
> (Participant 3)

However, equally common were obstacles to engaging, like time constraints and the priority for solving more urgent business issues.

> There's no time to engage! We have an informal advisory board but the focus is on not losing money – not ethical issues.
>
> (Participant 2)

Yet, the participants also understood the business benefits for engaging on social issues.

> Time is what holds us back from engaging more... But we realise the better we are at communicating with stakeholders the more the business will progress so we really need to do it.
>
> (Participant 12)

Over half the interviewees acknowledged the influence of stakeholders on their businesses.

> I think our partnership approach is also unique. We see everyone from our brewers to packagers to our accountants and lawyers as partners in the business which means we're all working toward a common goal.
>
> (Participant 9)

> Our values centre around creating long-term relationships with farmers. Actually we focus on lasting relationships with all our partners whether farmers, packaging producers, employees or even other associations – like our involvement with the Vegan Association.
>
> (Participant 11)

Some participants also expressed how stakeholders have a direct impact on their innovations.

> I think we'll continue to innovate by listening to the growers as a primary way of learning about the issues that we want to tackle. Of course new product development for us has to be kind of market driven so we need to be thinking about what people want to buy – and of course that's a benefit for the growers as well because if we can sell more chocolate we can buy more beans from them.
>
> (Participant 8)

> When we went to New York we met the guys from the New York City Parks department on their green roof. They gave us the crown jewels for learning how to set up a roof garden and now that we've paid our respects they're happy to collaborate with us. Any question I have on green roofs, bees and solar panels I now have a resource I can call on and that's tremendously valuable.
>
> (Participant 19)

These examples show how the foodpreneurial SMEs use stakeholder engagement as a resource for researching and implementing new ideas despite the lack of a formal engagement strategy.

4.4.3 Beyond profit maximisation, personal incentives and visions of change

In the type of SMEs studied here, business drivers centre on the personal values of the owner/manager. Visions of creating change and the owner/manager as change maker were evident, as were business incentives beyond profit maximisation. Spence and Rutherfoord's (2001) work helps to put these findings into perspective. They claim that SME entrepreneurs' business orientations can be described as profit maximisation priority, subsistence priority, enlightened self-interest priority or social priority. Briefly, profit maximisation refers to a drive for money, subsistence priority is based on long-term survival, enlightened self-interest refers to being active on social issues but in the knowledge that this could positively affect profit and social priority means that social values are embedded in business life and take priority over profit maximisation (ibid.). None of the participants in this research could be described by Spence and Rutherfoord's profit maximisation frame. Some participants fit their enlightened self-interest frame. However, most of the entrepreneurs can be described by the social priority frame, which embeds social issues into business life. For example:

> [Our product] contains no sugar or sweeteners, nothing artificial. We see this as an alternative [in the market]. We think too many people are consuming sugar without thinking about it – or even knowing about

> it. Our vision is to help improve the health of the country, increase health and happiness and decrease obesity by providing an alternative to sugary drinks.
>
> (Participant 5)

> We started the business to pursue our interest in beekeeping but we quickly realised there was an appetite for sustainable local food production in cities. So that's really the core of our business – to produce a local food source sustainably and make people aware of it and continue to promote it – that is finding the best ways to produce local food in an urban environment.
>
> (Participant 12)

That does not mean that the owner/managers discounted profit altogether, it is just that profit was not their top priority. This is evidenced here:

> I'm not sure about an ethical dilemma. I mean we're not about profit at the expense of the environment or sustainability but we're not a social enterprise either.
>
> (Participant 4)

When asked outright about their incentives for running their businesses responsibly the majority of subjects claimed personal moral conduct. This finding corroborates Jenkins' (2006) work that showed moral judgements outweighed business benefits as incentives for implementing CSR. Responses were generally quick and animated. For example:

> What's my incentive? Well – being able to look in the mirror!
>
> (Participant 14)

> I think the real incentive for running our business this way comes down to treating people how you would want to be treated. I could act like a cunt and I would probably make more money – but I don't want to act like a cunt!
>
> (Participant 10)

> Our incentive for running our business in a responsible way is our moral compass.
>
> (Participant 5)

These statements lead to another significant finding. None of the SMEs, not even the social enterprises, had a formal set of values or code of conduct yet half of the participants expressed values-driven decision-making. That means decision-making is ultimately made at the discretion of the owner/manager and inevitably according to his or her personal ethics.

> Right now we don't evaluate the future impacts of our work. We don't exactly know – it's impossible to predict where we're going but if our culture and ethics are intact then we know we'll head in the right direction and we'll make the right decision based on our values.
>
> (Participant 2)

> But, we've got a good reputation so far and we operate fairly. We hope if we live by the spirit of our values we can beat other companies at their own game!
>
> (Participant 9)

> The incentive for running the business this way is simply because I couldn't imagine doing it any other way. Between my upbringing, my education, my respect for the planet – this is just how I want to do business.
>
> (Participant 15)

Our findings suggest that foodpreneurial SMEs are driven by factors other than pecuniary reward, which creates a welcome platform for RI. However, it is possible that the personal incentives that drive decision-making could hinder RI processes if the results of the research process and stakeholder engagement conflict with the personal values and desires of the company's owner/manager.

4.5 Discussion

The extant literature on RI in industry has a tendency to concentrate on larger firms, relatively ignoring the SME context. Yet, SMEs form a significant part of the European and global economy, with the majority of firms falling into this category. Consequently, by analysing what these SMEs intuit about RI, the characteristics of their stakeholder engagement, their business motivation and their proclivity to innovate responsibly, this chapter enhances knowledge of how RI can be undertaken by this type of firm.

Our study corroborates previous findings (Chatfield et al., 2017) on RI awareness by demonstrating that, while many aspects of RI might be addressed by SMEs, the term itself is not well known. Thus, there might be cases where the subjects embrace RI principles without knowing it. This matters because, without using RI deliberately it will be coincidence if their innovation fulfils the needs of society. This is demonstrated in how many of the interviewees were not sure how to define a stakeholder or express how they engaged with them, but when probed could tell stories of how stakeholder influences have affected their business. Opposite to classic notions of stakeholder engagement that use a language of stakeholder salience based on legitimacy, power and urgency to prioritise stakeholder claims against the firm (Mitchell et al., 1997), our respondents told stories of building community, trust and reciprocity, which exemplify social capital. Perhaps these differences are the result of the foodpreneurial SMEs' stakeholder engagement process, which is informal, comes naturally and is even unintentional at times. Taken further, some of the interviewees expressed how they work with stakeholders to develop innovations. Due to the nature and size of their businesses, these SMEs are not engaged in high-level scientific research of the sort that dominates the RI discourse. Of course most collaborate with customers as a form of market research, but this is not necessarily innovative as most companies undertake customer feedback exercises. Still, many of the participants described more than simple customer research meant to improve their firm's offering and included stories that illustrated how engaging in research improves the community in which they operate, demonstrating in this way a propensity for RI.

Perhaps theories of stakeholder identification and salience are not necessary for these SMEs on account of their ability to network naturally within the communities they operate. While time and money constraints might impede a formal stakeholder prioritisation process, the informal networks they build as a process of product and business development are more appropriate to their size and structure. According to Pellé and Reber, "innovators are often driven by competition and by first-mover advantages that call for secrecy" (2013, p. 39). Our SMEs did not seem to suffer from this kind of competitive drive. Most of the participants expressed openness to sharing ideas, suggesting they would view the stakeholder engagement aspect of the RI process as complementary to their business objectives, rather than contradictory. Examples like Participant 8's cooperation with Ecuadorian farmers links to Hockerts et al.'s (2008) assertion that stakeholders in the SME setting are seen more like partners in the business than outsiders that need to be managed. Additionally, the stakeholder salience concept may actually be at odds with RI in that some researchers believe the idea of prioritising stakeholder claims disregards stakeholders that have legitimate claims on the firm (Derry, 2012). Furthermore, these marginal stakeholders may provide opportunity for innovation more in line with the principles of RI.

However, just because stakeholder identification and salience as described by Mitchell et al. (1997) do not fit within the context of RI in SMEs, it does not mean that the owner/managers should not take an active approach to stakeholder engagement. They already express how influential stakeholders are to their business and how they work with them to innovate so it would make sense for them to capitalise on the investments they are making to maximise the benefits of engagement. Doing so would result in an enhanced reputation and increase in confidence and loyalty, not to mention a more stable workforce and more sustainable business over the long term (Russo & Perrini, 2010). In fact, their preoccupation with the day-to-day running of the business is precisely the reason they should invest in strengthening their networks. In this way, they can compensate for their resource constraints by building ever-stronger networks.

Most subjects pointed to resource constraints as obstacles to engagement but their willingness to engage, particularly on societal challenges, provides fertile soil for RI to take hold. Maybe the greater obstacle to effective stakeholder engagement is the owner/managers' lack of knowledge for how to implement the deliberative aspect of the RI process. The owner/managers are the driving forces behind SMEs so choosing with whom they engage and how they engage is done very much in an image of themselves. That means that, for these SMEs, with their informal engagement processes, stakeholder salience is closely linked to the values and motivations of the owner/manager.

Part of what makes the businesses in this study unique is their vision for affecting change and motivations beyond profit maximisation. They already conduct their businesses responsibly, albeit on their own terms. Nevertheless, it would be a shame for RI to discourage this kind of thinking just because it might not fit within the RI framework. Perhaps RI policy would be more effective in tackling the reflective part of the RI process amongst SME owner/managers so that they become aware of how their decisions impact society from multiple perspectives, before they embark on a deliberative process that is perceived as harder and more resource-intense to implement.

Anticipating future impacts and reflecting on the underlying motivations driving research and innovation are important aspects of the RI process. Therefore, participants were asked about their motivations and incentives for running their businesses responsibly. The primary incentive the subjects revealed was the fulfilment of their own moral code. Decision-making and problem solving were based on the entrepreneurs' individual values and motivations, which indicates a high level of personal control over their organisations. This is further evidenced by the lack of formal values or codes of conduct made available to the public and open to scrutiny. Most of the owner/managers attributed this to their size. However, while personal values guided the business and could be changed on a whim due to their lack of formality, not one of the participants had what Spence and Rutherfoord (2001) would describe as a profit maximisation priority. Some fit the enlightened self-interest priority but the majority were oriented toward the social priority. Despite none of the subjects prioritising profit maximisation, only half could articulate what societal challenges affected their business or what challenges their business ideas could tackle. In other words, they work toward a personal goal that aligns with their own vision of what "grand challenges" are most important.

For RI to work, it needs to clearly target agreed desired outcomes through a research and innovation process that involves some inward reflection but also much outward deliberation and responsiveness to make sure outcomes meet the needs of society. Therefore, the extent to which responsible business practices are linked to personal values and motivations unearths a potential conflict with RI. It appears there is a gulf between the individualistic decision-making in these SMEs and the idea of meeting their responsibilities to all stakeholders that may make embedding the RI process difficult, even if the end result of their business practices is ultimately good for society.

4.6 Limitations and conclusion

This research is not exempt from limitations, which point towards possible directions for future research. The sample size of this study is not sufficient for gaining in-depth insights. More studies, with access to bigger samples, would provide a broader account of foodpreneurial SMEs' views on RI. Although our findings corroborate those from other sectors, more evidence is needed to improve the robustness of our claims. Furthermore, our interviews involved one subject who provided a single perspective on the drivers and challenges food SMEs face. Future studies could draw on additional sources of information to discuss these drivers and challenges. Another limitation stems from the fact that at times interviewees might want to appear altruistic or benevolent. It's possible that some of the subjects overstated their responsibility. They may have been apt to do this because of the nature of the study and because they are owner/managers trying to leave a good impression of their firm in the public.

Apart from its limitations, learning more about the subjects' incentives, motivations, stakeholder engagement and perceptions of RI provided scope for generating ideas on how RI could be implemented in SMEs of this kind. Even just establishing a better sense of what drives SMEs provides an opportunity for steering their responsible business practices toward tangible RI outcomes. More specifically, appealing to the SMEs' tendency to envision their businesses as agents of change by helping them build RI into their entrepreneurial story could make RI appealing to owner/managers, thereby proliferating the concept. This research suggests that the foodpreneurial SMEs in London are attuned to responsible business and that their business models could accommodate the proliferation of RI. Therefore, studying RI in the SME context has the potential for impacting dissemination of the concept to a wider audience.

References

Arsel, Z. & Thompson, C. J. (2011). Demythologizing Consumption Practices: How Consumers Protect Their Field-Dependent Identity Investments from Devaluing Marketplace Myths. *Journal of Consumer Research*, 37(5): 791–806.

Auer, A. & Jarmai, K. (2018). Implementing Responsible Research and Innovation Practices in SMEs: Insights into Drivers and Barriers from the Austrian Medical Device Sector. *Sustainability*, 10(1): 17.

Bogdan, R. B. & Biklin, S. K. (1998). *Qualitative Research for Education: An Introduction to Theory and Methods* (3rd ed.). Needham Heights, MA: Allyn and Bacon.

Chatfield, K., Iatridis, K., Stahl, B., & Paspallis, N. (2017). Innovating Responsibly in ICT for Ageing: Drivers, Obstacles and Implementation. *Sustainability*, 9(6): 971.

Creswell, J. W. (2009). *Research Design. Qualitative, Quantitative and Mixed Methods Approaches* (3rd ed.). Thousand Oaks, CA: Sage.

Derry, R. (2012). Reclaiming Marginalized Stakeholders. *Journal of Business Ethics*, 111(2): 253–264.

Dreyer, M., Chefneux, L., Goldberg, A., von Heimburg, J., Patrignani, N., Schofield, M., & Shilling, C. (2017). Responsible Innovation: A Complementary View from Industry with Proposals for Bridging Different Perspectives. *Sustainability*, 9(10): 1719.

European Commission. Entrepreneurship and Small and Medium-sized Enterprises (SMEs). http://ec.europa.eu/growth/smes/ (accessed 9 July 2018).

Fisher, T. (1993). The View from the Top: Chief Executives' Perceptions of Total Quality Management. *Australian Journal of Management*, 18(2): 181–195.

Fontana, A. & Frey, J. H. (2005). The Interview. From Neutral Stance to Political Involvement. In N. K. Denzin & Y. S. Lincoln (Eds.), *The Sage Handbook of Qualitative Research* (3rd ed.): 695–727. Thousand Oaks, CA: Sage.

Garst, J., Blok, V., Jansen, L., & Omta, O. (2017). Responsibility versus Profit: The Motives of Food Firms for Healthy Product Innovation. *Sustainability*, 9(12): 2286.

Genus, A. & Stirling, A. (2018). Collingridge and the Dilemma of Control: Towards Responsible and Accountable Innovation. *Research Policy*, 47(1): 61–69.

Gibbert, M., Hoegl, M., & Valikangas, L. (2007). In Praise of Resource Constraints. *MIT Sloan Management Review*, 48(3): 15–17.

Gurzawska, A., Mäkinen, M., & Brey, P. (2017). Implementation of Responsible Research and Innovation (RRI) Practices in Industry: Providing the Right Incentives. *Sustainability*, 9(10): 1759.

Halme, M. & Korpela, M. (2014). Responsible Innovation Toward Sustainable Development in Small and Medium-sized Enterprises: A Resource Perspective. *Business Strategy and the Environment*, 23(8): 547–566.

Healey, M. J. & Rawlinson, M. B. (1994). Interviewing Techniques in Business and Management Research. In V. J. Wass & P. E. Wells (Eds.), *Principles and Practice in Business and Management Research*: 123–146. Dartmouth: Aldershot.

Hemphill, T. A. (2016). Responsible Innovation in Industry: A Cautionary Note on Corporate Social Responsibility. *Journal of Responsible Innovation*, 3(1): 81–87.

Hockerts, K., Morsing, M., Eder-Hansen, J., Krull, P., Midttun, A., Halme, M., Sweet, S., Davidsson, P., Sigurjonsson, O., & Nurmi, P. (2008). *CSR-Driven Innovation: Towards the Social Purpose Business*. Frederiksberg: Center for Corporate Social Responsibility, CBS.

Holstein, J. & Gubrium, J. (1995). *The Active Interview*. Thousand Oaks, CA: Sage.

Iatridis, K. & Schroeder, D. (2016). *Responsible Research and Innovation in Industry: The Case for Corporate Responsibility Tools*. Cham, Switzerland: Springer.

Jenkins, H. (2006). Small Business Champions for Corporate Social Responsibility. *Journal of Business Ethics*, 67(3): 241–256.

Kowitt, B. (2018). Inside the trillion-dollar war on Big Food. http://fortune.com/2015/05/21/the-war-on-big-food/ (accessed 3 July 2018).

Lubberink, R., Blok, V., van Ophem, J., & Omta, O. (2017). Lessons for Responsible Innovation in the Business Context: A Systematic Literature Review of Responsible, Social and Sustainable Innovation Practices. *Sustainability*, 9(5): 721.

Martinuzzi, A., Blok, V., Brem, A., Stahl, B., & Schönherr, N. (2018). Responsible Research and Innovation in Industry – Challenges, Insights and Perspectives. Sustainability, 10(3): 702.

Mitchell, R. K., Agle, B. R., & Wood, D. J. (1997). Toward a Theory of Stakeholder Identification and Salience: Defining the Principle of Who and What Really Counts. *Academy of Management. The Academy of Management Review*, 22(4): 853–886.

Newsdesk (2018). Pip & Nut secures over £1 million in investment to drive next phase of growth. www.fdin.org.uk/2018/05/pip-nut-secures-over-1-million-in-investment-to-drive-next-phase-of-growth/ (accessed 30 March 2019).

Owen, R., Stilgoe, J., Macnaghten, P., Gorman, M., Fisher, E., & Guston, D. (2013). A Framework for Responsible Innovation. In R. Owen, J. Bessant, & M. Heintz (Eds.), *Responsible Innovation*: 27–50. London: John Wiley.

Pandza, K. & Ellwood, P. (2013). Strategic and Ethical Foundations for Responsible Innovation. *Research Policy*, 42(5): 1112–1125.

Pavie, X., Scholten, V., & Carthy, D. (2014). *Responsible Innovation: From Concept to Practice*. Singapore: World Scientific.

Pellé, S. & Reber, B. (2013). *Theoretical Landscape*. Paris: Governance of Responsible Innovation (GREAT) Project.

Putnam, R. D. (2001). *Bowling Alone: The Collapse and Revival of American Community*. New York: Simon and Schuster.

Russo, A. & Perrini, F. (2010). Investigating Stakeholder Theory and Social Capital: CSR in Large Firms and SMEs. *Journal of Business Ethics*, 91(2): 207–221.

Saunders, M., Lewis, P., & Thornhill, A. (2012). *Research Methods for Business Students* (6th ed.). Essex: Pearson Education.

Seale, C. & Silverman, D. (1997). Ensuring Rigour in Qualitative Research. *European Journal of Public Health*, 7(4): 379–384.

Silverman, D. (2006). *Interpreting Qualitative Data: Methods for Analysing Talk, Text and Interaction*. London: Sage.

Spence, L. J. & Rutherfoord, R. (2001). Social Responsibility, Profit Maximisation and the Small Firm Owner-manager. *Journal of Small Business and Enterprise Development*, 8(2): 126–139.

Stahl, B., Obach, M., Yaghmaei, E., Ikonen, V., Chatfield, K., & Brem, A. (2017). The Responsible Research and Innovation (RRI) Maturity Model: Linking Theory and Practice. *Sustainability*, 9(6): 1036.

Stilgoe, J., Owen, R., & Macnaghten, P. (2013). Developing a Framework for Responsible Innovation. *Research Policy*, 42(9): 1568–1580.

van de Poel, I., Asveld, L., Flipse, S., Klaassen, P., Scholten, V., & Yaghmaei, E. (2017). Company Strategies for Responsible Research and Innovation (RRI): A Conceptual Model. *Sustainability*, 9(11): 2045.

Voegtlin, C. & Scherer, A. G. (2017). Responsible Innovation and the Innovation of Responsibility: Governing Sustainable Development in a Globalized World. *Journal of Business Ethics*, 143(2): 227–243.

von Schomberg, R. (2013). A Vision of Responsible Research and Innovation. In R. Owen, J. Bessant & M. Heintz (Eds.), *Responsible Innovation*: 51–74. London: John Wiley.

5 Supporting RRI uptake in industry

A qualitative and multi-criteria approach to analysing the costs and benefits of implementation

Andrea Porcari, Daniela Pimponi, Elisabetta Borsella, Pim Klaassen, Maria João Maia, and Elvio Mantovani

5.1 Introduction

Motivated by the need to ensure that research and innovation (R&I) activities are societally desirable, ethically acceptable and sustainable, the European Commission and funding agencies in some European Union Member States have made Responsible Research and Innovation (RRI) a vital pillar of their research funding initiatives over the last decade. A growing body of literature is dedicated to the development of conceptual frameworks for RRI, generally either focusing on the process dimensions of RRI (e.g., Stilgoe, Owen, & Macnaghten, 2013; Klaassen, Rijnen, Vermeulen, Kupper, & Broerse, 2019) or on the European Commission-defined keys (European Commission, 2014) to RRI (Klaassen et al., 2019). More recently, increasing efforts have been devoted to investigating what adopting RRI entails for industrial actors (Dreyer et al., 2017). As the field is still relatively young, most contributions focus on one or a few finely delineated aspects of what RRI implementation requires (e.g., van Wezel et al., 2017 on safety and technology assessment; Gurzawska, Mäkinen, & Brey, 2017 on incentives for RRI implementation) or on features relevant to RRI's uptake in one particular field (Chatfield, Borsella, Mantovani, Porcari, & Stahl, 2017, on risk perception in the ICT industry (Lees & Lees, 2018 on the sheep dairy industry). Thus, in reviewing the literature, recurrent lessons and themes, knowledge gaps and gaps between knowledge and implementation stand out (Yaghmaei, Porcari, Mantovani, & Flipse, 2019).

The literature shows that companies have both positive and negative attitudes towards RRI (Brem et al., 2017). Specifically, while alignment with societal needs and conducting oneself in an ethically acceptable manner are generally seen as key aspects of any R&I activity, companies often question the added value of RRI compared to existing practices, such as quality management or corporate social responsibility (CSR) practices, and how best to embed RRI in company operations (Porcari, Borsella, & Mantovani, 2015).

The lack of (management) models to evaluate and showcase RRI's added value is one of the barriers to fostering corporate commitments to experiment with and use RRI approaches. The aim of this chapter is to address this gap. Based on and inspired by literature on RRI, CSR and multi-criteria decision analysis (MCDA) and by practical experiences with pilot companies within the Horizon 2020-funded PRISMA project, which focused on promoting RRI in industry, we propose a practical model to help companies identify RRI implementation strategies during product development – connecting goals, actions and impacts – and a simple methodology to perform qualitative evaluations of its impacts (benefits, barriers and costs).

We subscribe to the conviction implicit in the very notion of RRI that value is gained by integrally considering all of RRI's diverse aspects. This chapter therefore aims to help companies, especially small and medium-sized enterprises (SMEs), to identify RRI implementation strategies that fit within their unique realities and constraints.

5.2 Methodology and context

This study's methodology and primary sources of data come from the PRISMA[1] project, a coordination and support action dedicated to exploring and promoting RRI in industry. Central to this project were eight pilot projects involving companies working on transformative technologies (nanotechnologies, synthetic biology and biotechnology, internet of things, drones and autonomous vehicles), which allowed us to test the implementation of RRI principles in industrial settings.[2] The pilot projects involved close interactions with the participating companies, revolving around their motivations for, attempts at and successes in operationalizing RRI in an innovation trajectory or on another level of company functioning. These pilot projects were the primary data source for our study. We used a wide variety of data-gathering methods, which are summarized in Table 5.1. Activities took place between January 2017 and December 2018.

5.3 Background

This section presents the background for our study, particularly those elements from the literature that we considered in our analysis of RRI, RRI in industry and CSR.

5.3.1 Responsible Research and Innovation and corporate social responsibility

Various conceptualizations of RRI circulate in policy and academic circles, each emphasizing different aspects.[3] Elements almost universally agreed upon include the ideas that RRI entails the continuous alignment of 'research and innovation to the values, needs and expectations of society'

Table 5.1 Primary sources for the study: the PRISMA project's most significant activities and a literature review

Step	*Description*	*Methods used*	*Main outcomes*	*Further details*
1	Tools and methods inventory	Desk research; semi-structured interviews with experts (n = 11) and EC project officers (n = 5)	Selection of decision-support RRI tools tailored to industry needs	Klaassen et al., 2016; Klaassen, Smit et al. 2017
2	Pilots for RRI implementation in industrial R&I projects on transformative technologies	Close interaction with eight companies during a two-year period (at least six semi-structured interviews per company)	Selection of RRI actions tailored to industry needs	Nathan, 2017, 2018a, 2018b; Guelke, 2018
3	Selection and reflection on KPIs for RRI	Desk research and interviews with pilot companies (one focus group with all companies and one semi-structured interview with each company)	Input to identify criteria to analyse the impact of RRI actions	Yaghmaei et al., 2019
4	Dialogues with stakeholders on RRI aspects related to the pilot projects and their tech fields[14]	Five dialogues involving more than 100 stakeholders (companies, industrial organizations, public research organizations, non-governmental and civil society organizations and experts in social responsibility and RRI); plenary and interactive sessions using workshop, world-café and fish-bowl methods to stimulate discussion. Some discussions were recorded and transcribed verbatim for content analysis using MAXQDA software	Input and review to identify and select RRI actions and criteria to analyse the impact of RRI actions	Maia & Coenen, 2017, 2018
5	Literature review	Review of scientific and grey literature based on a specific set of keywords[15] and excluding studies not focusing explicitly on industry	Thematic framing	See references

EC, European Commission; KPIs, key performance indicators; R&I, research and innovation; RRI, Responsible Research and Innovation.

(European Union, 2014, p. 1) and that RRI practices should feature the process dimensions of anticipation, reflexivity, inclusion and responsiveness (Stilgoe et al., 2013). For our study, we use the latter definition and do not delineate RRI more distinctly than that.

As researchers have observed, it is still unclear how the current concept of RRI can fit within the business context (Blok, Hoffmans, & Wubben, 2015; Lubberink, Blok, van Ophem, & Omta, 2017; Gauttier, Søraker, Arora, Brey, & Mäkinen, 2017) and questions remain regarding how issues of motivation and practical implementation should be addressed. A major issue, which is also found with CSR, concerns the conflict between a company's aspirations for monetary profits and market growth, on the one hand, and the extra costs associated with addressing social objectives such as sustainability, ethics and well-being, on the other (Iatridis & Schroeder, 2016). Another important issue is the lack of guidance (e.g. action plans) regarding how to put RRI principles into practice. This lack of support makes it difficult for companies to assess the expected impacts of RRI and, thus, to understand whether and how to embed RRI in their strategies.

Such uncertainties are one reason why the implementation of RRI in industry is still in its infancy. In fact, initiatives to practically implement RRI in industry are still limited, with most being related to cooperative projects within EU framework programmes or national equivalents.[4]

To foster the alignment of RRI principles and objectives with corporate strategies, it could be useful to look at RRI in relation to socially beneficial processes and tools that companies already know or have already implemented – usually under the label of CSR (Porcari, Borsella, & Mantovani, 2015; Iatridis & Schroeder, 2016; Chatfield, Iatridis, Stahl, & Paspallis, 2017). However, like RRI, CSR is a broad concept. In the present chapter, we look at CSR as a management concept promoting forms of self-regulation businesses use to improve their impacts in a socially responsible way, conventionally with a focus on people, planet and profit (Graafland & Smid, 2019).[5]

Given that policies are not always implemented – and even when implemented, they might not have the impacts desired – we do not make any assumptions regarding CSR's success or failure in realizing societally, environmentally or financially beneficial impacts. Likewise, we do not make any a priori delineation regarding the scope of CSR policies (i.e. do they incorporate research and development (R&D) and thus innovation, or not?). However, although CSR usually applies to a company's overall conduct – for example, human resource management, waste management, stakeholder engagement or communication – rather than its innovation or R&D activities, CSR processes or tools may well be applicable to innovation too. That can be expected to be the case at least insofar as both CSR and RRI relate (business) conduct to 'the responsibility of enterprises for their impacts on society' (European Commission, 2011, p. 6). CSR approaches provide guidance on principles, procedures and actions for implementing

CSR in a company's business operations. In addition, reporting and certification mechanisms are available to measure and showcase CSR performance, providing both a strong incentive for companies to implement CSR as well as support in doing so (Gurzawska & Porcari, 2016).

Garst, Blok, Jansen and Omta (2017) identified three types of motives that push companies to introduce CSR in their organizations: instrumental, relational and moral. Examples include reduction of production costs, increasing sales, postponing of legislation, attracting employees and investors (instrumental); fulfilling stakeholder expectations and being recognized for moral relationship (relational); recognizing the intentionality behind a product's long-term impacts, knowing a product's long-term impacts or attempting to attain that knowledge (moral). Although innovation presents its own specific issues, particularly because of the uncertainty that comes with it (Collingridge, 1980), most of these motives are also relevant for RRI. However, this similarity in motives does not necessarily translate into similarities regarding the responsibilities companies take on in their innovation activities and other business functions.

5.3.2 Issues regarding RRI implementation in companies

While CSR initiatives generally apply to the overall conduct of a company, the RRI concept focuses on the earlier phases of a product's development and life cycle and, thus, on the R&I stages (Gurzawska & Porcari, 2016; Chatfield, Borsella, Mantovani, Porcari, & Stahl, 2017). Introducing RRI could provide ways to anticipate social needs, concerns and challenges, and it could offer opportunities to increase product desirability and positive social impacts (and reduce risks) starting at the early stages of development. It could therefore avert late interventions and reduce overall product development costs (in an attempt to address the well-known Collingridge dilemma; Collingridge, 1980).

However, responsibility for RRI implementation cannot be limited to just those people working in R&I; it falls on all areas of a company. Chatfield, Borsella, et al. (2017) emphasized that, although R&D departments in highly innovative enterprises are considered one of the key departments where societal risks and ethical issues should be addressed and relevant stakeholders should be engaged, other areas are expected to play strategic roles in implementing RRI principles, particularly top management, human resources, CSR and legal functions and marketing.

Many studies suggest that an RRI approach should be conceived as a 'holistic' framework, given that RRI should be implemented along the entire R&I value chain and that management should disseminate RRI principles among all the people working inside or close to the company. In this regard, Lubberink et al. (2017, p. 23) wrote:

> The conclusion can be drawn that responsible innovation does not only ask for new corporate practice in terms of innovation activities, but

> it also demands that companies reflect on their business models, leadership, and their roles and responsibilities for the political and socio-economic system in which they operate.

Chatfield, Iatridis et al. underlined that:

> if RRI is perceived as being something that is 'bolted on' or in some way separate from the core activity of the company, then it will be difficult to achieve. For effective RRI, it may be necessary for the whole company to be on board.
>
> (Chatfield, Iatridis et al., 2017, p. 14)

During the PRISMA project's stakeholder dialogues, discussed in the methods section, participants confirmed that the endorsement of RRI should primarily be a strategic decision made by higher levels of a company's hierarchy, one that is then put into practice by the whole organization. However, stakeholders also discussed how this top-down approach could, in some cases, not be enough. Although a strong commitment from management is needed, the existence of an RRI promoter at other levels in a company (bottom-up approach) could be beneficial and used as a complement to the top-down approach.

5.4 The PRISMA model

Several studies, mainly in previous European Commission-funded projects, have aided the development of benchmarks for RRI practices, which have led to RRI criteria (Kupper, Klaassen, Rijnem, Vermeulen, & Jacqueline, 2003; Wickson & Carew, 2014) or indicators (Woolley & Rafols, 2016; MoRRI Consortium, 2018). Yet, RRI suffers from a lack of widely accepted RRI models that provide *practical guidance* in relating RRI principles to company goals, strategies and attitudes, as well as to technology and product features. Our reflection on the experience of RRI pilots with innovative companies (described in the methods section) taught us that such guidance is needed to help companies recast abstract RRI dimensions into a set of (management) strategies, tools and actions that can support and motivate all company departments in endorsing and adopting RRI. Practical guidance could also help provide ways to measure the impacts of RRI on company operations, thereby enriching the ongoing work on RRI benchmarks and indicators. A model that could play this role would also be a useful starting point for investigating the extent to which RRI could benefit from existing CSR practices and whether RRI could be integrated into those practices.

The PRISMA project developed a conceptual model for RRI implementation (van de Poel et al., 2017) that provides guidance to address the issues to consider in the implementation of RRI in industry. It is inspired by three models developed in the context of the project Responsible Industry (Gauttier et al., 2017; Stahl et al., 2017) and provides a pathway for RRI implementation, represented graphically in Figure 5.1, that incorporates the following steps:

- *analysis* of RRI dimensions relating to the company's characteristics and values and the overall business and technology context to set the goals to be achieved through RRI
- *design* of the RRI strategy, defining the actions and tools needed to reach the goals, motivations and responsibilities and the desired level of RRI integration throughout the company's functions and the value chain
- *implementation* of those actions and tools in the different business areas and along the value chain
- *measurement and monitoring* of the impacts of RRI implementation to provide feedback for improving the strategy.

This set of steps is cyclic, meaning that it can be repeated to constantly improve the RRI strategy and actions to enhance the outcomes.

The PRISMA model follows a typical observe–plan–do–check–adjust (OPDCA) process (Smart, 2017). This is a variation of the plan–do–check–act/adjust (PDCA or Deming cycle) method already adopted in business contexts for the monitoring and improvement of processes and products.[6] The PDCA approach to RRI was previously proposed by other European Union-funded projects focused on RRI: SMART-MAP (Marschalek & Schrammel, 2018) and SATORI (*CWA 17145-2*, 2017). However, the OPDCA process emphasizes observation, a step often used in industrial contexts (e.g. in the literature on lean manufacturing, such as in the Toyota production system[7]).

Based on reflection on experiences with the eight pilot companies and the stakeholder dialogues, we aim to move beyond the PRISMA and Responsible Industry models in this study by distilling a set of RRI actions and criteria (and their relationships) that can support companies in assessing the impacts of RRI implementation on product development, at least qualitatively. Our goal is to provide a forward-looking approach to support RRI uptake in companies. In-depth analyses of PRISMA's activities and of the pilots in particular informed our results (for those analyses, see Maia & Coenen, 2017, 2018; Nathan, 2017, 2018a, 2018b; Guelke, 2018;).

5.5 The pathway for RRI implementation in companies

Defining suitable models for RRI implementation and methods for impact analyses needs to begin with an evaluation of the criteria for RRI uptake – the strengths and opportunities (benefits) and the barriers (risks, costs) of implementing RRI. This type of analysis (Section 5.5.1) provides information regarding the definition of an RRI action plan (Section 5.5.2) and the

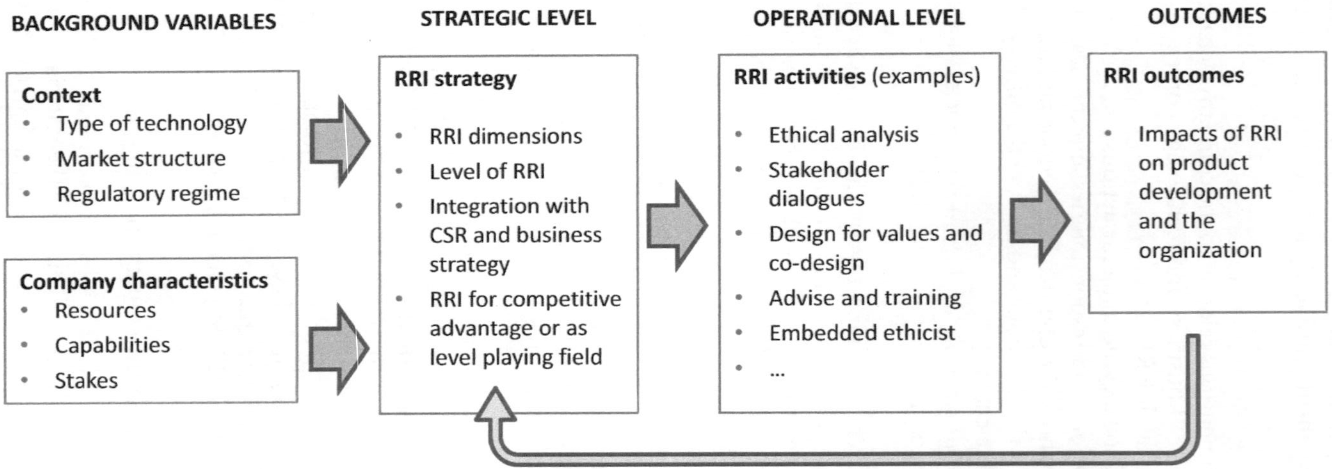

Figure 5.1 A conceptual model for Responsible Research and Innovation (RRI) implementation connecting the background variables, strategic level, operational level and final outcomes. CSR, corporate social responsibility. (Adapted from van de Poel et al., 2017.)

design of a multi-criteria and qualitative approach to analyse the impacts of RRI actions (Section 5.5.3).

5.5.1 Opportunities and barriers for implementing RRI

In analysing the experience of the eight pilot projects, we found that the costs related to RRI actions are generally perceived as being immediate, while most of the (positive) outcomes cannot realistically be expected until many months or even several years later (as in the case of company reputation). This perception is a major barrier to RRI implementation. However, when focusing on specific products, there are short-term RRI actions that can translate into immediate benefits, thus balancing the economic barrier of immediate costs versus deferred benefits. The PRISMA pilot projects provided examples of such actions, as reported in Guelke (2018) and Yaghmaei et al. (2019).[8]

The PRISMA pilots suggested that barriers to engaging with RRI exist at both strategic and operational levels and that these are not restricted to economic issues. In the PRISMA stakeholder dialogues, concerns over costs and resources associated with RRI were often mentioned by industry stakeholders, who raised issues about RRI being a difficult and bureaucratic process whose implementation would entangle internal and external resources. It was also clear that, once the companies became aware of the concept, RRI was perceived as potentially advantageous and as a possible way for them to save money. Most pilot companies agreed it would reduce several risks of product failure and would likely increase product acceptability. Other issues were also discussed:

- A company's 'maturity level' can influence its degree of interest in adopting RRI: a low awareness about responsibility or ethical issues could reduce this interest. Moreover, the pre-existing assumption of responsibilities within the CSR framework could reduce the perceived added value of RRI compared to current CSR practices.
- Identifying RRI approaches that fit a company's specific business case is not always straightforward, and thus, the related potential benefits might not be clearly visible. This calls for an interaction with experts or an in-depth reflection on company activities.
- Implementing RRI requires that a company opens its production processes and cooperates with different stakeholders in different phases of R&D. These actions could cause confidentiality problems and raise intellectual property rights issues that conflict with the company's usual management procedures.
- Developing a strategy for RRI adoption implies the internal agreement of several departments within the company and possibly a close cooperation between the R&D department and external partners along the R&I value and supply chains. This might be challenging.

- Specific skills for and experiences with RRI might be lacking within a company, and the use of external advisors and experts might be required. The more RRI is embedded in the existing process, the more this externalization could be resource demanding.

One influencing factor relates to the size of the company, which can strongly influence the possibility of investing in RRI activities. Large companies have several departments, each having different objectives and following specific formal procedures, and this complexity can challenge the introduction of new activities and procedures throughout the organization. However, big companies may already have some activities in place that support RRI or RRI-like principles at different stages of the value chain. Thus, it could be easier for them to integrate RRI concepts into existing procedures, possibly within their CSR framework. SMEs, on the other hand, have fewer resources to invest, but their organizational and decision-making processes are simpler: responsibilities are often shared across the company, which paves the way to an RRI implementation that involves the R&I value chain (at least within the company and with partners closer to the company). However, involving the entire value chain connected to product development might be more challenging.[9]

Regardless of their size, companies should monitor and evaluate the results of RRI adoption. The PRISMA project highlighted that, even if bigger companies can afford to use internal or external expertise for monitoring activities, both large and small companies would benefit from the availability of simple methodologies for self-assessment and evaluation that could provide constant feedback on the impact of implementing RRI.

Other factors influencing RRI implementation are closely related to the realities and constraints of the specific companies and sectors considered, including differences in the type of organization, the sector, the product and the technology considered. The main differences are found when comparing companies dealing with transformative technologies with those related to conventional technologies or traditional sectors. For instance, in one of the stakeholder dialogues, a participant from industry compared the example of nanomedicine, where transparency and openness are routine practices, with other sectors where the use of nanotechnology is sometimes hidden from product communication in order to avoid critical and opponent voices.

Innovative businesses are often more familiar with RRI concepts or activities. This can be due to a variety of factors:

- the need to foster customers' and society's acceptance of new technologies, which pushes companies to endorse specific values and adopt procedures for quality and social responsibility
- a strong willingness to intercept the public's needs and desires in order to facilitate access in the market of new technologies

- the uncertainties in normative requirements often associated with new technologies.

Some telling examples of RRI aspects already embedded in company operations were seen during the PRISMA experience, namely, the identification of companies' core values (such as quality, knowledge and sustainability) and companies' compliance with certifications for quality, environmental protection, health and safety, ethical aspects, sustainability and the like, in connection with R&I products.

Dialogue initiatives, performed on a regular basis with a wide range of stakeholders, were also mentioned as actions already implemented to build trust and relationships, improve transparency, develop common understandings and inform the political and societal debate. Finally, life cycle assessments, eco-design, sustainability strategies toward compliance with the United Nation's sustainable development goals (SDGs), responsible manufacturing and supply chain management were also identified as RRI-related activities already taking place in some companies.

Companies using transformative technologies have to deal with uncertainties and thus are looking for novel and advanced methods of risk and quality management, ones that are more responsive to users' needs and perspectives. RRI can be helpful in addressing these aspects.

One example of how RRI can improve responsivity comes from the PRISMA pilot projects involving nanomedicine – a rapidly transforming sector in which the emerging ethical and societal issues are critical to its eventual success. Because of the ethical and societal implications of new and challenging health therapies, which are sometimes personalized, companies require support from RRI approaches in dealing with patients and healthcare professionals and, more generally, with public opinion. A comment at one of the stakeholder dialogues attests to this: 'This is the responsibility for industry: to communicate in a proper way the benefits, but also the risks. ... This can be managed with a proper Responsible Innovation strategy'.[10] Thus, a therapy's acceptability could strongly benefit from RRI actions that encourage a process of product 'co-creation' that includes all the involved stakeholders and from transparency in communications about the real cost–benefit ratio for the patient.

Based on the experience with the pilots, and literature on RRI implementation in industry (e.g., Chatfield, Borsella, et al., 2017),[11] PRISMA developed a strengths, weaknesses, opportunities and threats (SWOT) analysis to help companies compare the human and economic efforts required for RRI implementation with the benefits and opportunities, such as profits, efficiency or public image (reported in Porcari, Pimponi, Borsella, & Mantovani, 2019). We used this SWOT analysis and the reflections discussed in this section to inform the definition of RRI actions (Section 5.5.2) and the selection for criteria for the assessment of RRI impacts (Section 5.5.3).

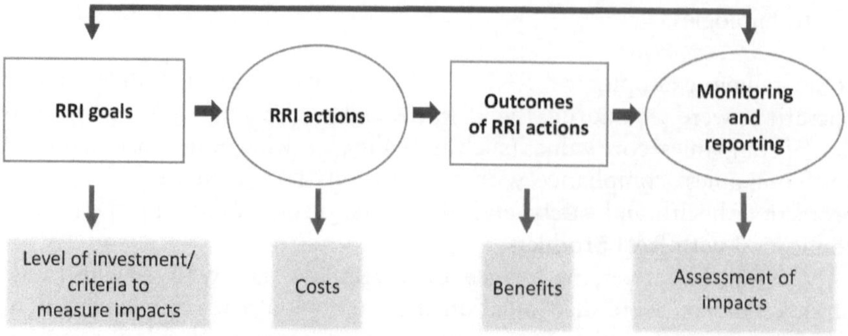

Figure 5.2 Responsible Research and Innovation (RRI) implementation phases and relative steps for impact evaluation (a cyclic approach).

5.5.2 Setting the strategy for RRI implementation

Based on the issues and models for RRI implementation discussed in the previous sections, we created a conceptual map to support development of an RRI implementation strategy. The map in Figure 5.2 includes and links the following elements:

- the actions required to reach the desired RRI 'maturity' or 'performance' level in every dimension of RRI
- the responsibilities needed for RRI deployment, with respect to the different company functions
- the organizational and financial impacts of these activities inside and outside the organization
- the relevant – qualitative and quantitative, tangible and intangible – criteria to assess the impacts related to RRI.

RRI impact analyses can be performed for specific products/projects or for overall R&D activities, and they can be done in a procedural/formal way or a more informative/informal way, depending on the conditions and needs of the company. Regardless of their target or method, impact analyses will likely include these main steps:

- defining the target (level of maturity/commitment/investment to achieve in RRI) and selecting criteria for monitoring costs and benefits based on the RRI goals and strategy
- estimating costs based on the RRI action plan and tools
- evaluating benefits based on (expected or actual) RRI outcomes
- monitoring RRI implementation based on selected criteria and using the results to continuously refine the RRI goals, strategy and indicators.

Given the broad and diverse impacts that RRI could have on an organization, as shown in Section 5.5.1, impacts should be evaluated on a case-by-case basis. Note that the overall impact of an RRI implementation strategy could actually exceed the sum of the specific benefits derived from each action.

A cornerstone for impact evaluation is the defining of a concrete action plan for RRI. Based on our experiences with the PRISMA pilot projects, we identified three key actions relating to the different RRI dimensions:

1. Reflection and anticipation ('observe/plan'): Integrate analysis of ethical, legal and social impacts (ELSI) beginning in the early stages of product development.
2. Inclusiveness ('do'): Perform stakeholder engagement to inform all phases of product development.
3. Responsiveness ('check/adjust'): Integrate monitoring, learning and adaptive mechanisms to address public and social values and normative principles in product development.

As indicated, the order of these actions follows an OPDCA cycling process. Tables 5.2–5.4 provide examples of how these three key actions can be used for implementing RRI in product development at the company level, with reference to expected benefits, the R&I value chain, the corporate functions, the stakeholders involved and the investment term. Further information on the strategy selected by each of the PRISMA pilot projects, including their RRI actions and expected benefits, is reported in Porcari et al. (2019).

5.5.3 Using a multi-criteria and qualitative approach to analyse impacts of RRI actions

Given the difficulties in defining and measuring relationships between actions and impacts, methods such as cost–benefit analysis and cost-effectiveness analysis seem difficult to apply to RRI, at least given the current level of RRI knowledge and practical experimentation in industrial contexts. Instead, MCDA (Linkov et al., 2006) may be more effective in evaluating the broad and diverse impacts related to RRI. MCDA is typically used to choose between different alternatives, considering multiple criteria on different scales or of various natures. With MCDA, impact categories (called lines of evidence) and criteria for each category are identified in accordance with the specific case. A scoring system is used to evaluate qualitative and quantitative impacts against the criteria, and weights can be applied to each criterion. This kind of analysis provides a qualitative or semi-quantitative comparison of impacts between different RRI adoption scenarios.

This section provides company project/product managers with a simplified MCDA model for self-evaluations of the positive impacts (benefits),

Table 5.2 Reflection and anticipation actions for integrating analysis of ethical, legal and social impacts (ELSI) throughout all stages of product development

Actions	*Benefits*	*R&I value chain phase*	*Corporate functions involved*	*Stakeholders involved*	*Investment term*
Conduct ethical analysis, through foresight, scenario analysis, social phenomena and trends evaluation, etc.	Improve product quality, desirability and acceptability Improve product sustainability, safety and reliability Address uncertainties, prevent and mitigate risks Motivate workers	Basic and applied research, engineering and testing	Management, R&D	R&I partners, end users, policymakers	Short, medium
Design for values, stakeholder and value inventory/scenarios (values hierarchy, conflicting values, etc.)			R&D	R&I partners, suppliers, end users	Short, medium
Include RRI principles in company's mission and vision			Management, R&D	Internal to the company	Medium
Hold internal meetings with R&D personnel to reflect on ethical issues			R&D	Internal to the company	Short
Solicit ELSI advice from (independent and external) experts as needed			R&D, CSR, legal	Internal to the company	Short

Table 5.2 (*Cont.*)

Actions	*Benefits*	*R&I value chain phase*	*Corporate functions involved*	*Stakeholders involved*	*Investment term*
Develop and introduce ethical frameworks, codes of conduct		All	Management, legal, R&D, CSR, quality	Internal to the company	Medium
Implement life cycle assessment (LCA) and social LCA		Applied research, engineering and testing	Management, R&D, quality, CSR	Suppliers	Long
Re-evaluate expected impacts prior to market launch				Internal to the company	Short

CSR, corporate social responsibility; R&D, research and development; RRI, Responsible Research and Innovation.

Table 5.3 Inclusiveness actions for stakeholder engagement that inform all phases of product development

Actions	*Benefits*	*R&I value chain phase*	*Corporate functions involved*	*Stakeholders involved*	*Investment term*
Set and implement a communication and dialogue strategy on ELSI	Strengthen relations and trust with all stakeholders, network building	Engineering and testing, go to market	R&D, CSR, marketing	All	Long
Work with business and social actors sharing values and create positive ethical networks	Reconcile opposing views and bridge opposing values New values creation Anticipate potential regulatory change Product quality, desirability and acceptability	All	CSR	All	Medium
Co-design product through dialogue with policy actors, authorities and normative bodies (EU, regional and local)		Applied research, engineering and testing	R&D, quality, legal	Policymakers, regulators	Short
Organize public dialogues, build/use public platforms for expressing needs and concerns			R&D, CSR	End users, consumers	Medium
Connect to or organize living labs and social experimentation using participatory methods			R&D, CSR	End users, consumers	Short, medium

Table 5.3 (Cont.)

Actions	*Benefits*	*R&I value chain phase*	*Corporate functions involved*	*Stakeholders involved*	*Investment term*
Build user-based communities of practice			R&D, CSR	End users, consumers	Medium
Promote initiatives for social inclusion, provide consumers with an official role in the innovation process		All	CSR	End users, policy makers	Medium
Promote capacity building with vulnerable actors in the value chain		Engineering and testing, go to market	R&D, CSR	End users	Medium

CSR, corporate social responsibility; ELSI, ethical, legal and social impacts; R&D, research and development; RRI, Responsible Research and Innovation.

Table 5.4 Responsiveness actions to integrate monitoring, learning and adaptive mechanisms that address public and social values and normative principles in product development

Actions	*Benefits*	*R&I value chain phase*	*Corporate functions involved*	*Stakeholders involved*	*Investment term*
Integrate user-centred design, user innovation, flexible and adaptive design, co-creation approaches	Create value, increase the social value/impact of R&D	Applied research, engineering and testing	R&D, management, legal, marketing	R&I partners, supply chain suppliers, end users, consumers	Long
Screen suppliers for positive practices	Build corporate image and reputation			Suppliers	
Put procedures in place for investigating reports of concerns or misconduct	Compliance with qualified norms and standards		Management, legal	Internal to the company	Medium
Employ adaptive risk management	Facilitate the access to financial support		Management, R&D, quality	Internal to the company	Medium
Embed ethicists in the R&I process		All	CSR, R&D	Internal to the company	Medium
Establish an ethical, social and legal (ELSI) monitoring board		All	R&D, management	R&I partners, suppliers, policymakers, end users	Long
Include ELSI criteria in internal procedures for R&D project quality monitoring		Applied research, engineering and testing	R&D, management	Internal to the company	Short
Perform regular ethical review and get ethical certification (by independent bodies)		Engineering and testing, go to market	CSR, quality	Certification bodies, regulators and authorities	Long

(*continued*)

Table 5.4 (*Cont.*)

Actions	*Benefits*	*R&I value chain phase*	*Corporate functions involved*	*Stakeholders involved*	*Investment term*
Obtain social accountability and quality certification at company and supply chain levels			CSR, quality	Certification bodies, regulators and authorities, investors	Long
Monitor post-marketing ELSI impacts		Go to market	R&D, quality	Regulators and authorities	Long
Include ELSI for R&D and innovation products in CSR/ sustainability reporting			CSR, marketing	All	Long
Support and invest in sustainable supply chains		All	R&D, management	Suppliers	Medium
Select funding mechanisms based on ethics/responsibility requirements		All	R&D, management	Funding bodies, investors	Short

CSR, corporate social responsibility; R&D, research and development; R&I, research and innovation.

negative impacts (barriers) and costs/resources resulting from RRI actions in product development.

As previously discussed, RRI aspects are connected to a broad spectrum of factors related to the type of company and its management style, the technology and products it works on, the sector and market, the pertinent regulatory frameworks and the stakeholders involved. Also, the conditions for RRI uptake relate to various issues, spanning from long-term strategic factors at the company level (e.g. company reputation) to short-term factors in product development (e.g. alignment with user needs and stakeholder values).

Analyses of RRI uptake should consider both tangible and intangible short- and long-term impacts and are therefore quite complex. We propose a qualitative methodology, focusing on individual projects or products, that looks at multiple criteria and is based on the RRI actions discussed in the previous section. The proposed self-evaluation procedure for companies contains four steps:

1. identification of a set of RRI actions for product development (RRI strategy)
2. evaluation of the impacts of the RRI actions based on selected criteria. Impacts could be positive (benefits of RRI uptake), irrelevant or negative (barriers to RRI uptake)
3. evaluation of the direct costs of performing each of the RRI actions (high, medium, low) compared to product development costs
4. mapping and analysis of the overall impacts of RRI uptake on an evaluation matrix.

This exercise can be used to assess the impact of actions taken or to evaluate different RRI strategies before implementation in order to identify the most advantageous and cost-effective actions for RRI uptake with respect to the selected criteria.

Tables 5.2–5.4 provide many examples of RRI actions that could be used in step 1 of the self-evaluation procedure. However, a company might decide to focus on a smaller or larger set of actions depending on its conditions and goals. Ideally, though, it should focus on having at least one action for each table in order to fulfil all the RRI dimensions. The final action plans in the PRISMA pilot projects included four to nine RRI actions per pilot company, with an emphasis on inclusiveness and responsiveness actions.

Table 5.5 provides a model questionnaire for a company's self-assessment of the impact of RRI actions. The questionnaire is structured as a set of five questions (Q1–5) that are based on lines of evidence (LoE), as described in Section 5.4, plus a sixth question (Q6) related to the direct costs of RRI actions. Each of the first five questions is detailed by a set of sub-questions (criteria) derived by criteria for RRI implementation that were discussed in Section 5.5.1. Though both Q5 and Q6 involve financial aspects, Q5 refers

Table 5.5 Questions and sub-questions (criteria) proposed for a company's self-assessment of the impacts (benefits, barriers, costs) of Responsible Research and Innovation (RRI) actions

Main questions (Q)	*Criteria*	*Impact of RRI action(s)*
Q1: Scientific and technological line of evidence (LoE)	Q1.1: Inspire technological innovation Q1.2: Feasibility of the technology solution Q1.3: Product quality (performance/efficiency) Q1.4: Product reliability Q1.5: Extend the product life cycle Q1.6: Trust with/avoid conflicts with business partners, suppliers, end-users	Positive Irrelevant Negative
Q2: Ethical and societal LoE	Q2.1: Product acceptability Q2.2: Product safety Q2.3: Product environmental sustainability Q2.4: Effect on quality of life and health of customers Q2.5: Product-related services and guidance (e.g., ethical protocols) Q2.6: Address users' needs and rights (e.g., privacy, data ownership, etc.)	Positive Irrelevant Negative
Q3: Strategic LoE	Q3.1: Competitive advantage Q3.2: Corporate image Q3.3: Transparency on product qualities Q3.4: Customer satisfaction, meeting new consumers' needs or requests Q3.5: Build legitimacy and gain consumer loyalty for the product Q3.6: Improve relationships with partners, suppliers and sub-suppliers Q3.7: Fulfil ethical and social requirements (e.g., for access to funding)	Positive Irrelevant Negative
Q4: Organizational LoE	Q4.1: Allocation and deployment of resources (e.g., human resources) Q4.2: Team cooperation and motivation for product development Q4.3: Address regulatory barriers Q4.4: Risk management, safety at the workplace Q4.5: Gender and diversity contribution to product development Q4.6: Avoid irresponsible behaviour	Positive Irrelevant Negative

(*continued*)

Table 5.5 (Cont.)

Main questions (Q)	*Criteria*	*Impact of RRI action(s)*
Q5: Economic LoE	Q5.1: Product cost Q5.2: Time to market Q5.3: Profit, market penetration Q5.4: (Favoured) access to financial support Q5.5: Use of human resources	Positive Irrelevant Negative
Q6: RRI action costs	Q6.1: Direct costs to perform the RRI action (compared to product development costs)	Low Medium High

to economic criteria related to product development (e.g. time to market), while Q6 refers to the costs of performing the RRI action (e.g. conducting stakeholder engagement activities, establishing an ethical and social advisory board, hiring an 'embedded' ethicist in product development).

The PRISMA pilot companies considered the most significant criteria for each LoE to be the following (Porcari et al., 2019):

- (Q1) inspire technological innovation, product quality and reliability
- (Q2) product acceptability and safety, address user's needs and rights, trust with/avoid conflicts with business partners, suppliers and end-users
- (Q3) build legitimacy and gain consumer loyalty for the product, meeting new consumers' needs or requests, transparency on product qualities
- (Q4) address regulatory barriers, risk management
- (Q5) market penetration, use of human resources.

RRI costs were a significant issue for only a few of the pilot companies.

To use the questionnaire, product/programme managers should evaluate the impact each RRI action would have on each criterion, using a three-score scale (positive, neutral/irrelevant or negative for Q1–5; low, medium or high for Q6). An example outcome is provided in the self-assessment matrix in Table 5.6 (based on a generic set of three possible RRI actions: A1, A2 and A3). Note that, instead of using scores, the evaluation could be visualized by using emoticons (a 'sentiment analysis'[12] technique), to facilitate qualitative comparisons of results.

In this example, RRI action 1 (A1) would have a positive impact from the scientific and technological (Q1) and ethical and societal (Q2) points of view; its impact would be irrelevant at the strategic (Q3) and organizational (Q4) levels, but negative at the economic (Q5) level. The direct cost (Q6) of the action would be low or negligible. A2 would provide the same beneficial impacts, but with an irrelevant impact at the economic level, whereas

Table 5.6 Example of a self-assessment matrix for the overall impact of possible Responsible Research and Innovation (RRI) actions, based on specific criteria for product development

Impact of RRI actions on criteria	*Impact of RRI A1*	*Impact of RRI A2*	*Impact of RRI A3*
Q1: Scientific and technological line of evidence (LoE)	Positive	Positive	Irrelevant
Q2: Ethical and societal LoE	Positive	Positive	Positive
Q3: Strategic LoE	Irrelevant	Irrelevant	Positive
Q4: Organizational LoE	Irrelevant	Irrelevant	Positive
Q5: Economic LoE	Negative	Irrelevant	Positive
Q6: Direct costs of the RRI action	Positive	Positive	Negative

A3 would have positive impacts on four lines of evidence, and no impact on the remaining line. However, the costs for its implementation would be quite high.

These results suggest that A2 should be implemented, while implementing A1 and A3 might be challenging due to their negative economic and cost impacts. However, A1's and A3's results are not necessarily as negative as they may seem at first. In fact, although the two actions have negative financial impacts, the mixed impact of the other categories may well outperform the financial impacts and therefore give a green light to the overall RRI strategy.

After beginning with this simplified methodology, companies should identify a more quantitative set of indicators to refine the scoring system and to develop and apply more complex and in-depth MCDA.

5.6 Discussion and conclusion

This chapter proposes a practical model to help companies to identify RRI implementation strategies during product development, and offers a simple methodology for a first, qualitative evaluation of RRI impacts along the lines of benefits, barriers and costs. Based on experiences from PRISMA pilots as well as other European Commission-funded projects on RRI in industry, it includes a selection of RRI actions that exemplify how RRI principles can be put in practice.

The PRISMA experience suggests that RRI could help companies that deal with disruptive technologies improve the societal impacts of their innovative products. But to do so, RRI needs to be implemented from the early stages of development onwards and should be considered as a medium- to long-term investment. The model proposed in this study offers a path for that implementation. Moreover, the PRISMA pilots also indicate that success of RRI uptake is strongly context-dependent and is affected by several factors, including company size and organizational complexity and the level of innovation and its associated risks. Our analysis suggests that the benefits of RRI can be both tangible and intangible and short-term and long-term. Prominent barriers include the immediate human and financial resources required to operationalize RRI.

The strengths of this model are that it provides a hands-on approach, that helps to ensure that RRI implementation fits as much as possible within the realities and constraints of individual companies. It is based on widely recognized management approaches (SWOT, Deming cycle) and is flexible and modular in order to accommodate differences across sectors, technologies and types of companies. The model makes use of the 'RRI maturity level' concept, thus giving the opportunity to integrate and consider any company activities that already address RRI aspects, including existing risk, quality and social responsibility actions ('de facto' RRI). Moreover, the model presented here provides a way to assess the impacts of RRI actions on product development, and it is complementary to the development of key performance indicators for monitoring RRI at the organizational level.

However, the model was derived through reflection on experiences with a limited number of pilot projects (and for a limited period), and although companies participated on a voluntary basis, their activities were driven and supported by the resources and targets of the PRISMA project. Testing on real cases, designed to fit the needs, requirements and timelines of a specific company or project (e.g. a task or work package within an industry-led R&I action), is necessary to further develop and refine the model, as well as ways of implementing it. Analysis of different types of companies (e.g. size, sector) and innovations (e.g. technologies, stage of development) would also be instrumental to making further progress in the field.

The approach presented in this study will become a part of a pre-standard document developed as a European Committee for Standardisation (CEN) Common Workshop Agreement (titled 'Guidelines to develop long-term strategies (roadmaps) to innovate responsibly'). This broader document, which will include experiences from other projects and initiatives, will be targeted towards any kind of organization dealing with R&I.[13]

Acknowledgement

This article is based on results of the PRISMA project, that has received funding from the European Union's Horizon 2020 research and innovation programme under grant agreement no. 710059.

Notes

1 www.rri-prisma.eu
2 The organizations involved in the PRISMA project were: Hub of All Things (UK), Aerialtronics (NL), Spectro (NL), RDM Group (UK), Colorobbia Consulting (IT), Laboratori Archa (IT) and Evolva (CH). The eight pilot projects were a public–private partnership linked to a cooperative project called Bisigodos (UK).

3 For instance, see Von Schomberg, 2012; Stilgoe et al., 2013; Van de Hoven, 2013; Klaassen, Kupper, et al., 2017; Gianni, Pearson, & Reber, 2019. Probably the most contested yet possibly most influential conceptualization is the one put forward by the European Commission (2014) in which RRI is defined in terms of the commission's six keys to RRI: ethics, gender, open access, governance, science education and public engagement. For criticism of this conceptualization, see, e.g., Klaassen, Rijnen, Vermeulen, Kupper, & Broerse, 2019.

4 For more information on such programmes, see www.nwo.nl/onderzoek-en-resultaten/programmas/maatschappelijk+verantwoord+innoveren (accessed 10 October 2018) and www.nwo.nl/onderzoek-en-resultaten/programmas/maatsch appelijk+verantwoord+innoveren (accessed 10 October 2018).

5 CSR can simultaneously be seen also as (1) a (cynical) discourse used to legitimize the status quo in industrial conduct and consolidate the power of big firms (Banerjee, 2008) and (2) an academic research field in which the social responsibility of firms is studied either descriptively (Ritala, Huotari, Bocken, Albareda, & Puumalainen, 2018) or normatively (Blok, 2019).

6 The Deming cycle is currently used in standardization: for example, in the ISO 9001 quality management system and the ISO 14000 series on environmental management. See www.iso.org

7 See Rother (2010) for more information on this example.

8 For example, in the pilots on nanotechnologies, introducing specific procedures for the safe use of nanomaterials in production phases increased the safety and sustainability of products; promoting stakeholder dialogues improved understanding of consumers' expectations and helped to address normative and market requests, such as alignment with distributors' requirements for 'premium' consumer products.

9 These considerations are strengthened by the outcomes of the Delphi study conducted in the Responsible Industry project and reported in Porcari et al. (2015).

10 Direct quotation (line 109) from a business and industry participant at the 1st PRISMA Stakeholder Dialogue, 'The Future of Technology: Putting Responsible Innovation into Practice', held in Brussels on 13 April 2017.

11 For example, the literature provided by previous RRI-related projects (such as Responsible Industry, MoRRI, Proso, Satori and COMPASS).

12 A 'sentiment analysis' aims to determine the subject's attitude regarding a specific topic or the emotional reaction to a document or an event. The attitude could be an emotional state but also a judgement or evaluation.

13 For more information, see www.cen.eu/news/workshops/Pages/WS-2019-010.aspx

14 See Maia and Coenen (2017) for detailed information on strategy and methodology for these events. See Maia and Coenen (2018) for a report on the events.

15 Keywords used in the literature search: RRI, CSR, industrial research, responsible innovation, business practices, social innovation, sustainable innovation, innovation management, governance, entrepreneurship, innovation process, R&D management, corporate shared values, ethical leadership, SDGs, business ethics, responsible investments, innovation ecosystem, business strategy, industrial pilots, social impact assessment.

References

Banerjee, S. B. (2008). Corporate social responsibility: The good, the bad and the ugly. *Critical Sociology*, *34*(1), 51–79. https://doi.org/10.1177/0896920507084623

Blok, V. (2019). Innovation as ethos. In C. Neesham & S. Segal (Eds.), *Handbook of Philosophy of Management* (pp. 1–14). Springer, Cham. https://doi.org/10.1007/978-3- 319-48352-8_19-1

Blok, V., Hoffmans, L., & Wubben, E. F. M. (2015). Stakeholder engagement for responsible innovation in the private sector: Critical issues and management practices. *Journal on Chain and Network Science*, *15*(2), 147–164.

Brem, A., Stahl, B., Schroeder, D., Martinuzzi, A., & Blok, V. (Eds.). (2017). *Special Issue "Responsible Research and Innovation (RRI) in Industry." Sustainability*. Basel: MDPI. Retrieved from www.mdpi.com/journal/sustainability/special_issues/RRI#info

Chatfield, K., Borsella, E., Mantovani, E., Porcari, A., & Stahl, B. C. (2017). An investigation into risk perception in the ICT industry as a core component of responsible research and innovation. *Sustainability*, *9*(8). https://doi.org/10.3390/su9081424

Chatfield, K., Iatridis, K., Stahl, B. C., & Paspallis, N. (2017). Innovating responsibly in ICT for ageing: Drivers, obstacles and implementation. *Sustainability*, *9*(6), 1–22. https://doi.org/10.3390/su9060971

Collingridge, D. (1980). *The Social Control of Technology*. London: Francis Printer.

CWA 17145-2 Ethics Assessment for Research and Innovation. Part 2: Ethical Impact Assessment Framework. (2017). Brussels. Retrieved from www.cencenelec.eu/research/CWA/Pages/default.aspx

Dreyer, M., Chefneux, L., Goldberg, A., von Heimburg, J., Patrignani, N., Schofield, M., & Shilling, C. (2017). Responsible innovation: A complementary view from industry with proposals for bridging different perspectives. *Sustainability*, *9*(10), 1–25. https://doi.org/10.3390/su9101719

European Commission. (2011). *A Renewed EU Strategy 2011–14 for Corporate Social Responsibility*. Brussels. Retrieved from www.europarl.europa.eu/meetdocs/2009_2014/documents/com/com_com(20 11)0681_/com_com(2011)0681_en.pdf

European Commission. (2014). *Responsible Research and Innovation: Europe's Ability to Respond to Societal Challenges*. https://doi.org/10.2777/95935

European Union. (2014). *Rome Declaration on Responsible Research and Innovation in Europe*. Retrieved from https://ec.europa.eu/research/swafs/pdf/rome_declaration_RRI_final_21_November. pdf

Garst, J., Blok, V., Jansen, L., & Omta, O. S. W. F. (2017). Responsibility versus profit: The motives of food firms for healthy product innovation. *Sustainability*, *9*(12). https://doi.org/10.3390/su9122286

Gauttier, S., Søraker, J. H., Arora, C., Brey, P. A. E., & Mäkinen, M. (2017). Models of RRI *in Industry (Responsible Industry Project Deliverable 3.3)*. Brussels.

Gianni, R., Pearson, J., & Reber, B. (Eds.). (2019). *Responsible Research and Innovation From Concepts to Practices*. London: Routledge Taylor and Francis.

Graafland, J., & Smid, H. (2019). Decoupling among CSR policies, programs, and impacts: An empirical study. *Business & Society*, *58*(2), 231–267. https://doi.org/10.1177/0007650316647951

Guelke, J. (2018). *Final Report on Pilots (PRISMA Project Deliverable 2.4)*. Brussels.

Gurzawska, A., & Porcari, A. (2016). *Models for Ethics Assessment and Guidance in Industry (SATORI Project Deliverable 4.1)*. Brussels.

Gurzawska, A., Mäkinen, M., & Brey, P. (2017). Implementation of Responsible Research and Innovation (RRI) practices in industry: Providing the right incentives. *Sustainability*, 9(10), 1759. https://doi.org/10.3390/su9101759

Iatridis, K., & Schroeder, D. (2016). *Responsible Research and Innovation in Industry. The Case for Corporate Responsibility Tools*. Cham: Springer International. https://doi.org/10.1007/978-3-319-21693-5

Klaassen, P., Kupper, F., Vermeulen, S., Rijnen, M., Popa, E., & Broerse, J. (2017). The conceptualization of RRI: An iterative approach. In L. Asveld, R. van Dam-Mieras, T. Swierstra, S. Lavrijssen, K. Linse, & J. van den Hoven (Eds.), *Responsible Innovation 3* (pp. 69–92). Cham: Springer International Publishing. https://doi.org/10.1007/978-3-319-64834-7_5

Klaassen, P., Rijnen, M., Vermeulen, S., Kupper, F., & Broerse, J. (2019). Technocracy versus experimental learning in RRI: On making the most of RRI's interpretative flexibility. In R. Gianni, J. Pearson, & B. Reber (Eds.), *Responsible Research and Innovation: From Concepts to Practices* (pp. 77–98). London: Routledge.

Klaassen, P., Smit, K., Tham, J., Groenewold, M., & Westra, J. (2017). *PRISMA's Responsible Industrial Research and Innovation Toolkit (PRISMA project Deliverable 1.1)*. Brussels.

Klaassen, P., Smit, K., Tham, J., Groenewold, M., Westra, J., Nathan, C., ... Porcari, A. (2016). *Workplans for PRISMA Pilots (PRISMA project Deliverable 1.2)*. Brussels.

Kupper, F., Klaassen, P., Rijnem, M., Vermeulen, S., & Jacqueline, B. (2003). *Report on the Quality Criteria of Good Practice Standards in RRI (RRI Tools Project Deliverable 3.1)*. Brussels.

Lees, N., & Lees, I. (2018). Competitive advantage through responsible innovation in the New Zealand sheep dairy industry. *International Food and Agribusiness Management Review*, *21*(4), 505–524. https://doi.org/10.22434/IFAMR2017.0013

Linkov, I., Satterstrom, F. K., Kiker, G., Batchelor, C., Bridges, T., & Ferguson, E. (2006). From comparative risk assessment to multi-criteria decision analysis and adaptive management: Recent developments and applications. *Environment International*, *32*(8), 1072–1093. https://doi.org/10.1016/j.envint.2006.06.013

Lubberink, R., Blok, V., van Ophem, J., & Omta, O. (2017). Lessons for responsible innovation in the business context: A systematic literature review of responsible, social and sustainable innovation practices. *Sustainability*, *9*(5), 721. https://doi.org/10.3390/su9050721

Maia, M., & Coenen, C. (2017). *Dialogue Strategy and Stakeholder Mapping (PRISMA Project Deliverable 4.1)*. Brussels.

Maia, M., & Coenen, C. (2018). *Final Report on the Stakeholder Dialogues (PRISMA Project Deliverable 4.2)*. Brussels.

Marschalek, I., & Schrammel, M. (2018). *Online Learning Material Accessible (SMART Map Project Deliverable 2.6)*. Brussels.

MoRRI Consortium. (2018). *The Evolution of Responsible Research and Innovation in Europe: The MoRRI Indicators Report. (MoRRI Project Deliverable 4.3.)* Brussels.

Nathan, C. (2017). *Report on Kick-off Meeting (PRISMA Project Deliverable 2.1)*. Brussels.

Nathan, C. (2018a). *Interim Report on Pilots (PRISMA Project Deliverable 2.2)*. Brussels.

Nathan, C. (2018b). *Report on Ethicists' Views (PRISMA Project Deliverable 2.3)*. Brussels.

Porcari, A., Borsella, E., & Mantovani, E. (2015). *A Framework for Implementing Responsible Research and Innovation in ICT for an Ageing Society*. Rome: Agra Editrice.

Porcari, A., Pimponi, D., Borsella, E., & Mantovani, E. (2019). *PRISMA RRI-CSR Roadmap (PRISMA Project Deliverable 5.2)*. Brussels.

Ritala, P., Huotari, P., Bocken, N., Albareda, L., & Puumalainen, K. (2018). Sustainable business model adoption among S&P 500 firms: A longitudinal content analysis study. *Journal of Cleaner Production*, *170*, 216–226. https://doi.org/10.1016/j.jclepro.2017.09.159

Rother, M. (2010). *Toyota Kata: Managing People for Improvement, Adaptiveness and Superior Results*. New York: McGraw-Hill Education.

Smart, J. M. (2017). Models foundations for organizational foresight. In The Foresight Guide *[electronic version]*. Foresight University. Retrieved from www.foresightguide.com/shewhart-and-deming/

Stahl, B., Obach, M., Yaghmaei, E., Ikonen, V., Chatfield, K., & Brem, A. (2017). The Responsible Research and Innovation (RRI) maturity model: Linking theory and practice. *Sustainability*, *9*(6), 1036. https://doi.org/10.3390/su9061036

Stilgoe, J., Owen, R., & Macnaghten, P. (2013). Developing a framework for responsible innovation. *Research Policy*, *42*(9), 1568–1580. https://doi.org/10.1016/j.respol.2013.05.008

Van de Hoven, J. (Ed.). (2013). *Options for Strengthening Responsible Research and Innovation*. Brussels: European Commission. https://doi.org/10.2777/46253

van de Poel, I., Asveld, L., Flipse, S., Klaassen, P., Scholten, V., & Yaghmaei, E. (2017). Company strategies for Responsible Research and Innovation (RRI): A conceptual model. *Sustainability*, *9*(11), 2045. https://doi.org/10.3390/su9112045

van Wezel, A. P., van Lente, H., van de Sandt, J. J., Bouwmeester, H., Vandeberg, R. L., & Sips, A. J. (2017). Risk analysis and technology assessment in support of technology development: Putting responsible innovation in practice in a case study for nanotechnology. *Integrated Environmental Assessment and Management*, *14*(1), 9–16. https://doi.org/10.1002/ieam.1989

Von Schomberg, R. (2012). Prospects for technology assessment in a framework of Responsible Research and Innovation. In D. Marc & B. Richard (Eds.), *Technikfolgen abschätzen lehren* (pp. 1–19). VS Verlag für Sozialwissenschaften. https://doi.org/https://doi.org/10.1007/978-3-531-93468-6_2

Wickson, F., & Carew, A. L. (2014). Quality criteria and indicators for responsible research and innovation: Learning from transdisciplinarity. *Journal of Responsible Innovation*, *1*(3), 254–273. https://doi.org/10.1080/23299460.2014.963004

Woolley, R., & Rafols, I. (2016). *Development of Metrics and Indicators for RRI Benefits (MoRRI Project Deliverable D6)*. Brussels.

Yaghmaei, E., Porcari, A., Mantovani, E., & Flipse, S. (2019). Monitoring the value of RRI in industrial nanotechnology innovation projects. In I. Eisenberger, A. Kallhoff, & C. Schwarz-Plaschg (Eds.), *Nanotechnology: Regulation and Public Discourse* (pp. 147–175). London: Rowman and Littlefield International.

6 Do voluntary standards support responsible innovation implementation and reporting in industry?

The case of the European food sector

Edurne A. Inigo, Jilde Garst, Vincent Blok, and Konstantina M. Pentaraki

6.1 Introduction

Over the last years, Responsible Research and Innovation (RRI) has strongly emerged as a way to tackle the grand challenges of our time. These grand challenges (e.g. climate change, socio-economic inequality or the obesity epidemic, among many others) require the intervention of several societal actors and the development of new, innovative solutions. Industry actors, as one of the major players of innovation, are presented with an opportunity to contribute to the solution of such problems while cultivating their competitive advantage (Von Schomberg, 2013). In order to do so, RRI proposes a transparent, democratic innovation process whereby stakeholders are mutually responsive and which aims to yield ethically acceptable, socially desirable and sustainable outputs that tackle the grand challenges of our time (Von Schomberg, 2013).

Responding to socio-ethical issues through transparent, democratic and mutually responsive innovation processes has often proven a struggle for industry. This has operationalised in several concepts as value-sensitive design, closely connected to responsible innovation (van den Hoven, 2013). This approach acknowledges that products and technologies are value-laden; that is, they reflect societal values in their design. A classic example is the 'racist bridges' in Long Island, NY, USA, which were designed deliberately too low to allow buses (the mode of transportations used by poor blacks and Puerto Ricans) to reach the beach. Therefore, values can be reflected in innovation, but transparent and democratic processes are necessary to identify which values are shared by society. However, there are certain difficulties to translate values in commercial innovation processes: apart from problems derived from information asymmetries and different degrees of responsibility among the intervening stakeholders (Blok & Lemmens, 2015), the complexity of introducing several (and often contradicting) voices in the deliberation process may conflict with the desirable time to market in competitive environments.

One of the governance mechanisms suggested to simplify the process of RRI to make it more compatible with commercial innovation is the use of voluntary standards for product development (Von Schomberg, 2013; Iatridis & Schroeder, 2016; Voegtlin & Scherer, 2017). Voluntary standards are a mechanism of soft law through which businesses bind themselves to certain criteria for the achievement of a common goal (Brunsson & Jacobsson, 2002; Kirton, Trebilcock, & Trebilcock, 2017). The rationale for the possible role of voluntary standards is that they provide companies with guidelines for responding to socio-ethical issues in the innovation process and its outcomes, while greatly reducing the procedural burden that a case-to-case response may entail.

In case of business-to-consumer products, voluntary product standards often come attached to a form of certification that can be easily recognised by the customer, the so-called front-of-pack (FoP) labels. Many studies have been conducted on the influence of voluntary standards on product innovation and their positive influence on the profitability of companies (Besen & Farrell, 1994; Acemoglu, Gancia, & Zilibotti, 2012). In particular, FoP labels could be beneficial for this purpose as they allow producers to communicate their certification to consumers and show their awareness and responsiveness towards particular socio-ethical issues, which might appeal to a certain customer segment (Nikolaeva & Bicho, 2011; Kleef & Dagevos, 2015). However, there is little (and inconclusive) evidence on the utility of voluntary product standards when directing innovation towards addressing grand challenges (Rennings, 2000). Thus, the value of voluntary product standards to integrate societal values in innovation practices needs to be further investigated and examined in practice. In this chapter we critically examine the potential of voluntary product standards as an RRI governing mechanism. The question that we are aiming to answer with the case study of the European food industry is: How do FoP labels based on voluntary product standards support RRI implementation and reporting in industry?

In order to do so, we first provide an overview of the literature that describes the supportive mechanisms of voluntary standards for RRI as well as possible drawbacks. Then we connect theory with industry practice by presenting a case study with data from seven European food companies on the adoption of FoP labels. We end this chapter with a comparison of the theory and our case study results, drawing preliminary conclusions and suggesting further research on the role of voluntary standards for target setting, monitoring and assessment in the governance of RRI in industry.

6.2 Voluntary standards as an RRI practice

6.2.1 *Classifications of voluntary standards*

The use of voluntary standards has greatly increased in industry in the last years (Hughes, Buttle, & Wrigley, 2007; Nadvi, 2008), since they respond to multiple needs of industry, from the absence of strong authorities to legislate or enforce laws and regulations in a globalised economy (Nadvi, 2008) to the will to pre-empt new, harder regulations from coming into place through the use of alternative standards (Okhmatovskiy & David, 2011). Typically, the socio-ethical issues addressed in these standards are not directly observable by customers and stakeholders, and for reasons of transparency and competitive advantage, these standards 'flag to customers and other stakeholders that producers or traders who adopt ... standards show a higher socioenvironmental performance than their uncertified counterparts' (Wijen, 2014, p. 303).

Voluntary standards can be categorised according to three characteristics: (1) the aspect of a company they assess; (2) the range of companies they assess; (3) the actor(s) that develop them and/or monitor their compliance (Wijen, 2014). Regarding which aspect of business voluntary standards assess, a distinction can be made between management standards and product standards. Management standards assess whether the business practices of the company are conducted in a way that addressed socio-ethical issues, such as protection of the natural environment, the protection of the human rights of employees or the mitigation of climate change (Wijen, 2014). Examples of voluntary management standards are ISO 14001 for international environmental management, ISO 9001 for international quality management and Fair Trade certification (Christmann & Taylor, 2006; Delmas & Montes-Sancho, 2011; Wijen, 2014). Product standards assess the final outcome of a company's business practices, also referred to as absolute performance requirements (Wijen, 2014). These standards assess the impact of a specific product on the socio-ethical issues. Examples of voluntary product standards are the standards of the Forest Stewardship Council (FSC) and the standards of the Marine Stewardship Council (Wijen, 2014; Moog, Spicer, & Böhm, 2015a). As one product of a company might comply with the product standard while another does not, the certification of product standards is often combined with a FoP label (Boström, 2006; Van Kleef & Dagevos, 2015). Concerning the range of standards, they can apply to one specific company, to a whole sector or even extend across sectors (Wijen, 2014). Furthermore, for their geographical range, scholars often make a distinction between international and national standards. Finally, the governance of voluntary standards can be categorised into structures with only one type of actor – such as trade organisations representing multiple companies – or a combination of multiple types of actors, referred to as a multi-stakeholder initiative or agreement (MSA) (Fransen & Kolk, 2007; Wijen, 2014).

6.2.2 *The potential role of voluntary standards in RRI*

The RRI framework calls for a transparent, democratic and mutually responsive innovation process (Von Schomberg, 2013; Blok & Lemmens, 2015). Following these principles, Stilgoe, Owen, and Macnaghten (2013) developed a framework of RRI that envisions four dimensions to be addressed: anticipation, inclusion, reflexivity and responsiveness. Anticipation refers to foresight tools and techniques that aim to predict, to the extent possible, the unintended consequences of the innovation, hence providing with governance mechanisms to understate future dynamics created by the innovation and in which the innovation will operate. Inclusion aims to remove traditional top-down streams in innovation goal-setting, integrating different stakeholders in the innovation process in order to better reflect and incorporate on societal values. Reflexivity is related to the ability of the intervening actors to be self-critical and mirroring the conclusions in the other dimensions in their actions. Moreover, second-order reflexivity can be built through standards or codes of conduct (Von Schomberg, 2013). Finally, responsiveness is related to the ability to respond and incorporate the requirements identified through anticipatory, inclusive and reflective measures in the innovation process, hence materialising in the final result.

Voluntary standards constitute a collaborative mechanism to address socio-ethical challenges through product innovation, and they are suggested to overcome the process of deliberation and facilitate the materialisation of socio-ethical objectives in the innovation process, by providing a soft-law governance system (Voegtlin & Scherer, 2017). In that sense, they level the competitive ground on socio-ethical issues, without the burden of engaging in costly deliberative processes individually, while promoting self-reflexivity (enhancing the capability of the company to reflect on its RRI performance as compared to the benchmark). The use of FoP labels, which are the result of compliance with product development standards, increases transparency towards consumers (Verbeke, 2005), hence constituting an element of direct reporting of RRI activities to consumers.

Moreover, multi-stakeholder-based standard setting is suggested to help overturn existing power structures, allowing non-traditional players to raise their voice in commercial innovation processes and have their interests considered further than they would be in other forms of stakeholder consultation (Mena & Palazzo, 2012; Ponte & Cheyns, 2013). Voluntary standards are often developed by MSAs, with the increasing participation of non-governmental organisations (NGOs) and other social stakeholders (Fransen & Kolk, 2007). In this sense, collaborative standard setting breaks business boundaries, by inviting different actors to the drafting of self-regulations. This is consistent with the inclusion dimension of RRI (Stilgoe et al., 2013), which was originally conceived for technology development in non-commercial environments (for instance, basic development of nanotechnologies). The progression down the innovation cycle to the competitive stage, where industry actors further develop these technologies for commercial purposes, changes the levelling ground and comes with additional challenges for stakeholder engagement. Hence multi-stakeholder-based drafting of voluntary standards makes stakeholder engagement feasible in an industry context: stakeholder involvement occurs in the early stage to develop criteria for innovation. Including stakeholder voices through standards can be a good way of overcoming the extra burden placed by inclusion measures on businesses, which often affects their ability to compete in the market (Blok & Lemmens, 2015). The development of these standards, and their adaptation to local markets and changing stakeholders' views is a time-consuming and complex effort also at the network level; therefore, MSAs are often constituted, providing a permanent governance structure that revises and adapts the standards over time: the success of the standard will also rest on the capacity (Ponte & Cheyns, 2013). Setting up a more permanent structure for dialogue allows for interaction and negotiation to revise standards and introduce elements of the collective deliberation advocated by the RRI framework, as well as providing mechanisms for certification and compliance (Gurzawska, Makinen, & Brey, 2017).

Blok and Lemmens (2015) indicate that such a process is at odds with actual commercial innovation practices. Voluntary product standards governed by MSAs have been suggested as a solution for overcoming each of these barriers (Von Schomberg, 2013; Blok, Hoffmans, & Wubben, 2015). First, for product innovation to contribute to the competitive advantage of a company, the innovation process is required to be relatively low-cost and have a short time to market to pre-empt competitors. Blok and Lemmens (2015) argue that the inclusive deliberation processes required for RRI are often lengthy and costly and will thus eliminate any competitive advantage of innovation. The second argument of Blok and Lemmens (2015) concerns the industry actors' need for information asymmetries to build a competitive advantage, which conflicts with the principle of transparency. Oversharing information and the risk of knowledge leaks to competitors through the deliberation process can be detrimental for the company's competitive advantage, particularly when no specific arrangements are made. In this regard, the standards frameworks can provide mechanisms for protection of intellectual property. Third, the power imbalances between companies – which carry the liability and financial burden of the innovation – and other stakeholders involved – which often do not financially contribute to innovation – trump the principle of mutual responsiveness on which RRI is built (Blok & Lemmens, 2015).

Voluntary standards can overcome the barrier regarding the time to market by targeting the whole sector at once. Thereby, any company that decides to bring a product to the market before the standards are completed runs the risk of having to reformulate the product at a later stage to make it comply. Potential benefits of being first to market are eliminated by the cost of reformulation. When it concerns re-accreditation after standard adjustment, the MSA has the option to allow a time lag between the publication and re-accreditation of the standards, providing companies time to adjust their products before losing their certification and thereby not interfering with the product innovation process of companies. The governance structure of the MSA can also soften the financial impact of inclusive deliberation processes by dividing the cost among the participating actors.

Consequently, voluntary standards for product development have an added value in the implementation of RRI in industry (Iatridis & Schroeder, 2016), since they perform a threefold mission: first, they support RRI reporting through certification and FoP labels, when compliance with the standards is audited and certified by an independent party or a permanent structure created ad hoc (Fulponi, 2006; Gurzawska et al., 2017). Second, they help to make the business case for RRI by helping to obtain a premium on sales price in the market, through the reporting of the RRI activities to consumers. Since certification and FoP labels report the efforts in the direction of including socio-ethical goals in the process, they have a direct effect on pricing (short-term advantage) and reputation (long-term advantage). Third, they help to set RRI targets that are incorporated in the innovation process. In fact, standardisation of product development can lead to more innovative outcomes through agenda setting and application of innovative methods (Inoue, Arimura, & Nakano, 2013; Amores-Salvadó, Martin-de Castro, & Navas-López, 2015). Studies in the environmental sustainability field have shown that the use of voluntary standards as part of environmental management systems has been positively related with the development of green innovation capabilities (Amores-Salvadó et al., 2015); with the development of green process innovations (Wagner, 2008); and with the profit maximisation of such innovations (Inoue et al., 2013).

In the case of standards for societal values it is argued that leaving standardisation to industry players may lead to socially undesirable results. The development of voluntary standards through MSAs supports companies in the daunting task of balancing the interests of different stakeholders and contributing to the inclusion of certain societal values in their products or services (King, Lenox, & Terlaak, 2005; York, Vedula, & Lenox, 2017), while providing compliance mechanisms(Ponte & Cheyns, 2013). The power imbalances can be softened by voluntary standards, as they can serve as a form of social contract between the involved stakeholders (Giovannucci & Ponte, 2005). Thereby, the non-commercial stakeholders will also connect their name to the initiative and the outcomes of the standards will thus impact their own legitimacy. That this impact can be both positive and negative can be seen with the Rainforest Alliance, an NGO that has set management and product standards in collaboration with several multinational enterprises and has received both praise and criticism for its role (Scherer, Palazzo, & Seidl, 2013). In this manner, the responsibility for developing the standards and assessing compliance lies with various societal actors, instead of businesses only. Nevertheless, as explored in the next sections, multiple shortcomings for voluntary standards as a mechanism for RRI in industry have been identified as well.

6.2.3 *The drawbacks of voluntary standards as RRI instrument*

Despite their potential contribution to responsible outcomes, the legitimacy and internal governance of voluntary standards are often questioned

(Vogel, 2010; Kleef & Dagevos, 2015). As previously mentioned, voluntary standards can support the implementation of RRI if they are based on multi-stakeholder engagement, including not only industry, but also NGOs, policy-makers and consumer associations. This allows for the multidimensionality that tackling complex problems entails (Schouten & Glasbergen, 2011), which includes the integration of social and technological concerns in innovation, or trade-offs between economic and social interests. However, the setting of voluntary standards has traditionally been considered a private-sector activity: a form of self-regulation in the absence of state regulation or as a corporate response to societal activism for a certain cause (Ponce del Castillo, 2010). When engagement of stakeholders is initiated, financed and conducted by industry players, it may lead to a bias in representation of unfavourable, marginal or overly critical societal actors (Fransen & Kolk, 2007; Roloff, 2008). Therefore, although these industry-led initiatives are prevalent, their moral legitimacy is highly questioned because of their selective representation of interests (Suchman, 1995; Fransen & Kolk, 2007), even when accreditation is left to third parties and appropriate measures for multi-stakeholder decision-making are in place (Bäckstrand, 2006; Thabrew, Wiek, & Ries, 2009). These issues of representativeness of different societal actors and interests in the development of voluntary standards can be particularly troubling when FoP labels are associated with them, since FoP labels grant an increased degree of trust from the consumer, although the standards may have not been evaluated and approved by a significant representation of societal actors.

To overcome these issues of illegitimacy and to contribute to responsible outcomes, voluntary standards need to be governed by MSAs that are characterised by: (1) balanced representation; (2) structures that assure deliberative communication and decision-making; and (3) systems for monitoring or verification of corporate behaviour (Scherer & Palazzo, 2008; Moog et al., 2015). In practice, however, difficulties with these three characteristics hinder the involvement of NGOs and government representatives in the development and monitoring of the standards (Moog et al., 2015). First, the balanced representation is hindered by the large organisational burden of MSAs, which leads to high resource demands for its participants. Membership is thus easier for larger organisations, as they have more resources to spare. In practice, this has led to under-representation of both small, local companies and small, local NGOs – e.g. underrepresentation of small farms in the Roundtable of Responsible Soy (García-López & Arizpe, 2010; Schouten, Leroy, & Glasbergen, 2012) and overrepresentation of WWF, Greenpeace and Amnesty International in MSAs (Fransen & Kolk, 2007). While RRI may open the innovation process for other stakeholders through deliberation, this deliberation may still disproportionately favour the larger players due to the resources required for participation. This underrepresentation not only leads to a less inclusive process, but also decreases the success of local implementation of the standards. To include local companies and NGOs, the larger organisations will be required to contribute more resources to the MSAs. However, this will create a power imbalance, as was the case with the FSC case in the long term (Moog et al., 2015).

Second, the procedural legitimacy of MSAs for voluntary standards is often criticised because of its lack of transparency (Auld & Gulbrandsen, 2010), despite this being a crucial element of RRI (Stilgoe et al., 2013). The reason for this lack of transparency can be found in the need to reach agreement across stakeholders about the details of the standards. For this purpose, the actors involved need to be encouraged to reflect on their views and possibly reconsider them to get closer to the views of the opposing actors (Owen et al., 2013). Such reflexive processes and reconsiderations are easier to achieve in closed-door negotiations because then 'negotiators can freely exchange ideas and thoughts more easily than in the public sphere where they have to stick to their guns' (Risse, 2004, p. 312). Furthermore, the act of reconsidering their views might also be easier for companies than for NGOs. While the reputation of companies is often positively influenced by MSA participation, an NGO's reputation can be threatened when they are seen to be compromising towards companies (Moog et al., 2015). The reputation of the NGO is based upon their ability to act upon a socio-ethical issue and business is often seen as the source of the issue at stake. The main purpose of companies is often portrayed as providing economic welfare and, although the social reputation of a company might be damaged, non-compliance with the standards can always be excused by the financial risk compliance brings. Although these conditions count for any MSA, the objective of RRI to govern innovation brings in an additional difficulty: the novelty of innovation creates uncertainty about future outcomes. Even if the standards are very detailed, the mere implementation of them can have unforeseen consequences due to the complex nature of grand challenges (Blok & Lemmens, 2015). The combination of this need for consensus and this uncertainty is shown to be a reason for NGOs not to participate, considering the multiple tensions that the involvement of stakeholders brings in the innovation process (see, also for a case in the food industry, Blok et al., 2015).

Third, the daunting task of monitoring and verification is often carried out by independent bodies (Albareda, 2013; Gurzawska et al., 2017). Due to the lack of impartiality of companies and the lack of resources of other actors, professional auditing organisations are tasked with monitoring the implementation of the standards in many MSAs. Although independent, the FSC case has shown that the influence of these auditors on the standard formulation increases over the years, often at the expense of the influence of non-industry representatives (Moog et al., 2015). Thereby, the different viewpoints that are necessary for the success of the voluntary standards in terms of integration of socio-ethical issues are often overlooked (Meybeck & Redfern, 2014). Additionally, to govern the uncertain nature of innovation, it is not only important to monitor compliance with the standards, but also the often unforeseen consequences of their implementation to the grand challenges. In the case of voluntary standards for RRI, this monitoring system could therefore become extensive and difficult to manage. The costs of developing such a monitoring system are in MSAs often covered by the initiating partners. Although inclusiveness is important for RRI, any actors that join the initiative at a later stage can be viewed as free-riders, making the MSA less open for new members and thus limiting its growth potential (Delmas & Keller, 2005).

The final potential drawback of voluntary standards lies in their effectiveness to result in more socially desirable, ethically acceptable and sustainable outcomes. Even if there is no previous empirical evidence focused primarily on RRI and voluntary product development standards, research on voluntary management standards for sustainability challenges shows that, while adherence to them is often high, the lack of enforceable sanctions in case of underperformance reduces their effectiveness (Ruysschaert & Salles, 2014). For product standards, however, it must be noted that MSAs can have strict rules for compliance, which result in removal of the FoP label (e.g. the label 'organic' or 'fair trade') (Boussalis, Feldman, & Smith, 2018). For RRI, the need to be responsive to the consequences of innovation requires the standards to be highly dynamic, focusing more on product *improvement* then meeting threshold requirements (see for instance the case of the Choices' logo for healthy food in the Netherlands: Garst, Blok, Jansen, & Omta, 2017). This adds another difficulty for consistent sanctioning practices. Without these sanctions, however, a company can enjoy the reputational benefits of engaging in a voluntary standards programme (Nikolaeva & Bicho, 2011) while keeping them at the periphery of their business operations (Vogel, 2010).

To summarise, literature theorises that voluntary product standards governed by MSAs and combined with FoP labels for public communication could potentially have a supportive role for RRI implementation in industry, but at the same time might not overcome all barriers and might even create new hurdles. Although some empirical evidence on the effectiveness of voluntary product standards and FoP labels exists (Brunsson & Jacobsson, 2002; Iatridis & Schroeder, 2016), their role in RRI implementation has not been empirically investigated. We critically examine how voluntary standards may support RRI reporting through a case study on the use of FoP labels in the food industry and compare these results with another illustrative example: the standards for the use of nanoparticles in cosmetics.

6.2.4 *Application of voluntary standards for RRI: the use of front-of-pack labels in the food industry*

To address many socio-ethical issues that it faces, the food industry has widely adopted voluntary product standards associated with FoP labels. One of the grand challenges that RRI aims to address is the increase of diet-related non-communicable diseases (e.g. obesity, cardiovascular disease and type 2 diabetes). In the context of the food industry, FoP labels are developed to increase transparency for the consumer through the provision of comparable information, to facilitate interaction through formal socialisation mechanisms and to enhance responsiveness via monitoring, assessment and objective target setting, which are important stakeholder engagement practices in RRI practice (Blok et al., 2015).While FoP labels are also used by the food industry to communicate about other societal values (e.g. fair-trade origin, animal welfare), the labels aimed at the health of the consumer are unique for this industry and have an increasing presence (Kleef & Dagevos, 2015). In the European context, policy-makers and scientific institutions as well as NGOs often collaborate in the development of the standards used for the certification of an FoP label (Brunsson & Jacobsson, 2002; Leipziger, 2010). Therefore, to investigate the use of standards for incorporating societal values in innovation processes, we draw upon examples from practice, using the food industry and their consumer-related health standards as an illustrative case.

6.3 Methods

To gather examples on the use of FoP labels in the food industry, we conducted in-depth interviews with innovation managers of seven European food companies of different sizes. To select the cases, we used a non-probabilistic purposive sampling method. To allow for verification of observations between companies and to identify contextual characteristics that influence observations, both similar and contrasting cases were selected (Miles, Huberman, & Saldaña, 2013). To ensure case similarity and relevance to the research question, the selection criteria were: companies (1) that are part of the food and beverage processing and manufacturing sector (for definition see: Lehtinen et al., 2016); (2) that are located in Europe and operating in the Netherlands; (3) that have shown activity in healthier product innovation; and (4) that have adopted FoP labels regarding the nutritional composition of their products. To provide contrast, the selected companies differed in size, geographical market and product range. The cases were selected from the membership list of the Dutch branch organisation for food manufacturers and their websites were used to determine their fit with the selection criteria. We selected the case of the Dutch food industry not only based on convenience sampling, but also because of the existence of FoP food labels for healthier products, and because of the prevalence of the food industry and the effect of its practices on health in the nationwide debate (de Vries, de Hoog, Stellinga, & Dijstelbloem, 2016). In total 21 companies were contacted, of which seven agreed to take part in the study. The main features of the respondent companies are summarised in Table 6.1. To ensure confidentiality, the names of the companies and their brands are anonymised.

The semi-structured interviews were conducted in English through Skype or face to face with an average duration of 40 minutes. The interviews

Table 6.1 Main features of sample companies

Case	*Geographical market*	*Product range*
Company A	Europe	Freshly preserved vegetables and fruits
Company B	Worldwide	Production, development and packaging of meat substitutes
Company C	The Netherlands	Bakery products
Company D	Worldwide	Fresh, frozen and dried potato products
Company E	The Netherlands	Chilled soups, sauces and meal components
Company F	Europe	Preserved fruits, vegetables and pulses
Company G	Worldwide	Pasta, ready-to-use sauces and bakery products

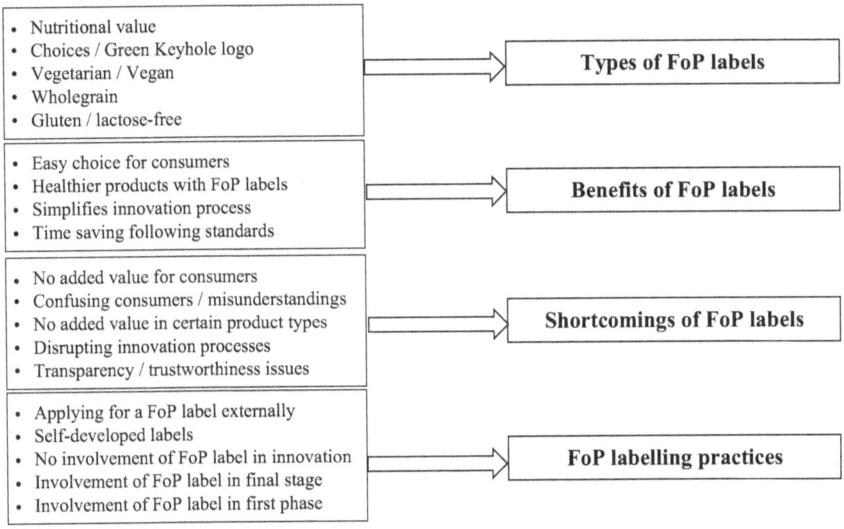

Figure 6.1 The coding tree used for data analysis.
Note: FoP, front-of-pack.

were semi-structured to examine specific topics, mainly the advantages and disadvantages presented in the previous section, but to still be open to for diverging topics that might extend existing theories (Bryman, 2011). To ensure the trustworthiness of the observations by the interviewees, the factual data were triangulated with secondary data, such as company websites, newspaper articles and industry reports. We transcribed the interviews verbatim and analysed the transcripts through inductive-deductive codification, data-driven but guided by the literature across coding cycles (Pratt, 2009) using the software Atlas.ti (ATLAS.ti Scientific Software Development GmbH, 2013; Gioia, Corley, & Hamilton, 2013). This allowed for the development of categories and the identification of patterns within the sample. The coding tree can be found in Figure 6.1.

6.4 Findings

The interviews partially confirmed that FoP labels played a significant role in translating the value 'improving health' into specific design requirements that could be directly implemented during product development, such as reduced sugar and fat levels. By taking over the search for consensus on the definition of health, the governance system behind FoP labels alleviated the companies from the need to navigate conflicting stakeholder views. In this manner, FoP labels seem to support the implementation of RRI. However, the participants nuanced this vision as well, showing that it also comes with attached hurdles. One of the difficulties of the integration of the societal value 'public health' in product innovation according to the companies in our case study is the abstract nature of the concept 'health' and the many views in society on how it relates to food products. Translating these abstract values was seen by the companies as a cumbersome process involving continuous stakeholder engagement, as illustrated by Company A:

> The different angles to look at healthiness of a product makes it more complex...So, what we are focusing on is telling people what it is, and we try to add as less salt as possible but it needs to be tasty because otherwise people don't eat it so you can make a very healthy product but if nobody eats it makes no sense.

These processes did not fit in their drive towards lean and efficient product development processes. However, the product standards helped them by providing clear guidelines useful in navigating these divergent views.

However, the use of voluntary product standards with FoP labels, while helpful in incorporating values in the innovation process, did not lead to companies directly participating in inclusive deliberation, since this task was mainly conducted by the MSA. Moreover, most of the companies in the sample doubted whether the FoP labels that they were currently deploying were developed in a sufficiently inclusive manner, considering the little diversity of stakeholders involved in the development of the standard, and the little space for direct stakeholder engagement beyond the application of the standard as a second-order reflexivity measure. The companies used the voluntary product standards (developed by the government or governmental agencies); only one of them additionally consulted a roundtable of scientific researchers throughout the three stages of the innovation process (conceptualisation, development and commercialisation). In cases when an external organisation was consulted – for instance, in order to be able to utilise the 'vegan' FoP label – these were only consulted in the final, validation phase of the project.

Therefore, the participation of stakeholders external to the innovating company was rather minor and limited to the final stages. In this manner, the deliberation and inclusion dimensions of RRI were somehow externalised to the MSA, hence reducing the active agency of the firm in incorporating RRI principles, but helping to overcome some of the barriers for RRI implementation in industry (Blok & Lemmens, 2015). Only companies B and F indicated that for some FoP labels they inquired about the standards in the ideation phase of their innovation process. The main argument given for not using the voluntary product standards earlier in the process was their limiting effect on the creativity in the product innovation. Thereby, in our sampled companies, the criteria of the FoP labels only acted as a validation system for the innovation outcome and had only limited effect on the target setting of the innovation process.

This observation is consistent with the various companies' perception that stakeholder inclusion in the innovation process is cumbersome and disturbed the usual pace of the innovation process. This was a major reason why they used voluntary product standards instead of consultation or co-creation processes. Some other setbacks identified by the companies concern the marketing value of the FoP label, as raised by company A, despite their ability to convey instant information to the consumer, because the FoP labels made the packaging less attractive. Interestingly, most companies in the sample doubted the effectiveness of voluntary standards to produce healthier food outcomes (see Table 6.2).

The ability of FoP labels to promote self-reflexivity in companies can also be doubted, based on our data. When asked about the effectiveness of the FoP labels to promote healthy innovation, several companies indicated that the labels were not particularly useful as their company already had its own nutritional guidelines in place and therefore the FoP labels did not change their innovation processes. Even though they acknowledged the usefulness of stakeholder engagement for the legitimacy of FoP labels, only one company involved external actors in their development of their internal guidelines, in the form of an advisory board of nutrition scientists. The other companies did conduct stakeholder engagement activities, but not directly related to setting health standards for their innovation process. Besides inefficiency, their main argument for not involving stakeholders was the fear of losing control over the end-product. The companies indicated that they are responsible not only for developing the innovation, but also diffusing it. Letting other actors be directly involved in the innovation process and thus have ownership of the process was for the companies not a guarantee that these actors would also take responsibility for the societal impact of the end-product. As this product always carried their brand name, the ownership of the product lies with the companies and thus also the responsibilities for its societal impact. As illustrated by company F, it is difficult to share aims, outcomes, rewards and responsibilities beyond a punctual occasion: 'We try to understand what their [other stakeholders'] horizon is, but it is never co-operational like sharing, we share insights, but we do not deliver results together'.

Besides a role in making the innovation process more responsible, FoP labels are also suggested to promote responsible consumption by transparently communicating the impact of the product on the socio-ethical issue at stake. On the one hand the companies in our case study indicated that the FoP labels indeed supported the communication of health messages to the consumer. On the other hand, they highly doubted whether the consumer understood that message and even suggested that the FoP labels might cause confusion. Three reasons were provided by the interviewees. First, although the FoP labels translated the abstract value to practical product requirements for the producers, when the label was printed on the package its simplicity did not communicate the standards it represented. Although a few FoP labels have now been developed that are more detailed, most labels do not allow consumers to compare their definition of health with label criteria. For instance, while the label 'organic' is relatively straightforward and criteria can be easily consulted, the criteria for the label 'healthy choice' are more blurred (e.g. is it healthy, or just healthier than an alternative?). Second, the number of FoP labels has increased over the years and the companies indicated that they perceived confusion among their consumers about the meaning of labels and the value they represent. Third, the increasingly globalised food market asks for global regulation of food labels. This is particularly challenging in this highly regulated sector, since degrees of food quality and safety vary from country to country, and consumer preference is very closely tied to local cultures. However, implementing such a system was indicated by the companies as a daunting task due to the many local differences, and limits the effectiveness of FoPs as a mechanism of RRI reporting.

A brief summary of the advantages and shortcomings raised in the literature and by the sample companies, accompanied by exemplar quotes, can be found in Table 6.2.

6.5 Discussion and conclusions: the future of voluntary standards as instrument for supporting RRI implementation and reporting in industry

Our study contributes to the study of the potential role of voluntary standards (and associated FoP labels) to support the implementation and reporting of RRI in industry. There are issues that emerge from the analysed cases that concern the value of voluntary standards for innovation and their effectiveness, and issues that are related to their value as mechanism for RRI assessment and reporting developed at the early stage of the innovation process.

Our results confirm the mixed trends in the literature, with some of the interviewees confirming the value of voluntary standards for the development of innovation – and most importantly, as a tool to integrate social values in innovation – and others noting how they can constitute obstacles for the innovation process. Albeit their value for objective target-setting and providing benchmarks (Vellema & van Wijk, 2015; York, Vedula, & Lenox, 2017), our research shows that voluntary standards are not per se sufficient for the operationalisation of RRI, and that the companies in our sample

Table 6.2 Benefits and shortcomings of front-of-pack (FoP) labelling practices as raised by the literature and the sample companies in the food industry

Advantages of FoPs identified by the sample companies	*Supported by*	*Exemplar quote*	*Main advantages of FoPs in the literature*
Facilitate the innovation process and save time	A, B	"… if those labels help us and they also help the consumers why not … we hope that it is positive and easier for people to select products…" (Company A)	Avoid costly individual (company-by-company) process of deliberation
Ease choice for consumers	A, B, C, D	"It is good to have criteria which state the health claim and then follow those on the development phase. It facilitates target setting" (Company B)	Increase transparency, providing comparable information to the consumer
			Fill in gaps where (often international) regulation is underdeveloped Facilitate stakeholder engagement Prevent power imbalances through the integration of various stakeholders in their development Enhance self-reflexivity Level the field for an industry
Shortcomings of FoPs identified by the sample companies	**Supported by**	**Proof quote**	**Main shortcomings of FoPs in the literature**

(*continued*)

Table 6.2 (*Cont.*)

Advantages of FoPs identified by the sample companies	*Supported by*	*Exemplar quote*	*Main advantages of FoPs in the literature*
Lack of transparency and / or trustworthiness	A, B, C, D, F	"yes, the trust of the people will increase, I think it will help. On the short term it might have a good effect. On the long term it would lead to less innovation because of its limits. A very controlled food system could lead to less innovation and a lot of bureaucracy, which is counterproductive" (Company F)	Lack of transparency during the development of the standards
Confuse customers and / or create misunderstandings	A, C, D, F, G	"in this moment the reality is very much fragmented between countries and as a company…we want to avoid increasing the complexity of reading all those FoPs from consumers" (Company G)	They can confuse consumers
Disrupt the innovation process	A, C, F	"I believe, that even a trustworthy labelling system would lead to less innovation because it limits always..." (Company F)	Their effectiveness has often been questioned
Make products less attractive	A	"so putting such labels on the products can be negative for people who just want to have a nice product… Logos are very ugly on the packaging" (Company A)	

perceive inclusion and deliberation as cumbersome in the innovation process despite the simplified mechanisms that FoP labels provide. Moreover, beyond taking the first steps in the inclusion of a certain value in the innovation process, complacency within the standards may lead to less innovative solutions for the challenge at hand.

The literature states certain benefits of engaging with voluntary standards in terms of establishing the business case of RRI. The companies in our sample confirmed this view, but also identified many shortcomings that show that the system of implementation of voluntary standards as a forum of RRI can be improved in order to adjust it to business realities. Companies see reputational gains as the major benefit of engaging with voluntary standards – rather than an increased capacity for RRI or a chance to remove obstacles for transparent and mutually responsive innovation process. These attitudes towards the voluntary standards show that the risk of their use by free-riders or as a greenwashing mechanism may be present (Risse, 2004; Moog et al., 2015), in line with the findings for other voluntary standards aiming to incorporate socially responsible or sustainable practices in business. Another major risk that has been identified in the cases, and which resonates with previous business engagement with social values – although not in the field of innovation (Delmas & Keller, 2005) – is that of the lack of transparency. The companies in the sample highlighted the remaining lack of access to information in the development process, which limits the trustworthiness towards the consumer and their value as an RRI reporting mechanism. The use of the label may veil the product under the category of 'healthier product', while there are no added health benefits to consuming that brand over others, creating a misleading 'fake transparency' effect that distorts the spirit of reporting of RRI activities to the consumer. This aspect was confirmed by our results, with several companies highlighting consumer confusion as a major shortcoming of voluntary standards. In addition, it was mentioned that the target-setting derived from the standards sometimes hindered or slowed down the innovation process. Even if this can be perceived as a problem from the perspective of RRI implementation in an ideal state, previous research shows that being able to be responsive to societal demands without needing to go through costly processes of stakeholder deliberation was a way in which industry has adopted RRI practices (Gurzawska et al., 2017).

The impact of these standards on the innovative capacity of a company or an industry is a controversial issue in the field of innovation. Despite the fact that values embedded in voluntary standards can span across business practices – e.g. the ISO 14000 standards for environmental management (ISO, 2009) – one of the most common practices targeted by standards is product development, which is the focus of this chapter. As noted by Wright, Sturdy, and Wylie (2012), standardisation implies working towards homogeneity and similarity in an industry. This intuitively goes against the narrative of status quo-defying developments that often characterise the innovation discourse. In that regard, codes of conduct that provide general behaviour guidelines instead of specific product requirements have also been developed for certain sectors, either promoted by policy-makers or industry players (Blok, 2017). Choosing to adhere to existing rules in product development is said to have a detrimental effect on the ability to develop disruptive innovations, since it limits the ability of the innovator to break away from existing dynamics by locking her to a set of behaviour connected with preceding practices.

In a nutshell, our research examines how voluntary standards and associated FoP labels can support RRI implementation and reporting. Examining the case of the European food industry, we observed that, while there is great potential value in the use of voluntary standards and associated FoP labels for RRI standard setting and direct reporting to the consumer, RRI principles of transparency and mutual responsiveness are not being incorporated in practice. The FoP labels can be considered an instrument of RRI reporting (once it had been implemented, it could be easily communicated to the consumer), while they have less value as a lever for RRI practices. In summary, our data suggest that FoP labels do not stimulate stakeholder engagement and self-reflexivity within the sample companies, which puts their validity for the purpose of understanding other stakeholders' views at risk. In other words, FoP labels may support RRI adoption because they circumvent the problems associated with stakeholder engagement (increased complexity, lengthening of innovation process); however, the price to pay for RRI in industry while sustaining competitiveness is then the loss of inclusion. This shows that new ways to approach the development of standards is needed. Examples from other cases include the collaboration with legislators as a stakeholder, as in the case of the use of nanocosmetics, or a move towards open global platforms for discussion that allow for a more inclusive approach.

6.6 Limitations of the study

While the case of the European food industry presented in this chapter is illustrative, the particularities of these sector limit the applicability of the results to other industries. Moreover, the geographical limitation and the small number of interviews may limit the transferability of the conclusions. However, the study is prone to analytical generalisation (Polit & Beck, 2010), hence contributing to the development of research around the usefulness of voluntary standards and FoP labels for the implementation and reporting of RRI. However, it must be noted that our research, limited to a country and region, shows inconsistencies in the literature that call for further research in the drivers and barriers of voluntary standards for RRI with larger samples. At this stage, we contribute with the identification of some variables explaining the mechanisms of implementation.

6.7 Future directions

Our case examines voluntary standards and FoP labels as a mechanism of assessment and reporting of RRI, with mixed results. While they show a great potential to measure and set targets for integration of social values in the innovation practices and reduce the burden of individual mutually responsive inclusion practices, their trustworthiness and transparency as an RRI reporting method are still questioned by the participating companies.

If this is the case, how can the drawbacks be overcome by governing voluntary standards differently, without losing the main benefits, as mentioned above? Standard development remains largely a task dependent on private actors, who decide for themselves or decide which stakeholders to involve in the MSAs (Fransen & Kolk, 2007; Moog et al., 2015). To that extent, it remains a power-imbalanced and not transparent process. In the light of this, in some sectors governmental agencies have taken the lead in the development of standards, in an approach that combines hard law (regulation) with soft law (voluntary standards) developed with industry, such as the case of nanotechnologies and nanoparticles in cosmetics. The European approach to nanoparticles in cosmetics combines supranational regulation, covering aspects related to the precautionary principle (e.g. REACH Regulation or the Cosmetics Regulation EC 1223/2009) with science-based standards supported by governmental organisations (Ponce del Castillo, 2010). While ISO, European Committee for Standardisation (CEN) or the Organisation for Economic Co-operation and Development (OECD) have focused on technical, nomenclature and safety-related standards, the European Union adopted a Code of Conduct for Responsible Nanosciences and Nanotechnologies Research, which complements regulation. The Code of Conduct deals with those aspects on which legislation would obstruct innovation because of scientific lack of consensus, allowing for further flexibility and timely revision of the Code of Conduct (Ponce del Castillo, 2010). This approach to standard development is effective and transparent; however, it does not fulfil principles of mutual responsiveness or inclusivity as proposed by the RRI framework. This illustrates the trade-off between opting for straightforward top-down designed regulations and codes of conduct or for voluntary standards developed in close collaboration with stakeholders, which provide legitimacy to private-sector decisions. The latter may be more appealing in terms of RRI reporting for user or consumer trust concerns.

In an interesting development in the field of ICT a global working group (named OCEANIS) aiming to discuss the very use of standards for the integration of ethical issues is emerging. OCEANIS was founded in 2018 by several national bodies for standardisation and stands for Open Community for Ethics in Autonomous and Intelligent Systems (OCEANIS, 2018). The aim is to integrate the view of the standardisation bodies (usually backed by governments) with that of businesses, scientists (including social scientists) and other organisations through an open and ongoing discussion about ethical concerns in the development of autonomous and intelligent systems. To that extent, it constitutes a governance innovation in the field of voluntary standards, whereby RRI principles of transparency, democracy and mutual responsiveness are incorporated before the innovation process commences. In this way, it addresses the issue of standardisation in a context of a globalised economy with different local ethical sensitivities.

While the involvement of governmental bodies for legitimatisation is not a necessary step for RRI assessment, it might provide a bonus of trustworthiness when reporting the RRI efforts to users and customers. However, a more nuanced approach where the discussion of criteria for voluntary standard setting is facilitated (instead of orchestrated) by governmental agencies in structures such as OCEANIS might show the future direction of voluntary standards for RRI. With direct involvement in the innovation process reported as too costly (Blok & Lemmens, 2015), and with private-sector-backed standards and labels suffering from a crisis of transparency and trustworthiness, new solutions are required. These solutions should allow for wider international participation, aiming to avoid power imbalances between the governing actor and the rest of the stakeholders, and in a spirit of open discussion where competitive information need not be shared yet.

References

Acemoglu, D., Gancia, G., & Zilibotti, F. (2012). Competing engines of growth: Innovation and standardization. *Journal of Economic Theory*, *147*(2), 570–601.e3. https://doi.org/10.1016/j.jet.2010.09.001

Albareda, L. (2013). CSR governance innovation: Standard competition–collaboration dynamic. *Corporate Governance: The International Journal of Business in Society*, *13*(5), 551–568. https://doi.org/10.1108/CG-06-2013-0076

Amores-Salvadó, J., Martin-de Castro, G., & Navas-López, J. E. (2015). The importance of the complementarity between environmental management systems and environmental innovation capabilities: A firm level approach to environmental and business performance benefits. *Technological Forecasting and Social Change*, *96*, 288–297. https://doi.org/10.1016/j.techfore.2015.04.004

ATLAS.ti Scientific Software Development. (2013). Atlas.ti (Version 7.1.0) [Windows]. Retrieved from https://atlasti.com/

Auld, G., & Gulbrandsen, L. H. (2010). Transparency in nonstate certification: Consequences for accountability and legitimacy. *Global Environmental Politics*, *10*(3), 97–119. https://doi.org/10.1162/GLEP_a_00016

Bäckstrand, K. (2006). Multi-stakeholder partnerships for sustainable development: Rethinking legitimacy, accountability and effectiveness. *European Environment*, *16*(5), 290–306. https://doi.org/10.1002/eet.425

Besen, S. M., & Farrell, J. (1994). Choosing how to compete: Strategies and tactics in standardization. *Journal of Economic Perspectives*, *8*(2), 117–131. https://doi.org/10.1257/jep.8.2.117

Blok, V. (2017). Bridging the gap between individual and corporate responsible behaviour: Toward a performative concept of corporate codes. *Philosophy of Management*, *16*(2), 117–136.

Blok, V., & Lemmens, P. (2015). The emerging concept of responsible innovation. Three reasons why it is questionable and calls for a radical transformation of the concept of innovation. In B.-J. Koops, I. Oosterlaken, H. Romijn, T. Swierstra, & J. van den Hoven (Eds.), *Responsible Innovation* 2 (pp. 19–35). Retrieved from http://link.springer.com/chapter/10.1007/978-3-319-17308-5_2

Blok, V., Hoffmans, L., & Wubben, E. f. m. (2015). Stakeholder engagement for responsible innovation in the private sector: Critical issues and management practices. *Journal on Chain and Network Science*, *15*(2), 147–164. https://doi.org/10.3920/JCNS2015.x003

Boström, M. (2006). Regulatory credibility and authority through inclusiveness: Standardization organizations in cases of eco-labelling. *Organization*, *13*(3), 345–367. https://doi.org/10.1177/1350508406063483

Boussalis, C., Feldman, Y., & Smith, H. E. (2018). Experimental analysis of the effect of standards on compliance and performance. *Regulation & Governance*, *12*(2), 277–298. https://doi.org/10.1111/rego.12140

Brunsson, N., & Jacobsson, B. (2002). *A World of Standards*. Oxford: Oxford University Press.

Bryman, A. (2011). *Business Research Methods*. Don Mills, Ont.: OUP Canada.

Christmann, P., & Taylor, G. (2006). Firm self-regulation through international certifiable standards: Determinants of symbolic versus substantive implementation. *Journal of International Business Studies*, *37*(6), 863–878. https://doi.org/10.1057/palgrave.jibs.8400231

Delmas, M., & Keller, A. (2005). Free riding in voluntary environmental programs: The case of the U.S. EPA WasteWise program. *Policy Sciences*, *38*(2), 91–106. https://doi.org/10.1007/s11077-005-6592-8

Delmas, M. A., & Montes-Sancho, M. J. (2011). An institutional perspective on the diffusion of international management system standards: The case of the environmental management standard ISO 14001. *Business Ethics Quarterly*, *21*(01), 103–132. https://doi.org/10.5840/beq20112115

de Vries, G., de Hoog, J., Stellinga, B., & Dijstelbloem, H. (2016). *Towards a Food Policy*. The Hague: Netherlands Scientific Council for Government Policy (WRR).

Fransen, L. W., & Kolk, A. (2007). Global rule-setting for business: A critical analysis of multi-stakeholder standards. *Organization*, *14*(5), 667–684. https://doi.org/10.1177/1350508407080305

Fulponi, L. (2006). Private voluntary standards in the food system: The perspective of major food retailers in OECD countries. *Food Policy*, *31*(1), 1–13. https://doi.org/10.1016/j.foodpol.2005.06.006

García-López, G. A., & Arizpe, N. (2010). Participatory processes in the soy conflicts in Paraguay and Argentina. *Ecological Economics*, *70*(2), 196–206. https://doi.org/10.1016/j.ecolecon.2010.06.013

Garst, J., Blok, V., Jansen, L., & Omta, O. S. W. F. (2017). Responsibility versus profit: The motives of food firms for healthy product innovation. *Sustainability*, *9*(12), 2286. https://doi.org/10.3390/su9122286

Gioia, D. A., Corley, K. G., & Hamilton, A. L. (2013). Seeking qualitative rigor in inductive research: Notes on the Gioia methodology. *Organizational Research Methods*, *July 24*, 1094428112452151. https://doi.org/10.1177/1094428112452151

Giovannucci, D., & Ponte, S. (2005). Standards as a new form of social contract? Sustainability initiatives in the coffee industry. *Food Policy*, *30*(3), 284–301. https://doi.org/10.1016/j.foodpol.2005.05.007

Gurzawska, A., Makinen, M., & Brey, P. (2017). Implementation of Responsible Research and Innovation (RRI) practices in industry: Providing the right incentives. *Sustainability*, *9*(10), 1759. https://doi.org/10.3390/su9101759

Hoven, J. van den. (2013). Value sensitive design and responsible innovation. In R. Owen, J. Bessant, & M. Heintz (Eds.), Responsible Innovation (pp. 75–83). https://doi.org/10.1002/9781118551424.ch4

Hughes, A., Buttle, M., & Wrigley, N. (2007). Organisational geographies of corporate responsibility: A UK–US comparison of retailers' ethical trading initiatives. *Journal of Economic Geography*, 7(4), 491–513. https://doi.org/10.1093/jeg/lbm011

Iatridis, K., & Schroeder, D. (2016). Responsible research and innovation in industry: the case for corporate responsibility tools. In *Springer Briefs in Research and Innovation Governance*. Cham: Springer.

Inoue, E., Arimura, T. H., & Nakano, M. (2013). A new insight into environmental innovation: Does the maturity of environmental management systems matter? *Ecological Economics*, *94*, 156–163. https://doi.org/10.1016/j.ecolecon.2013.07.014

ISO. (2009). *Environmental Management: The ISO 14000 Family of International Standards*. Geneva, Switzerland: ISO.

King, A. A., Lenox, M. J., & Terlaak, A. (2005). The strategic use of decentralized institutions: Exploring certification with the ISO 14001 management standard. *Academy of Management Journal*, *48*(6), 1091–1106. https://doi.org/10.5465/amj.2005.19573111

Kirton, J. J., Trebilcock, M. J., & Trebilcock, M. J. (2017). *Hard Choices, Soft Law : Voluntary Standards in Global Trade, Environment and Social Governance*. https://doi.org/10.4324/9781315253633

Kleef, E. V., & Dagevos, H. (2015). The growing role of front-of-pack nutrition profile labeling: A consumer perspective on key issues and controversies. *Critical Reviews in Food Science and Nutrition*, *55*(3), 291–303. https://doi.org/10.1080/10408398.2011.653018

Lehtinen, U. E. E., Ahokangas, P., & Lu, J. (2016). The role of intermediaries in food export: Case evidence from Finland. *British Food Journal*, 118(5). https://doi.org/10.1108/BFJ-03-2015-0116

Leipziger, D. (2010). *The Corporate Responsibility Code Book* (2nd ed.). Sheffield: Routledge.

Mena, S., & Palazzo, G. (2012). Input and output legitimacy of multi-stakeholder initiatives. *Business Ethics Quarterly*, *22*(3), 527–556. https://doi.org/10.5840/beq201222333

Meybeck, A., & Redfern, S. (2014). Voluntary standards for sustainable food systems: Challenges and opportunities. A workshop of the FAO/UNEP programme on sustainable food systems. Rome: FAO.

Miles, M. B., Huberman, A. M., & Saldaña, J. (2013). *Qualitative Data Analysis: A Methods Sourcebook* (3rd ed.). Thousand Oaks, CA: SAGE Publications.

Moog, S., Spicer, A., & Böhm, S. (2015). The politics of multi-stakeholder initiatives: The crisis of the Forest Stewardship Council. *Journal of Business Ethics*, *128*(3), 469–493. https://doi.org/10.1007/s10551-013-2033-3

Nadvi, K. (2008). Global standards, global governance and the organization of global value chains. *Journal of Economic Geography*, *8*(3), 323–343. https://doi.org/10.1093/jeg/lbn003

Nikolaeva, R., & Bicho, M. (2011). The role of institutional and reputational factors in the voluntary adoption of corporate social responsibility reporting standards. *Journal of the Academy of Marketing Science*, *39*(1), 136–157. https://doi.org/10.1007/s11747-010-0214-5

OCEANIS – Open Community for Ethics in Automation and Intellegent Systems. (2018). Retrieved October 11, 2018, from OCEANIS website: https://ethicsstandards.org/

Okhmatovskiy, I., & David, R. J. (2011). Setting your own standards: Internal corporate governance codes as a response to institutional pressure. *Organization Science*, *23*(1), 155–176. https://doi.org/10.1287/orsc.1100.0642

Owen, R., Stilgoe, J., Macnaghten, P., Gorman, M., Fisher, E., & Guston, D. (2013). A framework for responsible innovation. In R. Owen, J. Bessant, & M. Heintz (Eds.), *Responsible Innovation* (pp. 27–50). https://doi.org/10.1002/9781118551424.ch2

Polit, D. F., & Beck, C. T. (2010). Generalization in quantitative and qualitative research: Myths and strategies. *International Journal of Nursing Studies*, *47*(11), 1451–1458. https://doi.org/10.1016/j.ijnurstu.2010.06.004

Ponce del Castillo, A. (2010). European Trade Union Institute (ETUI) – The EU approach to regulating nanotechnology / Working Papers / Publications / Home. Retrieved September 12, 2018, from www.etui.org/Publications2/Working-Papers/The-EU-approach-to-regulating-nanotechnology

Ponte, S., & Cheyns, E. (2013). Voluntary standards, expert knowledge and the governance of sustainability networks. *Global Networks*, *13*(4), 459–477. https://doi.org/10.1111/glob.12011

Pratt, M. G. (2009). From the editors: For the lack of a boilerplate: Tips on writing up (and reviewing) qualitative research. *Academy of Management Journal*, *52*(5), 856–862. https://doi.org/10.5465/AMJ.2009.44632557

Rennings, K. (2000). Redefining innovation – eco-innovation research and the contribution from ecological economics. *Ecological Economics*, *32*(2), 319–332. https://doi.org/10.1016/S0921-8009(99)00112-3

Risse, T. (2004). Global governance and communicative action. *Government and Opposition*, *39*(2), 288–313. https://doi.org/10.1111/j.1477-7053.2004.00124.x

Roloff, J. (2008). A life cycle model of multi-stakeholder networks. *Business Ethics: A European Review*, *17*(3), 311–325. https://doi.org/10.1111/j.1467-8608.2008.00537.x

Ruysschaert, D., & Salles, D. (2014). Towards global voluntary standards: Questioning the effectiveness in attaining conservation goals: The case of the Roundtable on Sustainable Palm Oil (RSPO). *Ecological Economics*, *107*, 438–446. https://doi.org/10.1016/j.ecolecon.2014.09.016

Scherer, A. G., & Palazzo, G. (2008). Globalization and corporate social responsibility. In A. Crane, A. McWilliams, D. Matten, J. Moon, D. Siegel (Eds.), *The Oxford Handbook of Corporate Social Responsibility* (pp. 413–431). Oxford: Oxford University Press.

Scherer, A. G., Palazzo, G., & Seidl, D. (2013). Managing legitimacy in complex and heterogeneous environments: Sustainable development in a globalized world. *Journal of Management Studies*, *50*(2), 259–284. https://doi.org/10.1111/joms.12014

Schouten, G., & Glasbergen, P. (2011). Creating legitimacy in global private governance: The case of the Roundtable on Sustainable Palm Oil. *Ecological Economics*, *70*(11), 1891–1899. https://doi.org/10.1016/j.ecolecon.2011.03.012

Schouten, G., Leroy, P., & Glasbergen, P. (2012). On the deliberative capacity of private multi-stakeholder governance: The Roundtables on Responsible Soy and Sustainable Palm Oil. *Ecological Economics*, *83*, 42–50. https://doi.org/10.1016/j.ecolecon.2012.08.007

Stilgoe, J., Owen, R., & Macnaghten, P. (2013). Developing a framework for responsible innovation. *Research Policy*, *42*(9), 1568–1580. https://doi.org/10.1016/j.respol.2013.05.008

Suchman, M. C. (1995). Managing legitimacy: Strategic and institutional approaches. *The Academy of Management Review*, *20*(3), 571–610. https://doi.org/10.2307/258788

Thabrew, L., Wiek, A., & Ries, R. (2009). Environmental decision making in multi-stakeholder contexts: Applicability of life cycle thinking in development planning and implementation. *Journal of Cleaner Production*, *17*(1), 67–76. https://doi.org/10.1016/j.jclepro.2008.03.008

Vellema, S., & van Wijk, J. (2015). Partnerships intervening in global food chains: The emergence of co-creation in standard-setting and certification. *Journal of Cleaner Production*, *107*, 105–113. https://doi.org/10.1016/j.jclepro.2014.03.090

Verbeke, W. (2005). Agriculture and the food industry in the information age. *European Review of Agricultural Economics*, *32*(3), 347–368. https://doi.org/10.1093/eurrag/jbi017

Voegtlin, C., & Scherer, A. G. (2017). Responsible innovation and the innovation of responsibility: Governing sustainable development in a globalized world. *Journal of Business Ethics*, *143*(2), 227–243. https://doi.org/10.1007/s10551-015-2769-z

Vogel, D. (2010). The private regulation of global corporate conduct: Achievements and limitations. *Business & Society*, *49*(1), 68–87. https://doi.org/10.1177/0007650309343407

Von Schomberg, R. (2013). A vision of responsible research and innovation. In *Responsible Innovation.* Retrieved from http://tethys.eaprs.cse.dmu.ac.uk/rri/sites/default/files/obs-approach-to-rri/2013-von%20Schomberg%20RRI%20-owenbookChapter.pdf

Wagner, M. (2008). Empirical influence of environmental management on innovation: Evidence from Europe. *Ecological Economics*, *66*(2–3), 392–402. https://doi.org/10.1016/j.ecolecon.2007.10.001

Wijen, F. (2014). Means versus ends in opaque institutional fields: Trading off compliance and achievement in sustainability standard adoption. *Academy of Management Review*, *39*(3), 302–323. https://doi.org/10.5465/amr.2012.0218

Wright, C., Sturdy, A., & Wylie, N. (2012). Management innovation through standardization: Consultants as standardizers of organizational practice. *Research Policy*, *41*(3), 652–662. https://doi.org/10.1016/j.respol.2011.12.004

York, J. G., Vedula, S., & Lenox, M. (2017). It's not easy building green: The impact of public policy, private actors, and regional logics on voluntary standards adoption. *Academy of Management Journal*. https://doi.org/10.5465/amj.2015.0769

Part III

Responsible innovation assessment

7 Monitoring Responsible Research and Innovation in the European research area

The MoRRI project

Ingeborg Meijer and Wouter van de Klippe

7.1 Introduction

Responsible Research and Innovation (RRI) as a policy initiative exists to more closely align research and innovation (R&I) processes and products with societal needs. RRI draws on a diversity of theoretical work which has in the past sought for ways to more intentionally incorporate societal reflection within R&I processes (Stilgoe, Owen, & Macnaghten, 2013). While drawing on a rich history of efforts to develop more open and democratized R&I systems (for a discussion on the historical context of the discourse of RRI, see Rip, 2014), the explicit operationalization of RRI within European Commission (EC) R&I policy can be traced back to the 2014 Rome Declaration of RRI.

The 2014 Rome Declaration of RRI described the concept as "the ongoing process of aligning research and innovation to the values, needs and expectations of society" (Italian Presidency of the Council of the European Union, 2014). Within the declaration, there were calls for the integration of RRI within R&I policy, which have since been followed. The EC currently has an operational definition of RRI which draws on a diversity of theoretical elaboration (Owen, Macnaghten, & Stilgoe, 2012; Pandza & Ellwood, 2013; Stilgoe, Owen, & Macnaghten, 2013; von Schomberg, 2013; Rip, 2014), and is as follows:

> a process where all societal actors (researchers, citizens, policy makers, business representatives, third sector organisation representatives, etc.) work together during the whole R&I process in order to better align R&I outcomes with the values, needs and expectations of European society.

This operational definition of RRI contains six guiding "keys" of RRI, including: gender equality, science literacy and science education, public engagement, ethics, open access and governance. The aspirations of RRI are bold and will likely benefit from the development of tools which will further support the creation, evaluation, implementation, and adaptation of R&I policy to accomplish these goals (Owen, Macnaghten, and Stilgoe, 2012).

The Monitoring the Evolution and Benefits of Responsible Research and Innovation (MoRRI) project (2014–2018) was the first large-scale effort to develop, implement, and assess an RRI monitoring system for the European research area (ERA). MoRRI was a three-and-a-half-year-long project which consisted of initial conceptual work, data collection, data analysis, reflections, adaptations, and findings. In order to identify and measure both the scope, benefits, and pitfalls of RRI within the ERA, the MoRRI project developed and operationalized a conceptual framework and methodology according to the EC's current definition and framework of RRI. Subsequently, the project tested this methodology for its ability to monitor the current state of RRI, based around various metrics which were developed and elaborated throughout the project. Finally, MoRRI conducted various case studies to assess the realized socio-economic and democratic impacts that the implementation of RRI might have had.

The monitoring system developed by the MoRRI project (Peter et al., 2018) consisted of 36+ indicators which were based on a review of literature which covered the EC's six key areas of RRI. The process of developing this monitoring system included examining existing data sources, analysing their appropriateness, locating any potential gaps in coverage, and the iterative development of new indicators which required primary data collection. Throughout this chapter, primary data will refer to the collecting of new data through tools (surveys, questionnaires, etc.) that were created throughout the MoRRI project.

After the various indicators were developed and potential shortcomings were identified, an extensive data collection process covering all European Union (EU) Member States was then carried out. This data collection process consisted of a large-scale survey-based data collection, a set of case studies critically examining the effects of the implementation of RRI, collection and analysis of bibliometric and patent databases, and secondary analyses of existing data at both individual- and country-level data. The survey was distributed among researchers at European institutions, research-funding organizations (RFOs), research-performing organizations (RPOs), societal stakeholder organizations, and representatives in industry.

In this chapter, we continue with a description of the challenges of the MoRRI project (Section 7.1.2), describe the EC's six RRI keys (Section 7.1.3), discuss the development of the indicators (Section 7.1.4), highlight emerging patterns at the country level (Section 7.1.5), and discuss the identification of benefits of RRI (Section 7.1.6) and the way forward (Section 7.1.7).

7.2 Challenges

The task of creating a monitoring and evaluation system for the implementation of RRI and its benefits can be understood through the lenses of being a technical, scientific, practical, and irreducibly political challenge.

- The challenge is technical: The growth of quantitative measures to control and evaluate a diversity of swathes of life is so pernicious and widespread that it has been dubbed a "veritable metric tide" (Wilsdon, Doubleday, & Stirling, 2015). Indicators to assess countries, organizations, institutions, and even individuals within R&I landscapes have become ubiquitous and controversial. Despite this growth of indicator-led life, few scholars have attempted to develop indicators of RRI, and even fewer have gone beyond the theoretical step of their creation and sought to populate these indicators with data. While some laudable efforts for the creation of such indicators have been developed (Wickson & Carew, 2014; Strand & Kaiser, 2015; Heras & Ruiz-Mallén, 2017; Monsonís-Payá et al., 2017), each shows considerable (and understandable) humility towards the task. An emphasis on the need for a diversity of context-sensitive indicators has been repeatedly articulated. This importance was not lost throughout the MoRRI project, as it is vital to ensure that the creation of these indicators does not serve to foreclose potential notions of responsibility.
- The challenge is scientific: There exist gaps in knowledge in the efficacy of RRI policies, the practices that they ostensibly support, and the benefits that these aim to have. The MoRRI project sought to remedy these gaps in knowledge through the development of a theoretical frame to understand the links between policy, practice, and effect, and additionally provided empirical work to support this theory. This required an integration of work being conducted in various disciplinary contexts; most importantly, theoretical work elaborating RRI and quantitative social science work on monitoring and evaluation.
- The challenge is practical and political: Throughout the MoRRI project, there existed a myriad of technical and theoretical decisions. These choices, made according to a variety of different value systems which were continuously negotiated throughout the project, inevitably have impacts on what vision of RRI is monitored, evaluated, and supported. The successor project of SUPER MoRRI (acronym for Scientific Understanding and Provision of an Enhanced and Robust Monitoring system for RRI) will provide RFOs, RPOs, and other parties with tools and knowledge which may be needed for their own assessments and action plans for the implementation of RRI. This novel framing of a monitoring and evaluation system is to be thought of as an opportunity to facilitate change rather than a system of domination and control. Indeed, consortia members caution against RRI (and the evaluation and monitoring systems seeking to aid in its implementation) becoming a "box-ticking" exercise or bureaucratic hurdle for researchers and innovators. Proponents of RRI envision RRI to be a novel opportunity for the democratization of scientific work and a space for opportunity rather than hindrance. In recognition of the risk of the "bureaucratization" of RRI, the monitoring and evaluation system within MoRRI and the follow-up project of SUPER MoRRI seek to be as open, context-dependent, and amenable to the needs of users as possible.

Nevertheless, we acknowledge that there were irreducibly political decisions made throughout the project (whom to enlist for surveys, which questions to pose, how to pose those questions, how to depict their results, which value systems to adopt in the framing of all of these actions, etc.) which concomitantly open and close potential visions of what RRI is and will become. One framing decision that was made throughout the MoRRI project was the use of the RRI framework articulated by the EC which uses six RRI keys to orient work. The next section will describe in detail the ways in which these keys were used throughout MoRRI.

7.3 RRI keys

The first phase of the MoRRI project required a review of the theoretical literature which underlies the six RRI keys which make up the EC's operationalization of RRI. Subsequent to this literature review, a stocktaking and assessment of potential sources of data for each key were conducted. Below, we continue with a brief discussion of each of the six keys and the theoretical work which has been developed around them.

7.3.1 Public engagement (PE)

The definition of the key public engagement in RRI that is featured on the EC's website is as follows: "[Public engagement] is about co-creating the future with citizens and civil society organizations, and also bringing on board the widest possible diversity of actors that would not normally interact with each other, on matters of science and technology" (EC, 2019b). The work that has investigated public engagement with science and technology is diverse, multifaceted, and often contentious.

The Royal Society published a report entitled "The public understanding of science" (1985) which is often held as a point of orientation for the history of the development of the field. Within this report, increasing the public understanding of science was advocated primarily for two reasons. First, due to the ways in which science and technology have permeated contemporary society, citizens benefit from a general understanding of science when navigating their lives. Second, a higher level of public understanding of science contributes positively to a nation's innovation system, and thus has positive implications for national prosperity.

More recent work in the field of public engagement with science and technology has articulated various problematic assumptions which were contained within the Royal Society's 1985 report. For a discussion on the developments of the field up until a decade after the report's publication, see Wynne (1995). Central to the development of the field were the identification and critique of what has been labelled the "deficit model" of public engagement with science and technology (Wynne, 2006). In short, the deficit model assumes that whatever tensions may exist within the relationship between the scientific community and "lay" publics is due to the "lay public" being ignorant of science, thus exemplifying a knowledge "deficit" (Bucchi, 2008). The task then, according to this model, is to remedy this tension by informing this supposedly ignorant public. What is made invisible within this model of public engagement with science and technology is the very real possibility that, when tensions exist between publics and science, this may be due to differences in underlying value commitments which are not due to any particular lack of knowledge by "lay" publics (Wynne, 2006). If this different interpretation of the cause of tension between these communities is accepted, then the role of public engagement work changes. Rather than providing platforms to inform (ostensibly) ignorant publics, instead what is needed are venues and forms of engagement which encourage the active articulation of value disputes between publics and the scientific community. In recognition of this alternative function for public engagement with science, the following categories of public engagement work were used:

- *Public communication* – communicating and disseminating scientific work to public audiences
- *Public activism* – informing decision-makers of public concerns to inform and mobilize action according to publics' needs
- *Public consultation* – informing decision-makers of public opinions regarding certain topics
- *Public deliberation* – providing venues to facilitate group deliberation policy-related scientific work
- *Public participation* – providing opportunities for publics to contribute to, steer, and co-construct scientific work.

7.3.2 Science literacy and scientific education (SLSE)

The second RRI key operationalized within MoRRI of science literacy and scientific education is intimately related with the key of public engagement, but places more emphasis on critically inspecting the various forms through which citizens are able to "comprehend and express opinions about science, as well as the ability to contribute to 'doing science'" (Talmon-Gros & Teichler, 2015). The underlying theoretical grounding for this key was also informed by the theoretical work conducted for the public engagement key, most notably the shift away from a deficit model of science literacy and scientific education towards one of co-production. This resulted in three foci within this key:

1. Science education (specifically for young audiences) covering contemporary scientific work (textbook knowledge) and norms and values of science
2. Science communication work to generate awareness among citizens of all ages about science and science-related issues

3. Science communication work with the aim of encouraging and investigating practices of co-production of scientific work with citizens.

Similar to the model used within the public engagement key, the form of science literacy and scientific education work envisioned to be monitored within MoRRI is one that enables a more inclusive R&I environment which aims to ensure the public as an active participant and steward of science and innovation work.

7.3.3 Gender equality (GE)

The gender equality key within the EC's operationalization of RRI contains three objectives (EC, 2019c):

1. Ensuring that an equal gender balance is found within research teams. This is to address systemic inequalities in the makeup of research teams which have historically contained more men than women. Importantly, this objective not only emphasizes the general balance, but also attends in particular to the gender balance at different institutional and managerial levels within the R&I environment.
2. Decision-making environments must also be more equitable in terms of gender balance. This means that, when panels, work groups, and any other decision-making space are created, these must be representative in terms of gender equality.
3. The integration of the gender dimension in R&I content. This will extend the relevance of R&I outputs and increase the overall quality of these outputs. Furthermore, this integration will address the systemic lack of attention paid to gender-specific concerns within the production of knowledge.

Within the MoRRI project, the functional definition of the gender equality key of RRI was theoretically extended beyond that of the EC. MoRRI understood gender through a social constructivist lens, emphasizing that it is a social construct enacted through performative practices (West & Zimmermann, 1987; Butler, 1990). Recognizing that gender is a social construct requires attending to the ways in which gender categories are replicated in social practices. Additionally, this forces us to attend to other structural categories which are inseparably linked to gender, such as age, socio-economic status, race / ethnic background, disability, sexual preference, etc. (Harding, 2006). Consequently, MoRRI operationalized gender through an intersectional approach (Hancock, 2007).

7.3.4 Open access (OA)

Within the EC, the RRI key open access refers to "the global shift towards making research findings available free of charge for readers" (EC, 2019a).

The underlying logic supporting the move to make research findings available to all, free of charge for access, is often justified through claims that this will facilitate the development of a more efficient R&I system. This is envisioned through facilitating access to research findings for actors and institutions within the public and private sector who would, presumably, benefit from having access to these research findings. The EC put forth a recommendation to all Member States that all publicly funded research results should be made accessible to the public as early as 2012 (European Commission, 2012). The EC has since moved to an even more ambitious model of open science beyond only envisioning the findings of research to be made open to the public. This would entail, when appropriate, that both research processes and products be made more open and transparent.

Within MoRRI, the concept of open access was grounded upon the general policy concept of open science and sought to focus on the lack of available policies, frameworks, institutional tools, and institutional conditions, to encourage the opening of research processes and products (van den Eynden & Bishop, 2014). In particular, the indicators that were developed focused on the following:

- Open-access instruments for publications: these included indicators focusing on measuring publication practices within open-access journals (gold open access) and efforts for self-archiving in repositories (green open access), and thus focused predominantly on making *outputs* accessible;
- Developments in open-research data: These indicators used the FAIR (findable, accessible, interoperable, and reusable) framework to provide measures of the extent to which data produced within research processes were made available and usable according to ways in accordance with open-science principles. When compared to the research findings made available through open-*access* principles, data and datasets which have been made available and usable according to these principles are relatively scarce (Costas et al., 2013; Farhan et al., 2013).

7.3.5 Ethics

Within the MoRRI project's analytical report on the RRI key of ethics, the following working definition was operationalized: "Ethics as a scientific discipline is concerned with normative rules for everybody, which other than moral should be used to evaluate action and not to guide actions", which was taken from Griessler and Littig (2006, p. 134; translated by the author). However, an alternative definition supplemented this one as it was found to be more applicable within the context of R&I: "a common platform of right and wrong, [which] is influenced by cultural norms, and aims at informing policy making" (Ladikas et al., 2015, p. 3). Note that, according to this latter definition of ethics, importance is placed on the local specificity and need for context-specific considerations when deliberating norms and notions of what is right or wrong.

In order to draw out what normative considerations to take into account, while attending to culturally specific norms, an importance was placed on providing platforms to express ethical concerns held by different stakeholder groups. While this can be done according to formal deliberation in institutionalized settings (such as in policy-making at global, European, and national levels), this can also be done within more informal levels as well (Felt et al., 2009). This more informal engagement with ethics is often done through, for example, engaging diverse groups of individuals who may have a stake in the development of R&I in participatory and open public engagement exercises. There must be an emphasis placed on openness, transparency and accountability of processes, explicitly considering the needs of marginalized stakeholder groups, and other considerations within these efforts.

The operationalization of ethics according to this understanding was further categorized as the following:

- Ethical governance: the institutionalization of ethics through, for example, the development and implementation of standards in R&I policies (Brom et al., 2015), often done by ethics commissions, ethical codes, and soft law
- Ethical deliberation: the institutionalization of ethics through the facilitation of debate, among a diversity of different stakeholder groups, on issues pertaining to the development of science and technology, often taking place in advisory bodies
- Ethical reflection: the institutionalizing of ethics through providing venues for the reflection. This often takes the form of collective academic and societal reflection of ethical issues.

7.3.6 Governance

The MoRRI project's operational definition of governance (Stilgoe & Lindner, 2015) emphasized various aspects of the governance of science and innovation. First, "governance" was understood as consisting of "any form of control or management", recognizing that a broader understanding of governance includes the establishing of goals, setting up means, and assessing the performance of the undertaken steps to accomplishing the established goals. The project recognized that governance exists beyond institutions of the state, emphasizing that it can and does exist at the level of individual institutions (including non-profits, universities, RFOs and RPOs, and more). Furthermore, it is emphasized that governance need not take the form of explicit rules for guidance or regulation and is also performed in other social contexts which may be more or less explicit (Gluckman & Wilsdon, 2016).

The way that the notion of governance was understood throughout the MoRRI project also recognized that systems of governance impact each of the other keys which are included in RRI. As a consequence, governance in this sense should be understood as having implications on the other keys. The tools of governance which are recommended for supporting RRI are broadly in tune with a form entitled "anticipatory governance" (Barben et al., 2008, Guston, 2014) and include tools such as public deliberation methodologies, lay membership within expert committees, transparency guides, and forms of encouraging multidisciplinary collaboration.

7.4 Development of indicators

After these operational definitions of the EC's six keys were developed, the next task for the MoRRI project was the development of indicators which would provide insight into the ways in which these various aspects of RRI were developing across the European research area. For a more detailed discussion on the development of these indicators, please see Woolley and Rafols (2016).

Ultimately, the monitoring framework developed within the MoRRI project consisted of 36+ indicators. The indicators are listed in Table 7.1. Thirteen of the 36 indicators used readily available secondary data sources (emphasized with bold font), while 23 of the indicators required primary data collection. Many of the indicators which were developed throughout the MoRRI project required data collection at the institutional level and were subsequently aggregated at the national level. Table 7.2 provides an overview of the various methods of primary data collection which were used in the MoRRI project; these methods ranged from surveys of RPOs, RFOs, and societal stakeholders; bibliographic data and – to a lesser degree – data based on qualitative approaches (case studies). The secondary data set applied and collected in the MoRRI project include data from Eurostat, SheFigures, Eurobarometers, OpenAIRE, and specific cross-European projects such as MORE2, MASIS, EPOCH, and SATORI.

An important note should be made regarding the fact that there exists considerable overlap between the different RRI keys. The MoRRI project undertook the assumption that, while these keys can be distinguished as a useful heuristic for the purposes of implementation, they are intimately related and not independent of one another. As a consequence of the overlap between the keys, there are likely relationships between the different indicators which were developed to measure the different keys. An illustration of the potential links and overlaps between the different keys and the indicators developed to measure them is provided in Figure 7.1. The arrows with full lines within Figure 7.1 indicate that these links were directly assessed in the development of the indicators, whereas the arrows with dashed lines indicate links which require additional theoretical exploration. The direction of the arrows indicates the hypothesized direction of influence between the links, and also shows whether or not there is theorized uni-directionality or bi-directionality in the relationship between the indicators / keys.

Table 7.1 Indicators capturing aspects of Responsible Research and Innovation (RRI)

Key	*Indicator*	*Indicator title*
Gender equality	GE1	Share of RPOs with gender equality plans
	GE2	**Share of female researchers by sector (secondary data) – four sub-dimensions**
	GE3	Share of RFOs promoting gender content in research
	GE4	**Dissimilarity index (secondary data) – two sub-dimensions**
	GE5	Share of RPOs with policies to promote gender in research content
	GE6	**Glass ceiling index (secondary data)**
	GE7	**Gender pay gap (secondary data) two sub-dimensions**
	GE8	Share of female heads of RPOs
	GE9	Share of gender-balanced recruitment committees at RPOs
	GE10	Number and share of female inventors and authors – two sub-dimensions
Science literacy and education	SLSE1	Importance of societal aspects of science in science curricula for 15–18-year-old students (desk research)
	SLSE2	RRI-related training at higher education institutions
	SLSE3	**Science communication culture (secondary data)**
	SLSE4	Citizen science activities in RPOs – two sub-dimensions
Ethics	E1	Ethics at the level of RPOs – two sub-dimensions
	E2	**National ethics committees index (secondary data)**
	E3	RFO ethics index – two sub-dimensions
Public engagement	PE1	**Models of public involvement in science and technology decision making (secondary data)**
	PE2	**Policy-oriented engagement with science (secondary data)**
	PE3	**Citizen preferences for active participation in science and technology decision making (secondary data)**
	PE4	**Active information search about controversial technology (secondary data)**
	PE5	Public engagement performance mechanisms at the level of research institutions
	PE6	Dedicated resources for public engagement – indicator dropped
	PE7	Embedment of public engagement activities in the funding structure of key public research-funding agencies
	PE8	Public engagement elements as evaluative criteria in research proposal evaluations
	PE9	Reseach and innovation democratization index
	PE10	National infrastructure for involvement of citizens and societal actors in research and innovation
Open access	OA1	Open access literature – two sub-dimensions
	OA2	Data publications and citations – indicator dropped
	OA3	Social media outreach /take-up of open access literature – two sub-dimensions
	OA4	**Public perception of open access (secondary data)**
	OA5	**Funder mandates (secondary data)**
	OA6	RPO support structures for researchers as regards incentives and barriers for data sharing
Governance	GOV1	**Composite indicator: governance for responsible research and innovation (secondary data)**
	GOV2	Existence of formal governance structures for RRI within RFOs and RPOs
	GOV3	Share of RFOs and RPOs promoting RRI

Notes: RFOs, research-funding organizations; RPOs, research-performing organizations.

Table 7.2 Overview of primary data collection methods

Methods for collecting primary data	*Indicators*
Science in Society actor survey	PE9, PE10
Higher education institutions and research-performing organizations survey	GE1, GE5, GE8, GE9, SLSE2, SLSE 4, PE5, PE6, OA6, E1, GOV2, GOV3
Research-funding organizations survey	GE3, PE7, PE8, E3, GOV2, GOV3
Register-based data	GE10, OA1, OA2, OA3
Qualitative data, interviews/desk research	SLSE1

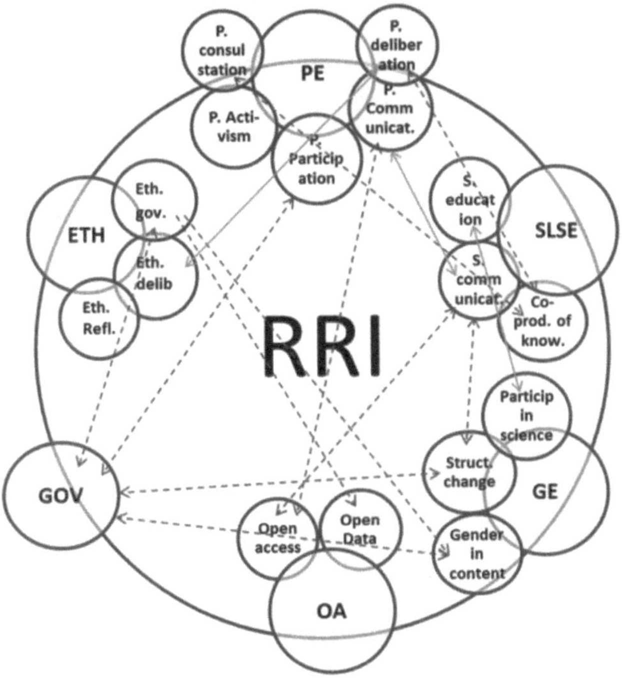

Figure 7.1 Existing and potential interlinkages/overlaps between Responsible Research and Innovation (RRI) (sub-)categories (Taken from Meijer et al., 2016 and Peter et al., 2018.)

7.5 Emerging patterns at the country level

"The Evolution of Responsible Research and Innovation in Europe: The MoRRI Indicators Report" (MoRRI Consortium, 2018; Peter et al., 2018) presents the results of individual indicators and the monitoring of developments. The methodological background to that analysis is described in more detail in Mejlgaard et al. (2019). The data collection aimed to populate the 36+ indicators covering the six keys of RRI. Subsequently, the empirical structure of the data collected was analysed with the purpose of understanding the appropriateness of the indicators towards the "key" dimensions that they were intended to be indicators of. For the year 2016 a cross-sectional dataset was available covering all 28 EU countries. A validation procedure was conducted to test four main aspects: the quality of indicators based on survey-questions (removal of indicators based on items with high non-response); the internal consistency of composite measures; the robustness of indicators on the basis of testing the effects of small adjustments in indicator specifications on ranking; and the degree of within-country variance compared to cross-country variance. While the coverage of the indicators across EU countries is comprehensive, the majority of indicators have a few missing cases, i.e. countries for which no observed value has been obtained. Also, on country level, there are more indicators than observations, which presents a challenge to any integrated statistical models drawing on multiple indicators. For the examination of the empirical patterns of the indicators, these limitations have been accommodated by imputing data points to replace missing values, and applying factor analyses to subsets of indicators rather than the full set of 36+ indicators. Principal component factor analyses were conducted for each key area separately. The aim of the imputation was to examine how the selected indicators relate to each other and to the dimensions one would expect them to be indicators of, in order to address integrated empirical findings.

The purpose of the factor analysis was to determine the number of distinct dimensions for each key. A second aim was identification of the indicators that are most relevant in capturing these dimensions. Interestingly, the parallel principal component analyses revealed two distinct and interpretable factors for all keys except governance, for which only one factor was retained. In Table 7.3, the 11 retained sub-dimensions are presented, along with the indicators most relevant to the respective sub-dimension.

Table 7.3 Responsible Research and Innovation (RRI) keys and core indicators

Dimension	*Core indicators*
GE action	GE1, GE5
GE status	GE2.3, GE10.1
SLSE training	SLSE1, SLSE2
SLSE culture	SLSE3, SLSE4
PE participation	PE1, PE4, PE9
PE in assessment	PE7, PE8
Ethics in RPOs	E1a, E1b
Ethics in RFOs	E3a, E3b
OA status	OA1.1, OA1.2
OA action	OA3, OA4, OA6
Governance	GOV1, GOV2, GOV3

Source: Peter et al. (2018).

Interestingly, the two dimensions for each key represent two aspects. One aspect is related to policies, action, activities, or planning to support or enhance an RRI key. The other one relates to the actual status, situation, or culture of the RRI keys. For example, for the key gender equality, the first dimension that materialized from the principal component analyses was labelled *GE actions*. This combines GE1, which measures the share of RPOs with gender equality plans, and GE5, measuring the share of RPOs with policies promoting gender content in research. The second sub-dimension of gender equality represents the actual *GE status* and is strongly interrelated with GE2.3 and GE10.1. They measure the share of female researchers in the higher education sector and the share of female authors of scientific papers, respectively. GE status is not about policies to promote gender equality but rather about the status observed concerning female representation in science.

Science literacy and science education was also divided into two dimensions. One relates to formal training activities around issues of responsibility in secondary education (SLSE1) and in higher education institutions (SLSE2) and can be called *SLSE training*, representing the "action". The other – the actual situation – is about the broader national science culture, indicated by SLSE3 capturing aspects of science communication culture and SLSE4 signalling the importance of citizen science activities in RPOs (thus, *SLSE culture*).

The "action" in public engagement revolves around citizens' active *participation*. Three indicators of public involvement in science and technology decision-making (PE1), citizens' search for information about controversial technologies (PE4), and the level of democratization of R&I (PE9) all pertain strongly to this sub-dimension. The "situation" sub-dimension is mainly concerned with the extent to which public engagement is included in *assessment* exercises. This dimension relates strongly with PE7, which measures the inclusion of public engagement activities in the activities of RFOs, and PE8, which is about the extent to which public engagement is used as evaluation criteria in the assessment of research proposals.

For the ethics key the first sub-dimension is concerned with planning, i.e. the existence of (E1a) and degree of importance attributed to (E1b) research ethics committees and research integrity offices at higher education institutions and other public RPOs (*ethics in RPO*). The second dimension (*ethics in RFO*) is more concerned with the culture around ethics in RFOs. It is informed by two indicators relating to the use of an ethics review in relation to funding decisions (E3a) and a composite indicator based on a set of questions relating to the importance of such assessment for funding decisions (E3b).

Almost equivalent to the pattern from gender equality, open-access indicators seem to support the existence of two dimensions capturing "status" and "action". There is a sub-dimension concerning activities promoting open access (*actions*), such as OA3 on social media outreach or uptake of open-access literature, OA4 on public support for open access to scientific information, and OA6 on support structures promoting data sharing within RPOs. They all relate to "open-access activities" which could be expected to push a Member State towards higher levels of open access. The OA sub-dimension concerning the *state of play* consists of indicators OA1.1 and OA1.2 measuring the share of open-access publications as proportions of all publications in a country.

Finally, the three indicators in the governance key end up in one single factor. GOV1 concerns the use of science in policymaking at the national level, while GOV2 addresses RRI-related formal governance mechanisms within RPOs and RFOs. GOV3 is a composite measure of institutional support for RRI among employees within these organizations. The empirical structure of the governance indicators thus supports retaining only one RRI *governance* dimension, which is in line with the actual interlinkage of governance with all other keys.

Thus, from the 36+ original indicators, 25 proved to be strong indicators for the11 RRI sub-dimensions that emerged empirically from the data. For those 25 indicators, a 0–1 normalized index was subsequently created, which was then used to characterize individual countries for the 11 sub-dimensions. In turn, the 11 indexes helped to explore similarities and differences between and within clusters of countries. The approach to cluster analysis of countries was a hierarchical, weighted average linkage cluster analysis.

Figure 7.2 presents the clusters, based on country scores on the 11 indices. The dendrogram reveals a somewhat heterogeneous set of countries;

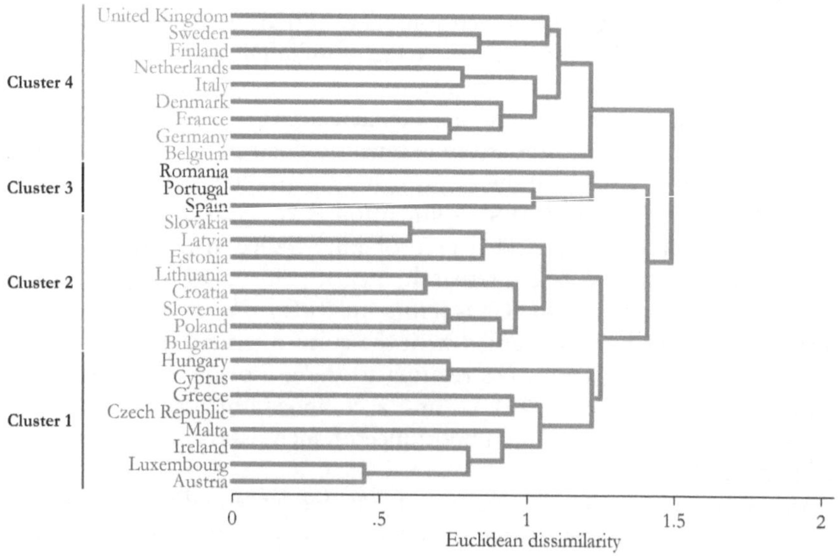

Figure 7.2 Dendrogram of country clustering. (Taken from Peter et al., 2018.)

looking horizontally from left to right, each Member State initially forms its own cluster but progressively Member States cluster together based on the average linkage algorithm, resulting in continuously fewer clusters. The four-cluster solution seems most useful from an interpretation perspective. Looking from the bottom up, Austria, Luxembourg, Ireland, Malta, the Czech Republic, Greece, Cyprus, and Hungary form a first cluster. This means that their individual country profiles are alike, and that they as a group are distinct from the other groups. The second group includes Bulgaria, Poland, Slovenia, Croatia, Lithuania, Estonia, Latvia, and Slovakia. The third and smallest group includes Spain, Portugal, and Romania. The fourth and last group includes Belgium, Germany, France, Denmark, Italy, the Netherlands, Finland, Sweden, and the United Kingdom.

In Figure 7.3, the characteristics of the four clusters are displayed as a radar plot showing the profile of the 11 RRI dimensions for each cluster of Member States. For each of the 11 sub-dimensions it captures the range from minimum to maximum possible effort within the sub-dimensions. The use of a clustering algorithm to group countries does not mean that countries within a cluster have exactly the same RRI profiles. Rather, it means that the profile of a country within a cluster is more alike to other members of that cluster than to countries belonging to a different cluster. However,

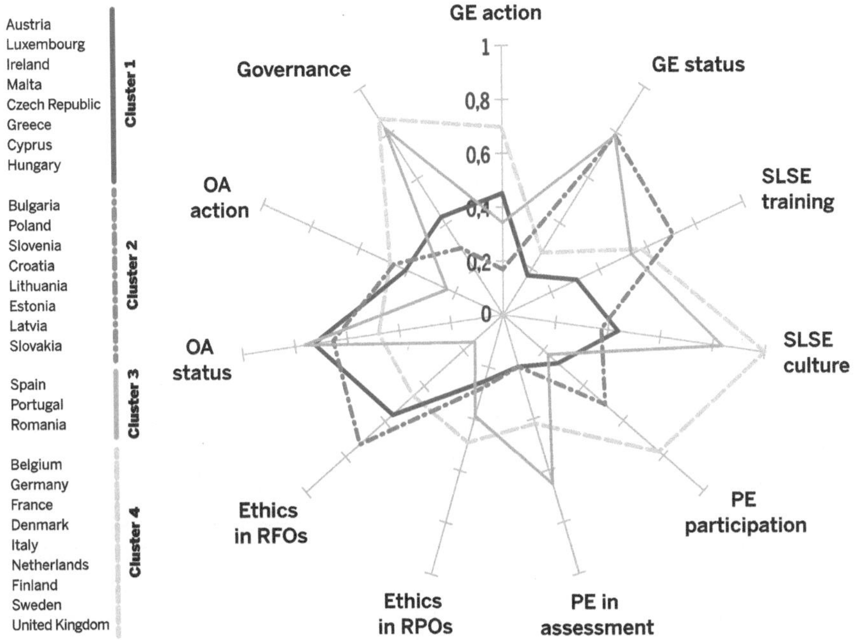

Figure 7.3 Responsible Research and Innovation (RRI) characteristics of four clusters of EU Member States. (Taken from Peter et al., 2018.)

there can be significant differences in the country profiles, even within the same cluster.

The first cluster of Member States is characterized by having below-average scores on most of the 11 RRI dimensions apart from "OA status" and "ethics in RFOs", where this cluster is performing relatively well. Within this cluster, there is a rather moderate level of accomplishment concerning RRI in general.

The second cluster of Member States performs particularly well on "GE status", "SLSE training", and "ethics in RFOs", and also rather well on both sub-dimensions of open access. At the same time, the average score of this cluster on "GE action" and "governance" is considerably lower than for the other clusters.

The third cluster resembles the shape of a star due to its high scores on nearly every second dimension and fairly low scores on the other half. On one dimension, "PE in assessment", this cluster is doing particularly well. Additionally, Member States within this cluster tend to score very highly on both gender equality status and open-access status.

The fourth cluster is generally performing above average. Exceptions include the dimensions of "GE status" and "OA status" where the average score of Member States within this cluster is low. Sub-dimensions related to inclusivity and co-creation of R&I with civil society (PE participation and SLSE culture) perform particularly well.

Interestingly, for both clusters 2 and 3, there is a negative relation between gender equality status and action. Countries with a high level of observed gender equality in science (according to our indicators) are seemingly less prone to be highly active regarding gender equality policies and action plans at the institutional level. In countries in which the historical labour market trajectories have been more conducive to gender equality in science, there may be less need for action. However, these conclusions are tentative and these results must be examined with the broader context of the history of gender equality in science.

Remarkably, three clusters (1, 2, and 3) have similarly high scores for OA status, which indicates that the use of open access for scientific publication relative to all publications is high across these clusters. This is however not the case in cluster 4, where open-science policies are prominent, such as the UK, the Netherlands, and Finland, but where the use of open access for publications is lower than in the other three.

On the one hand, the results of the cluster analyses show that there is significant diversity in the European RRI landscape, with unequal distribution across the 11 sub-dimensions and across Member States. Clearly, the origins of such diversity cannot be read from the graphics, but require further investigation of historical trajectories in the relationship between science and society, R&I policy approaches, as well as political and civic culture. Some distinctions, e.g. between the primarily north-western European cluster 4 and the primarily eastern European cluster 2, seem to be in line with earlier findings concerning science's role and responsibilities in society.

On the other hand, and despite the careful imputation of missing data, findings may not capture the diversity between the responding institutes across and within Member States. Research institutes have other priorities than higher education institutes with a significant teaching load; large general and traditional universities with a broad spectrum of research fields have different needs than specialized or technical universities; and the location of an institute in an urban or rural setting may induce various local or regional contexts. Especially when the number of observations does not exceed the number of indicators, a country profile may be the result of the (local) context of the responding institutes. If a realistic perspective of RRI is the aim, it makes sense to take the institutional level more centrally in further monitoring of RRI. Eventually, the institutional context is leading in responding to and implementation of policies that support the anticipation, reflection, responsiveness, and inclusivity of R&I.

7.6 Exploring the benefits of RRI

The MoRRI project also sought to investigate benefits which could be attributable, to some extent, to the implementation of changes in line with the aims of RRI. For a more detailed discussion on these potential benefits, see Wuketich et al. (2016).

Before describing the benefits which were observed, a theoretical and methodological note should be made. While the indicators which were developed throughout the MoRRI project were developed with a relatively conventional intervention logic (inputs, outputs, outcomes, and impacts), it was recognized that conceptualizing the benefits of RRI in this same logic would likely result in missing some of the eventual effects of its implementation.

The recognition of the fact that benefits cannot be immediately envisioned within this inputs-to-impacts intervention logic led to two elements which were of particular note when trying to attend to the benefits of RRI. The first element of the benefits of RRI which was recognized was the observation that frequently RRI benefits manifest themselves in the transformations of existing processes. For example, if RRI-informed policy-making resulted in the institutionalizing of a public engagement platform within an RPO, with the explicit desire to empower previously disenfranchised populations within research processes, this can be understood as a democratic benefit of RRI. The novel space for the enfranchisement of these actors within this institutionalized engagement exercise can be thought of as a potential RRI benefit irrespective of other additional substantive outcomes. As this example highlights, when one seeks to investigate the potential benefits of RRI, it is crucial to attend to forms of institutional change.

A second element which requires extending the interpretation of the benefits of RRI beyond an input impact model is that the benefits of RRI should be understood as being normative in character. Indeed, the question of how to align R&I with the needs of society is context-dependent, and requires asking detailed questions such as: Which actors in society do we mean? Whose needs are excluded? among other irreducibly political questions. Therefore, observing the benefits of RRI cannot be understood through a framework of being universally "positive", "net positives", or by measuring accumulated effects (Peter et al., 2018). Consequently, the MoRRI project used detailed case studies to identify potential benefits of RRI, and focused on understanding institutional change and attending to the different actors for whom these changes benefitted.

Throughout the case studies, the identified benefits were categorized according to four potential types. This typology consisted of *societal benefits*, *democratic benefits*, *economic benefits*, and *scientific benefits*. Although these four different categories of RRI benefits were identified, these are to be understood as a useful typology rather than a prescriptive set of benefit categories; the benefits of RRI are unlikely to be neatly distributed among these distinct categories and will likely transcend or evade the categories of this typology. For a detailed description of the case studies that were completed throughout the MoRRI project, please see Wuketich et al. (2016).

A principal *democratic benefit* that was identified in the case studies was the increased participation of a diversity of actors within the R&I system. By institutionalizing public engagement exercises within R&I processes, actors who otherwise would not have been able to express their values and contribute to the steering of R&I were able to do so. One case study exemplifying the institutionalization of public engagement was a large research project entitled "Building Adaptive Strategies for Environmental Change with Rural Land Managers" which adopted a participatory action research approach (Pain, Whitman, & Milledge, 2011). This project enlisted land managers from the outset and ensured that they played an active role in shaping the research project and contributed to knowledge production and tool development which would subsequently be used within policy-making contexts. In this example, the production of knowledge itself was done through a more democratic and inclusive way, highlighting the democratic benefits for the participating land managers.

Of the *societal benefits* which were identified throughout the MoRRI project, one of the most evident was the inclusion of gender-specific concerns within research processes. One of the case studies conducted throughout MoRRI involved researching the Institute of Gender in Medicine at the Charité University of Medicine in Berlin. The head of the institute led efforts to actively address the fact that women experience significantly worse outcomes regarding the diagnosis and treatment of cardiovascular diseases (Regitz-Zagrosek, 2012). One of the outcomes of this initiative was the creation of novel courses to train clinicians to identify gender-specific concerns and address inequalities in the ways in which diagnosis and treatment were performed. Additionally, research projects were developed which explicitly sought to investigate the different ways in which cardiovascular diseases manifest depending on gender.

With regards to the *economic benefits* of RRI, these were particularly difficult to measure. However, the MoRRI project investigated the growth of open-access publications within Denmark, the Netherlands, and Switzerland between the years of 2000 and 2012. It was found that the proportion of publications which were published open-access increased in each country during this time period. Although the direct economic benefits of this change are difficult to assess, it can be assumed that, when these publications were utilized by firms, there was an economic benefit for firms when they no longer had to pay access fees. Lowering the barrier to being able to access these publications may facilitate the transfer of knowledge between the academic and industrial sectors, having eventual economic benefits.

Finally, *scientific benefits* were observed in a number of case studies throughout MoRRI. The term "scientific benefits" is vague, and a rather open-ended operationalization of the term was used throughout the description of the case studies. With regard to scientific benefits observed through the implementation of efforts to consider gender-specific concerns in research, it was noted that the knowledge produced because of these institutionalized changes was beneficial in that it addressed concerns and gaps which otherwise would have been ignored. Additionally, the knowledge which was co-produced with diverse populations through the institutionalization of public engagement exercises benefitted from the diverse forms of expertise which were able to voice their needs within the R&I system. One example of the scientific benefits of the institutionalizing of public engagement exercises was evident in the analysis of the GAP2 project. The GAP2 project enlisted fishers within a research project investigating sustainability issues in fisheries. As a consequence of the inclusion of these actors within the research process, the participating researchers quickly found that it was not a process of informing fishers of novel developments within research, but instead that there was a benefit in efforts to "engage fishers as active agents in the knowledge production process". Evidently, the forms of knowledge and expertise which the fishers held would not have contributed to the production of scientific knowledge if these public engagement opportunities had not been created (see Wuketich et al., 2016).

As mentioned above, the separation of potential benefits that the implementation of RRI had into these categories is merely a useful typology. The benefits which occur due to RRI are likely to be intimately linked and extend well beyond the discussion above. The interlinked nature of these benefits, and several benefits which were not discussed, can be observed in Figure 7.4.

7.7 The future of monitoring RRI

RRI is not the first (policy) concept with the lofty aims of creating a more socially responsive R&I system and one can reasonably state that its implementation has been unsuccessful in comprehensively changing R&I practices. Indeed, hearing innovation myths which are antithetical to the underlying philosophy of RRI, such as R&I being politically neutral, or the continued use of linear models of innovation is still commonplace (for a rich discussion

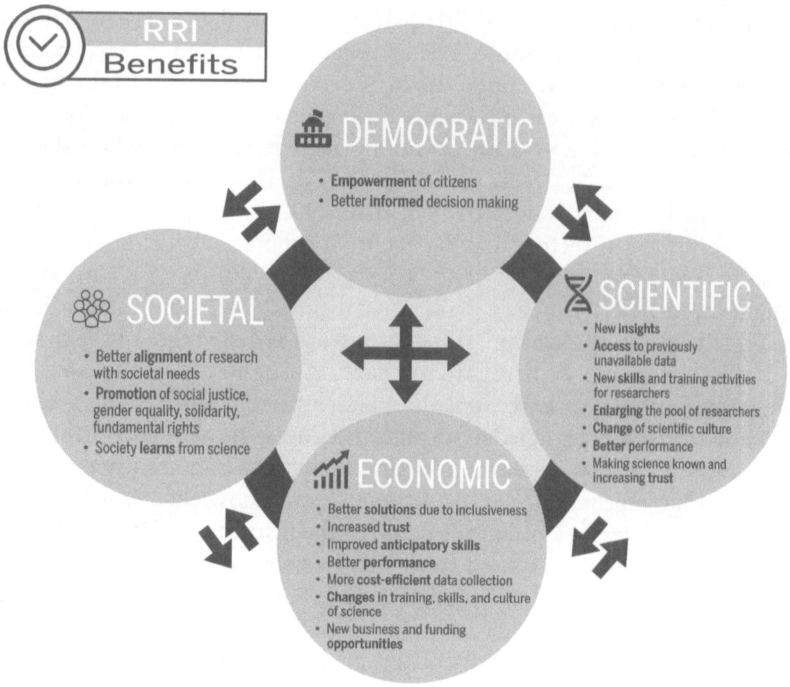

Figure 7.4 Societal, democratic, economic and scientific benefits of Responsible Research and Innovation (RRI). (Taken from Peter et al., 2018.)

of other oft-repeated innovation myths, see Joly, 2019). Even more problematic is the concern that RRI risks serving as "responsible washing" because of its voluntary and highly flexible character.

Nevertheless, the concept of RRI has considerable momentum and political weight behind it, and is at present a prominent device employed throughout a diversity of international research efforts, notably within the context of EC funding schemes (Smallman et al, 2015; EC, 2018). The growth of RRI within the R&I system is evinced by the many projects funded through the Science with and for Society (SwafS) programme of Horizon2020, many placing key tenets of RRI, such as co-creation environments, central (Mejlgaard et al., 2018). Furthermore, a wealth of RRI literature, much of it within the *Journal of Responsible Innovation*, and tools guiding RRI change (such as those present on the www.rri-tools.eu web page) further highlights the vast work which is being done under the heading of RRI (see also Von Schomberg & Hankins, 2019). It is crucial to recognize, however, that the swathe of efforts which are informed by RRI needs adjustment considering the shortcomings which have been articulated.

The question of how to continue working in the space of RRI, while acknowledging its many shortcomings, makes evident the need for continued reflection and negotiation of the concept of RRI itself, which has direct consequences on what a monitoring system for RRI should resemble. This question, and its multiplicity of deeply normative, political, and context-dependent answers are focal within the successor project of MoRRI: SUPER MoRRI. While the primary output of MoRRI was the creation of a monitoring system highlighting changes at the national level, the SUPER MoRRI project envisions an alternative future for the monitoring and evaluation of RRI which is much more intimately local and context-dependent.

It was through recognizing the importance of changes at the institutional level that SUPER MoRRI has shifted emphasis towards providing a monitoring framework which aims to provide knowledge and monitor changes at the institutional level rather than monitoring the state of the implementation at the national level and benchmarking nations, or clusters of nations, against each other. The case studies conducted within MoRRI highlighted that the implementation of RRI can often best be understood within an institutional context. Indeed, shifting towards a monitoring system seeking to aid in institutional change may have the benefit of empowering individuals to articulate their own values within their institutions by providing them with intellectual resources to do so. This is a considerably different view of the function of a monitoring framework – one that shifts away from simply measuring and actively moves towards intervening and changing practices.

RRI takes as an assumption that fact that science, technology, and society mutually co-construct each other. Against the backdrop of entrenched social inequalities, environmental degradation, and socio-political unrest, the R&I system has a duty to become more responsive to societal needs. Despite the limitations of RRI, its political weight is considerable, and the opportunity that it has to reconstruct an R&I system which is more just is yet to be determined. RRI may indeed be a venue for opening up a space within the R&I system for asking deeply normative questions and allowing for the enfranchisement of voices to elicit their needs within R&I. The task of the SUPER MoRRI project is to develop a monitoring system to better understand how and whether this space is indeed being opened.

Acknowledgement

This chapter builds upon and summarizes work in the Monitoring the Evolution and Benefits of Responsible Research and Innovation (MoRRI) project 2014–2018 that was funded by the European Commission, contract number RTD-B6-PP-00964-2013, duration 09/2013–03/2018. We are greatly indebted to our project partners, who are equally responsible for the work carried out and described in more detail in the MoRRI deliverables. The partners are: Viola Peter, Frederic Maier (editors) (Technopolis Group); Niels Mejlgaard, Carter Bloch, Emil B. Madsen (Aarhus University); Erich Griessler, Milena Wutekich (IHS); Ingeborg Meijer (CWTS, Leiden University); Richard Woolley (Ingenio CSIC-UPV); Ralf Lindner, Susanne Bührer, Angela

Jäger (Fraunhofer ISI); Lena Tsipouri (University of Athens); Jack Stilgoe (UCL). Figures in this chapter are directly taken from MoRRI deliverables with permission of the EC. Figures 7.1–7.4 and Table 7.3 appeared in the EC report: Monitoring the evolution and benefits of responsible research and innovation in Europe; Summarising insights from the MoRRI project, May 2018 (Peter et al., 2018); Tables 7.1 and 7.2 are included in the conference paper for the OECD Blue Sky 2016 paper (Meijer et al., 2016).

References

Barben, D., Fisher, E., Selin, C., & Guston, D. H. (2008). Anticipatory governance of nanotechnology: Foresight, engagement, and integration. *The Handbook of Science and Technology Studies*. London: MIT Press, p. 979.

Brom, F. W. A., Chaturvedi, S., Ladikas, M., & Zhang, W. (2015). Institutionalizing ethical debates in science, technology, and innovation policy: A comparison of Europe, India and China. In: Ladikas, M., Chaturvedi, S., Zhao, Y., & Stemerding, D. (eds.), *Science and Technology Governance and Ethics*. Cham: Springer, pp. 9–23. https://doi.org/10.1007/978-3-319-14693-5_2

Bucchi, M. (2008). Of deficits, deviations and dialogues: Theories of public communication of science. In: Bucchi, M. & Trench, B., *Handbook of Public Communication of Science and Technology*. New York: Routledge, pp. 71–90.

Butler, J. (1990). *Gender Trouble. Feminism and the Subversion of Identity*. New York: Routledge.

Costas, R., Meijer, I., Zahedi, Z., & Wouters, P. (2013). The value of research data: Metrics for datasets from a cultural and technical point of view. *A Knowledge Exchange Report* (pp. 1–48). Available at: www.knowledge-exchange.info/event/value-reseach-data-metrics

European Commission (EC). (2012). Commission Recommendation on Access to and Preservation of Scientific Information. Available at: http://ec.europa.eu/research/science- society/document_library/pdf_06/recommendation-access-and-preservation-scientific-information_en.pdf

European Commission (EC). (2018). Responsible Research and Innovation. Available at: https://ec.europa.eu/programmes/horizon2020/en/h2020-section/responsible-research-innovation

European Commission (EC). (2019a). Open Science (Open Access). Available at: https://ec.europa.eu/programmes/horizon2020/node/1031

European Commission (EC). (2019b). Public Engagement in Responsible Research and Innovation. Available at: https://ec.europa.eu/programmes/horizon2020/node/766.

European Commission (EC). (2019c). Promoting Gender Equality in Research and Innovation. Available at: https://ec.europa.eu/programmes/horizon2020/node/797

Farhan, H., Alonso, J., Davies, T., Tennison, J., Heath, T., & Berners-lee, T. (2013). Open Data Barometer 2013 Report, 1–45. Available at: http://opendatabarometer.org/doc/1stEdition/Open-Data-Barometer-2013-Global-Report.pdf

Felt, U., Fochler, M., Müller, A., & Strassnig, M. (2009): Unruly ethics: on the difficulties of a bottom-up approach to ethics in the field of genomics. *Public Understanding of Science*, 18(3), 354–371.

Gluckman, P., & Wilsdon, J. (2016). From paradox to principles: Where next for scientific advice to governments? *Palgrave Community*, 2, 16077.

Griessler, E. and Littig, B. (2006). Neosokratisch Dialoge zu ethische Fragen der Xenotranspantation. Ein Beitrag zur Bearbeitung ethische Probleme in partizipatiever Technikfolgenabschützung. In: E. Buchinger & U. Felt (eds.), *Technik und Wissenschaftssoziologie in Österreich. Stand und Perspektiven.* Österreichische Zeitschrift für Soziologie, Sonderheft. 8: S.131–157.

Guston, D. H. (2014). Understanding 'anticipatory governance'. *Social Studies of Science*, 44(2), 218–242.

Hancock, A. (2007). Intersectionality as a normative and empirical paradigm. *Gender and Politics*, 3(2), 248–254.

Harding, S. (2006). *Science and Social Inequality: Feminist and Postcolonial Issues.* Urbana, IL: University of Illinois Press.

Heras, M., & Ruiz-Mallén, I. (2017). Responsible research and innovation indicators for science education assessment: How to measure the impact? *International Journal of Science Education*, 39(18), 2482–2507.

Italian Presidency of the Council of the European Union. (2014). *Rome Declaration on Responsible Research and Innovation in Europe.* Brussels, Belgium: European Commission. https://ec.europa.eu/research/swafs/pdf/rome_declaration_RRI_final_21_November.pdf.

Joly, P-B. (2019). Reimagining innovation. In: S. Lechevalier (ed.), *Innovation Beyond Technology. Science for Society and Interdisciplinary Approaches.* doi.org/10.1007/978-981-13-9053-1_2

Ladikas, M., Chaturvedi, S., Zhao, Y., & Stemerding, D. (2015). Introduction: Embedding ethics in science and technology policy – a global perspective. In: M. Ladikas, S. Chaturvedi, Y. Zhao, & D. Stemerding (eds.), *Science and Technology Governance and Ethics.* Cham: Springer. doi.org/10.1007/978-3-319-14693-5_1

Meijer, I., Mejlgaard, N., Lindner, R., Woolley, R., Rafols, I., Griesler, E., Wroblewski, A., Bührer, S., Stilgoe, J., Tsipouri, L, Maroulis, N., & Peter, V. (2016). Monitoring the evolution and benefits of Responsible Research and Innovation (MoRRI) – a preliminary framework for RRI dimensions and indicators. OECD Blue Sky Forum 2016. www.oecd.org/sti/105%20-%20Meijer_MoRRI_paper_%20OECD_bluesky_16ul16_def.pdf

Mejlgaard, N., Bloch, C., & Bargmann Madsen, E. (2019). Responsible research and innovation in Europe: A cross-country comparative analysis. *Science and Public Policy*, 46, 198–209. DOI: 10.1093/scipol/scy048

Mejlgaard, N., Woolley, R., Bloch, C., Bührer, S., Griessler, E., Jäger, A., Lindner, R., Bargmann Madsen, E., Maier, F., Meijer, I., Peter, V., Stilgoe, J., & Wuketich, M. (2018). Europe's plans for responsible science. *Science*, 361(6404), 761–762. DOI: 10.1126/science.aav0400

Monsonís-Payá, I., García-Melón, M., & Lozano, J. F. (2017). Indicators for Responsible Research and Innovation: A methodological proposal for context-based weighting. *Sustainability*, 9(12), 2168.

MoRRI Consortium. (2015). Synthesis Report on Existing Indicators Across RRI Dimensions, Task 3, Progress Report, Deliverable 3.2. Available at: http://morri-project.eu/reports/2015-05-01-d3.1

MoRRI Consortium. (2018). The Evolution of Responsible Research and Innovation in Europe: The MoRRI Indicators Report. Monitoring Report (D4.3). Available at: http://morri-project.eu/reports/2018-02-21-the-evolution-of-responsible-research-and-innovation-in-europe-the-morri-indicators-report-d4-3

Owen, R., Macnaghten, P., & Stilgoe, J. (2012). Responsible research and innovation: From science in society to science for society, with society. *Science and Public Policy*, 39(6), 751–760.

Pain, R., Whitman, G., & Milledge, D. (2011). Participatory Action Research Toolkit. http://studentsatthecenterhub.org/wp-content/uploads/2015/04/Engaging-Youth-Researchers-partoolkit.pdf

Pandza, K., & Ellwood, P. (2013). Strategic and ethical foundations for responsible innovation. *Research Policy*, 42(5), 1112–1125.

Peter, V., Maier, F. J., Mejlgaard, N., Bloch, C., Madsen, E., Griessler, E., Wuketich, M., Meijer, I., Woolley, R., Lindner, R., Bührer, S., Jäger, A., Tsipouri, L., & Stilgoe, J. (2018). Monitoring the evolution and benefits of responsible research and innovation in Europe: Summarising insights from the MoRRI Project D13. Available at: www.technopolis-group.com/wp-content/uploads/2020/02/Final-report-%E2%80%93-Summarising-insights-from-the-MoRRI-project-D13.pdf

Regitz-Zagrosek, V. (2012). *Sex and gender differences in health. EMBO Reports*, 13(7), 596–603.

Rip, A. (2014). The past and future of RRI. *Life Sciences, Society and Policy*, 10(1), 17.

Royal Society. Council. (1985). *The Public Understanding of Science*. London: The Royal Society.

Smallman, M., Lomme, K., & Faullimmel, N. (2015). D 2.2. *Report on the Analysis of Opportunities, Obstacles and Needs of the Stakeholder Groups in RRI Practices in Europe*.

Stilgoe, J., & Lindner, R. (2015). Analytical Report on the Dimension of Research and Innovation Governance. Available at: https://www.technopolis-group.com/report/analytical-report-dimension-research-innovation-governance-d2-4-2/

Stilgoe, J., Owen, R., & Macnaghten, P. (2013). Developing a framework for responsible innovation. *Research Policy*, 42(9), 1568–1580.

Strand, R., & Kaiser, M. (2015). *Report on Ethical Issues Raised by Emerging Sciences and Technologies*. Bergen, Norway: Centre for the Study of the Sciences and the Humanities, University of Bergen.

Talmon-Gros, L., & Teichler, T. (2015). Analytical report on the dimension of science literacy and scientific education. Available at: http://morri-project.eu/reports/2015-04-01-d2.2

Van den Eynden, V., & Bishop, L. (2014). *Incentives and Motivations for Sharing Research Data: A Researcher's Perspective*. Colchester, UK: University of Essex, 19, p. 2016. Available at: http://www.knowledge-exchange.info/event/sowing-the-seed.

Von Schomberg, R. (2013). A vision of responsible research and innovation. In: Owen, R., & Bessant, J. (eds.) *Responsible Innovation: Managing the Responsible Emergence of Science and Innovation in Society*. London: John Wiley, pp. 51–74.

Von Schomberg, R., & Hankins, J. (eds.). (2019). *International Handbook on Responsible Innovation: A Global Resource*. Cheltenham, UK: Edward Elgar Publishing.

West, C., & Zimmerman, D. H. (1987). Doing gender. *Gender and Society*, 1, 125–151.

Wickson, F., & Carew, A. L. (2014). Quality criteria and indicators for responsible research and innovation: Learning from transdisciplinarity. *Journal of Responsible Innovation*, 1(3), 254–273.

Wilsdon, J., Doubleday, R., & Stirling, A. (2015). *Future Directions for Scientific Advice in Europe*. Cambridge: Centre for Science and Policy.

Woolley, R., & Rafols, I. (2016). Development of metrics and indicators for RRI projects, Available at: https://www.technopolis-group.com/wp-content/uploads/2017/09/D6.pdf

Wuketich, M., Lang, A., Griessler E., & Polt, W. (2016). In-depth case studies on the benefits of RRI across the scientific disciplines and industrial sectors. Available at: www.technopolis-group.com/report/depth-case-studies-benefits-rri-across-scientific-disciplines-industrial-sectors-d5-2/

Wynne, B. (1995). Public understanding of science. In: S. Jasanoff, G. E. Markle, J. C. Peterson, & T. Pinch (eds.), *The Handbook of Science and Technology Studies*. London: SAGE Publications, pp. 361–388.

Wynne, B. (2006). Public engagement as a means of restoring public trust in science – hitting the notes, but missing the music? *Community Genetics*, 9(3), 211–220.

Best practice I The B Impact Assessment

Joey van den Brink

B Lab, the non-profit organisation supporting the B Corporation movement, has developed the B Impact Assessment as a holistic framework for impact performance measurement and provides its user, typically for-profit companies, with various analysis and improvement options. The first version of the B Impact Assessment was launched in 2007 in line with the launch of the B Corporation movement. Over the course of 12 years (2007–2019), the B Impact Assessment has had multiple updates to maintain credible standards and provide a better user experience. In January 2019, version 6 was launched.

The B Impact Assessment provides companies the opportunity to evaluate their impact performance through a holistic framework that covers all aspects of the company's operations. The company's impact performance is captured in five key impact areas: governance, workers, community, environment and customers. There are indicators that examine the operational practices of the company as well as indicators that look at the core business activities or purpose of the company, referred to as impact business models. Companies gather a score across all topics to come to a final B Impact Score. An internal benchmark is provided to compare this score to similar companies using the B Impact Assessment.

The B Impact Assessment is a free, online tool available to any company interested in evaluating its impact performance. Companies can self-assess their impact performance and use the tools available to learn and improve. As a result, the B Impact Assessment is an accessible means for a large number of companies (approximately 80,000 in the database) to get started in the field of impact measurement and reporting. In addition, the holistic framework (as well as linked resource library) will provide them with a set of comprehensive indicators which can be included in the company's practices or operational strategy. Those companies that reach a B Impact Score that meets or exceeds the bar for B Corporation certification have the option to obtain a verified score that is made transparent through public channels. This provides the company with legitimacy on the credibility of their overall impact performance, as well as building trust and transparency with those interested in buying from or working with the company.

The success of the B Impact Assessment has been its accessibility and relevance for a wide range of companies. No matter their engagement with responsible business practices, each company has the option to get started and expand their knowledge on the topic. In addition, the B Impact Assessment has a focus that is not just based on evaluation, but also on improvement. The user is not penalised for negative practices, but incentivised to improve its positive practices. As a result, engaged users return to the assessment year after year to evaluate their progress, creating a dynamic focused on continuous improvement towards impactful, successful business practices.

8 The COMPASS self-check tool

Enhancing organizational learning for responsible innovation through self-assessment

Adele Tharani, Katharina Jarmai, Norma Schönherr, and Patricia Urban

8.1 Introduction

In an increasingly dynamic, global and complex business environment, companies are continuously challenged to learn and adapt. The capacity to create and absorb new information as well as the ability to translate it into organizational knowledge and tangible outcomes are key strategic issues for companies of all sizes (Balbastre *et al.* 2003; Bennet and Shane Tomblin 2006).

This is particularly visible in the area of innovation management. As a key business function of many companies, innovation is a requirement to stay competitive in light of ongoing digitalization, globalization and rapidly changing markets (Dess and Picken 2000; Crossan and Apaydin 2010; Mone *et al.* 1998). At the same time, the current speed of innovation, in combination with a general drop in trust in societal institutions (Pirson *et al.* 2019), leads many people to be wary of new technologies.[1] This requires companies to develop and maintain internal knowledge and skills, which enable them to anticipate impacts of their actions, respond effectively to concerns of other societal stakeholders and adapt to changes in the business environment.

The concept of responsible innovation provides a framework for companies to balance these competing demands (Iatridis and Schroeder 2016; Martinuzzi *et al.* 2018). It originates in discourses on research ethics in contested emerging technologies (Owen *et al.* 2012) and was initially developed for research organizations and research projects funded through public money. The European Commission has been promoting the concept under the term "Responsible Research and Innovation" (RRI) by supporting research on, and integration of, ethics, gender and diversity, public engagement, open access and science education in research projects through the previous and current European Framework Programmes for Research and Innovation, "FP7" and "Horizon 2020" (see Nwafor *et al.* 2017 for an overview of projects). In the academic literature, agreement about the meaning and key aspects of responsible (research and) innovation has developed in the form of the four dimensions of anticipation, reflection, inclusion/deliberation and responsiveness (Stilgoe *et al.* 2013). In the business context, the use of the abbreviated term "responsible innovation" has come to prevail.

Based on insights from the discourses on responsible innovation and organizational learning (Fortis *et al.* 2018), this chapter introduces a self-assessment tool tailored to evoke organizational learning for responsible innovation in a corporate setting. The COMPASS self-check tool takes a diagnostic approach and aims to enable learning by translating the concept of responsible innovation into concrete corporate practices and policies, ascribing them to specific business functions. This facilitates the assimilation of information about specific responsible innovation practices as well as the reflection about company strengths and weaknesses in terms of the responsible innovation approach.

The remainder of this chapter is organized as follows: Section 8.2 starts with a brief introduction to organizational learning and then describes the relationship between organizational learning and each of the two components that make up the concept of responsible innovation; namely corporate responsibility and innovation processes and output. Section 8.3 elaborates on the specific role of self-assessment for organizational learning within this field and documents the development process of the COMPASS self-check tool. Section 8.4 presents the structure and contents of the tool and gives a detailed account of how it supports organizational learning. The chapter concludes with an outlook on possible future research as well as on the next steps toward enhancing organizational learning and self-assessment for the establishment of responsible innovation in corporate practice.

8.2 Organizational learning and responsible innovation

The literature on organizational learning builds on concepts of learning at the individual level from the realms of psychology and sociology,[2] and applies them to learning at the level of organizational practices and routines. In this sense, organizational learning can be understood as the ongoing process of interpreting and assimilating information that takes place in an organization, and the output of which is organizational knowledge (e.g. Balbastre *et al.* 2003; Lloria and Moreno-Luzon 2014).

Information from within and outside the organization functions as prerequisite for organizational learning by providing new points of view, making connections visible or shedding light on previously invisible meanings. Information can be understood as the input to organizational learning processes, while knowledge is the output. In contrast to information, the knowledge that is created in organizational learning processes is connected to a particular perspective, connected to intention, context-specific and relational (Nonaka and Takeuchi 1995). Organizational learning thus enables the generation of organizational practices and routines, which in turn facilitate the achievement of organizational goals (Balbastre *et al.* 2003).

Argyris and Schön (1978, 1996) coined a distinction between two types of organizational learning that is useful in the context of this text; as will become evident later on. They distinguish between "single-loop" and "double-loop" learning.[3] Single-loop learning is described as the integration of new information into established routines or practices. Learning in this form can increase the effectiveness and efficiency of established modes of action. It is useful for solving problems without questioning underlying mechanisms or assumptions. Double-loop learning occurs when there is a critical review of assumptions, principles or operational procedures. It involves reflection, adaptation and possible abandoning of established concepts in the organization. For this reason, organizations may try to avoid double-loop learning unless they are facing an acute problem, which cannot be solved within existing patterns of action. At the same time, however, double-loop learning creates the possibility for new patterns of behaviour and organizational routines to emerge.

Both types of organizational learning can be beneficial for companies that wish to engage with responsible innovation. Single-loop learning would, for example, allow the company to adapt their stakeholder engagement activities to include additional groups of stakeholder. Double-loop learning would enable a different company to adapt their innovation processes in a way that would make the engagement of external stakeholders a prerequisite in the first place.

8.2.1 Organizational learning in innovation processes

Innovation is frequently cited as one of the key drivers for long-term business success (Schwab 2016), especially in highly dynamic environments. In such environments, companies must innovate to survive, maintain competitive advantage and adapt to the expectations of internal and external stakeholders (Christensen *et al.* 2015; Stata 1994). In order to do this, it is of essential importance for companies to develop and maintain internal knowledge and skills to adapt to these, sometimes rapid, changes (Crossan and Apaydin 2010). This process of continuous organizational learning depends on the effective assimilation of information and creation of knowledge – which is why knowledge is considered a key strategic resource for organizations (Balbastre *et al.* 2003). Referring to the centrality of knowledge in Western modern "knowledge society", Nonaka (1994) emphasizes the need for a shift in thinking about innovation that specifically considers how organizations process and create knowledge.

Jiménez-Jiménez and Sanz-Valle (2011) conclude that organizational learning can enable an organization to develop capabilities that enhance innovation. Tamayo-Torres *et al.* (2016) likewise describe how organizational learning can be conducive to innovation by enhancing an organization's capacity to generate new ideas, products, services and processes.

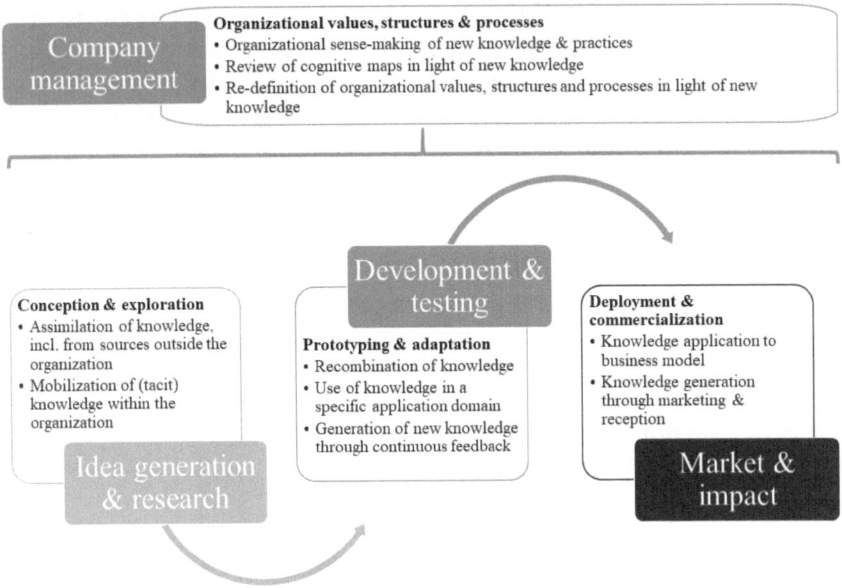

Figure 8.1 Organizational learning in company management and the innovation process.

The relevance of organizational learning for innovation can be illustrated by reviewing the specific types of knowledge-gathering, assimilation and generation activities that happen within each phase of the innovation process (Figure 8.1):

- In the idea generation and research phase (sometimes also called exploration or conception phase), innovators explore innovation opportunities by gathering knowledge on trends, emerging technologies and customer needs. In addition, creativity and problem-solving activities within organizations are used to mobilize existing knowledge (including tacit knowledge) of the innovators to generate ideas for potential new products, services, business models or management innovations.
- Ideas to be pursued further advance to the development and testing phase. In this phase, innovations are prototyped and knowledge on what works and what does not is being generated through continuous testing and adaptation. In this phase, a recombination of the knowledge gathered in the previous phase takes place in a specific application domain. In addition, the continuous feedback and adaptation of prototypes required for development of an innovation is a process of knowledge creation, in which innovators engage in learning by doing and re-doing.

- Finally, mature innovations are deployed in the market where they create impacts on users, generate revenue for the organization and may over time even effect changes in the wider socio-economic system in which they are deployed. Knowledge on the viability of the business model underpinning the innovation as well as the reception of the innovation in the market and in society is fed back to the organization and expands the knowledge base for future innovations.

8.2.2 Organizational learning for responsibility

Organizational learning has been shown to be an important factor in the adoption of responsibility within companies (Wicki and Hansen 2019). In other words, the process of assimilation of existing knowledge and the generation of new knowledge are key prerequisites for establishing more responsible corporate practices within existing organizational processes. Building on the (sometimes tacit) knowledge already present in companies presents a chance to leverage responsible innovation for organizational learning in two areas that are key to the long-term success of companies, notably innovativeness and responsibility. Regarding the former, responsible innovation offers the potential to bring already existing knowledge in corporate sustainability from the margins into core strategic decision processes, thus achieving a stronger integration of the creation of social value in addition to economic returns (Gallego-Álvarez *et al.* 2011). This may open up previously untapped resources and avenues for innovation. Regarding the latter, responsible innovation connects core business concerns to overarching societal challenges and is therefore considered a promising way for companies to leverage their core competences for the well-being of individuals, communities, countries, regions and global society (Antal and Sobczak 2004; Goodman *et al.* 2017).

When knowledge on responsibility and responsible practices is injected into the innovation process based on existing paradigms and ways of doing things, this corresponds to what is called single-loop learning in the organizational learning discourse. Responsibility is treated as just another issue to consider within the established innovation process. Outcomes of single-loop learning may entail changes to procedures and practices, e.g. wider and more inclusive engagement with stakeholders, yet central tenets of how and why an organization innovates remain unchanged.

However, responsible innovation is concerned with more than the innovation process itself – it adds an additional layer of organizational learning by also addressing the organizational structures and processes underpinning innovation within organizations. In other words, it also addresses company management. Since issues concerning responsibility are often complex and multi-layered, internal organizational sense-making processes are central for giving meaning to responsible innovation and developing a common framework of understanding that can be integrated into company management.

In this interpretation process, existing cognitive maps are reviewed and new ones can be created (Basu and Palazzo 2008; Richter and Arndt 2018; Zietsma *et al.* 2002). Sonenshein (2005) further argues that organizational sense-making enables the constructive voicing of internal social criticism, which helps companies to define their moral standards and ascribe meaning to information and practices regarding responsibility (Fortis *et al.* 2018). This is what is conceptualized as double-loop learning.

8.2.3 The contribution of self-assessment to organizational learning

Self-assessment can be defined as the process of evaluating an organization along with its achievements, improvements and processes, which is undertaken by members of an organization themselves (Hillman 1994). Similarly, Balbastre and Luzón (2003) conceptualize self-assessment as the comprehensive, systematic and regular review of activities and results of an organization that is contrasted with best practice or organizational objectives. Given their high diagnostic capability (Conti 1997), self-assessments are generally leveraged for organizational learning in emerging or highly dynamic fields to identify improvement potential across the organization. Such benefits are particularly likely when self-assessments are regularly applied and results are shared within an organization, but may remain limited when applied on a one-off basis by individual members of an organization without being widely communicated (Balbaster Benavent *et al.* 2005).

Research into the institutionalization of responsibility in organizations has shown that self-assessment can enable organizational learning (Balbastre and Luzón 2003). However, the type and purpose of self-assessment tools determine the effectiveness of such learning processes in terms of knowledge assimilation and change in organizational practices. Whereas external reporting-focused corporate social responsibility (CSR) assessment tools have been found to yield limited organizational learning effects (Gond and Herrbach 2006; Mitchell *et al.* 2012), self-assessments have been applied as promising tools for organizational learning in the field of quality management (Tarí 2008).

In contrast to third-party assessments, which are generally employed in an audit-like function and with the aim of rating or ranking organizations according to pre-defined quality criteria, the use of self-assessments can be considered improvement-oriented (Conti 1997). The focus on identifying strengths, weaknesses and underlying causes (Conti 1997) makes it possible to reflect on organizational routines and structures in the fashion of what has been called double-loop learning (see above).

In the specific context of innovation, self-assessment can enhance organizational learning with a view to fostering an organizational culture and organizational capabilities for innovation that in turn have a positive effect on innovation performance measures. Through self-assessment, an organization can gather the knowledge needed to build innovation capability

(Lau *et al.* 2010). When self-assessment focuses on responsible innovation, it provides the opportunity to trigger double-loop learning in an organization that might lead to the emergence of organizational routines that support responsible innovation.

8.3 Developing a self-assessment tool for responsible innovation

First evidence from the business realm suggests that some companies have implemented practices that already operationalize some aspects of responsible innovation (Auer and Jarmai 2018; Lubberink *et al.* 2017). However, responsible innovation as a concept, in its entirety, or its specific aspects, remains difficult to implement in corporate practice. This is because the principles of responsible innovation have not yet been systematically aligned with corporate processes and innovation management practice (Blok *et al.* 2015; Dreyer *et al.* 2017). Despite a growing corpus of literature on *why* companies (should) commit to responsible innovation, there is still limited knowledge on *how* companies assimilate, institutionalize and translate responsible innovation principles into concrete behaviours and practices.

A few attempts have already been made to operationalize responsible innovation in learning tools for different organizational settings. Lubberink *et al.* (2017) take stock of corporate practices that operationalize anticipation, reflexivity, inclusion, deliberation, responsiveness and knowledge management aspects of responsible innovation. Stahl *et al.* (2017) propose a maturity model to investigate where a company stands with regard to responsible innovation management. In addition, van de Poel *et al.* (2017) also consider company-external factors and firm strategy to assess the contextual, strategic, operational and outcome practices across different product development and life-cycle phases. However, currently proposed models of responsible innovation in a business context do not yet offer an overview of responsible innovation considerations across company management and the innovation process in the form of a self-assessment tool. Concurrently, the discourse on implementing responsible innovation in companies has attracted substantial criticism of the concept. Existing conceptualizations are criticized for using language that is irrelevant to business organizations, for lack of consideration for the specific characteristics of innovation processes (as opposed to research processes) and for misalignment with corporate practices and cognitive frames (Blok *et al.* 2015; Dreyer *et al.* 2017).

The COMPASS self-check tool was developed to address this gap by creating a self-assessment tool that would support organizational learning while avoiding the caveats addressed in the literature. Intended as a self-consulting resource for innovative companies, it aims to facilitate self-diagnosis and monitoring of responsible innovation practices without the need for expert consultation or advice. Its ambition is to go beyond existing learning tools by systematically deconstructing responsible innovation into observable practices, and thus contributing to better alignment of responsible innovation with business realities and company processes.

The concept, self-assessment questionnaire and accompanying tools were developed in the context of a European Union-funded collaborative project. COMPASS consortium partners involved in the development of the self-assessment included academics with expertise in corporate sustainability, organizational learning, innovation ethics and responsible innovation (Vienna University of Economics and Business, University of Central Lancashire Cyprus, De Montfort University), interactive process design experts (Strategic Design Scenarios), as well as business and innovation support organizations (European Business and Innovation Centre Network, B Lab Europe, La Caixa Foundation).

The COMPASS self-check tool was developed in a four-step process.[4] In the first step, an extensive stocktaking of the elements that make up responsible innovation was conducted, drawing on responsible innovation academic literature (with a focus on responsible innovation in industry), grey literature, existing responsible innovation tools, CSR tools and standards. Based on this stocktaking, an initial scope, concept and structure of the tool were devised by the team of tool authors at the Institute for Managing Sustainability, Vienna University of Economics and Business.

In the second step, the tool was further refined in two workshops involving the COMPASS consortium partners. The first of these two workshops, which took place in Leicester, UK, in June 2017, was dedicated to developing a suitable tool architecture and to defining the contents of the different sections of the tool. At this point in time, an early version of the tool architecture connected the elements of RRI as defined by the European Commission (2012), supplemented by an element focusing on social and environmental effects, with the three main phases of an innovation process, namely idea generation and research, development and testing and market entry and diffusion (Godin 2006; Hansen and Birkinshaw 2007), as well as with two sections concerning the company as a whole. Table 8.1 displays this early tool architecture. In a group exercise, consortium partners identified responsible innovation practices for each section (i.e. each cell of Table 8.1). It was agreed that all practices should have observable, factual actions or policies that are implemented in an organization to operationalize elements of responsible innovation. In this fashion, responsible innovation practices break down the concept of responsible innovation into actionable parts which, in their entirety, represent all elements that constitute the concept of responsible innovation.

At the end of the workshop, individual experts from within the consortium were selected to complement and further develop identified practices into sub-sections for each of the five columns, in collaboration with the team of tool authors.

The second workshop, which took place in Brussels, Belgium, in October 2017, was used to review the tool architecture after work conducted by the

Table 8.1 Early draft of tool architecture

	Company as a whole		*Innovation process phases*		
	Governance	*Employees*	*Idea generation and research*	*Development and testing*	*Market and impact*
Gender equality and diversity					
Public engagement					
Science education					
Open access					
Ethics					
Social and environmental effects					

different experts selected from within the consortium. It was also used as an opportunity to cross-check that all elements of responsible innovation as defined by the European Commission (2012) and the so-called AREA framework (anticipate, reflect, engage, act) (Stilgoe *et al.* 2013) were sufficiently represented by the different sub-sections. The different sections and sub-sections (Figure 8.2) were presented to the consortium partners and discussed in plenary. One of the major structural decisions taken at this point was to integrate the two sections covering sub-sections that concerned the whole company.

The tool authors used the input gathered to develop an initial set of questions and answer options. This initial draft consisted of 65 questions in five sections: "company governance" (integrating sub-sections listed under "governance" and "employees" in Figure 8.2), "idea generation and research", "development and testing" and "market and impact".

This first complete draft was then shared and discussed with the experts of the COMPASS project's high-level advisory board in February 2018. The main purpose of this exercise was to receive an external expert assessment of the contents of the tool, ensure its completeness in terms of the responsible innovation concept and to eliminate redundancies and unnecessary components. A content-related change that was implemented based on the advisory board's advice was, for example, the integration of gender analysis and gender responsibility in all innovation process sections dealing with anticipating impacts, testing products, monitoring innovation effects and stakeholder involvement.

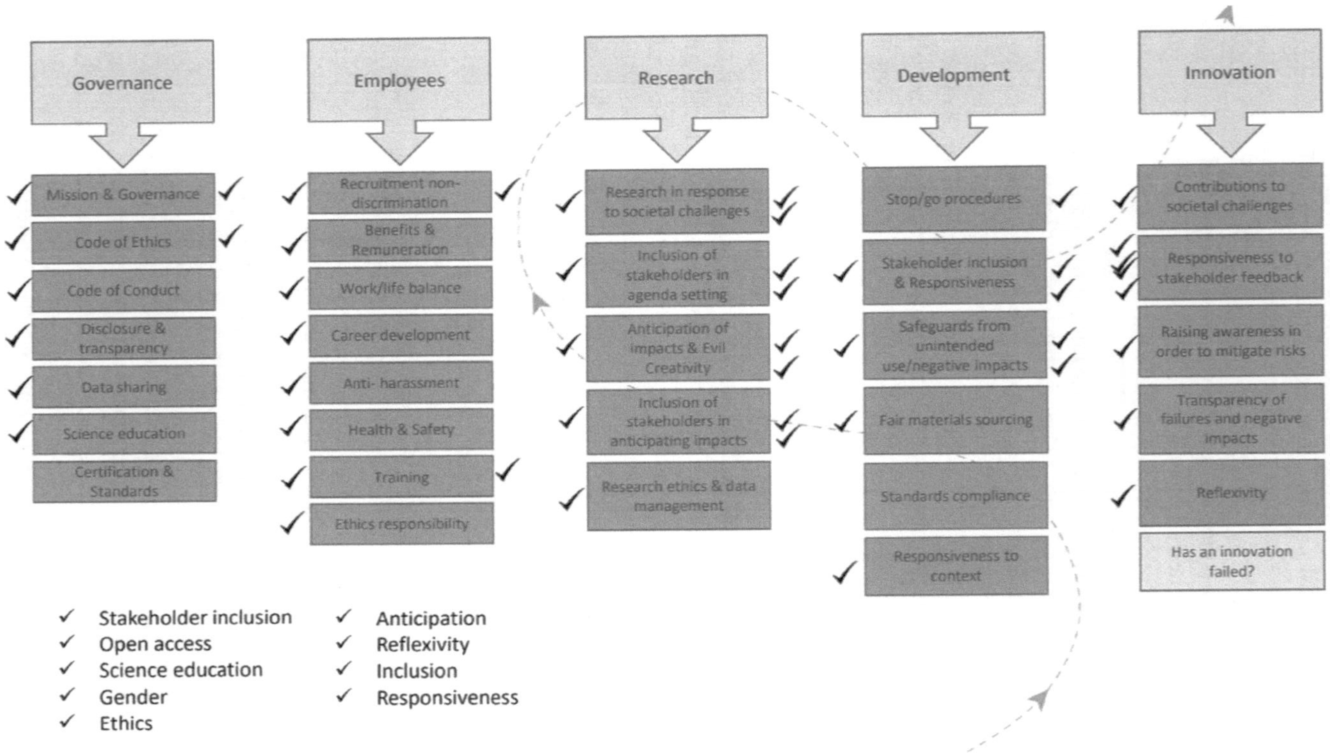

Figure 8.2 Advanced draft of tool architecture.

In the third step, an offline prototype of the envisioned online tool was tested in bilateral interviews or group discussions with 84 individuals representing small and medium-sized enterprises (SMEs), civil society organizations, business support organisations, consultancies, a funding agency and a research organization in the second half of the year 2018. The objective of this round of testing was to receive feedback on general usability and comprehensibility of the questionnaire and to, subsequently, improve wording and terms to maximize understandability and usability for the target group. The collected feedback was diverse and covered comments on the modular approach (which was considered to be a highly valuable and useful approach), to the amount of text in the tool (which was recommended to be reduced) to functional advice (such as e.g. awarding points for each of the four sections in the online tool). The feedback was evaluated by the tool authors in collaboration with the lead testing and piloting consortium partner, European Business and Innovation Centre Network (EBN). Changes were implemented to finalize the questionnaire and develop it into an interactive online tool.

As the fourth and last step of development, a beta version of the online self-check tool was published on the COMPASS project website in February 2019. Another 30 individuals representing SMEs from across Europe provided their feedback on usability of the online tool. Their feedback again concerned diverse subjects, from the length of questions and answer options to the perceived "maleness" of the user icon. Feedback was again evaluated by the tool authors and changes consequently implemented in the final version of the COMPASS self-check tool, which has been available online and free of charge at https://innovation-compass.eu/self-check/ since March 2019.

8.4 Introducing the COMPASS self-check tool

The COMPASS self-check tool[5] is a self-assessment tool that guides a user through a questionnaire on responsible innovation practices, explains why these practices are important and illustrates how they can be implemented. It has the purpose of facilitating learning about what responsible innovation is, how the concept applies to company management and innovation processes, as well as how to implement it. As a user goes through the questionnaire, he or she will discover the elements and practices that constitute responsible innovation. Concurrently, users are able to apply these elements to their own company situation by answering the questions. This helps companies to understand responsible innovation and to reflect on their own company practices. At the same time, it provides them with an analysis of their strengths and areas for improvement to instigate change.

For the purpose of the tool, the concept of responsible innovation was operationalized in the broadest sense, utilizing both the AREA framework prevalent in the academic discourse (Stahl *et al.* 2017; Stilgoe *et al.* 2013) as well as the European Commission's conceptualization (European Commission 2012). Consequently, aspects such as science education or open access that do not play a major role in the AREA framework but are key components of the European Commission's understanding of responsible innovation have been addressed in the tool to ensure a comprehensive approach to the assessment of responsible innovation practices in companies.

The primary target group of the COMPASS project consisted of companies in industries that are characterized by high levels of innovation. The tool has therefore been developed with involvement of nanotechnology, cybersecurity and biomedicine companies; however, it was designed to be applicable by other organizations and in other sector contexts without any adjustments.[6] The COMPASS self-check tool can be implemented by anyone in the company who has a good understanding of how the company operates as well as of the policies and procedures that govern the company's management and innovation process. For larger companies it can be that a collaboration among more people is needed to answer all questions, as one person may not have all the information at hand. A collaboration among a few persons from the same company can also be recommended for smaller organizations to facilitate a discussion. Implementing the tool together and discussing the issues it addresses may yield immediate learning effects and help prioritizing issues and defining actions for integrating responsible innovation in company and innovation management and setting up an improvement plan.

8.4.1 Aligning company practice and responsible innovation

The COMPASS self-check tool is structured along key phases of the innovation process (idea generation and research, development and testing, and market and impact) in conjunction with a general company management section that invites reflection about wider organizational structures and practices (Figure 8.3). The focus of the tool is on an organization, primarily a private company and its processes as a whole, rather than an individual innovation project (although the tool's modular approach also allows for using sections of it for a specific innovation project without assessing the whole company).

Each section comprises a number of responsible innovation elements that are applicable to each phase of the innovation process or company management, respectively. Each element is deconstructed into questions and answer options that point to good responsible innovation practice. Keeping in mind that responsible innovation in company settings is still an evolving concept, most questions have an open comment option where users can enter their own approach of how they tackle that specific issue, if this particular approach is not presented as a possible answer option in the multiple choice answers.

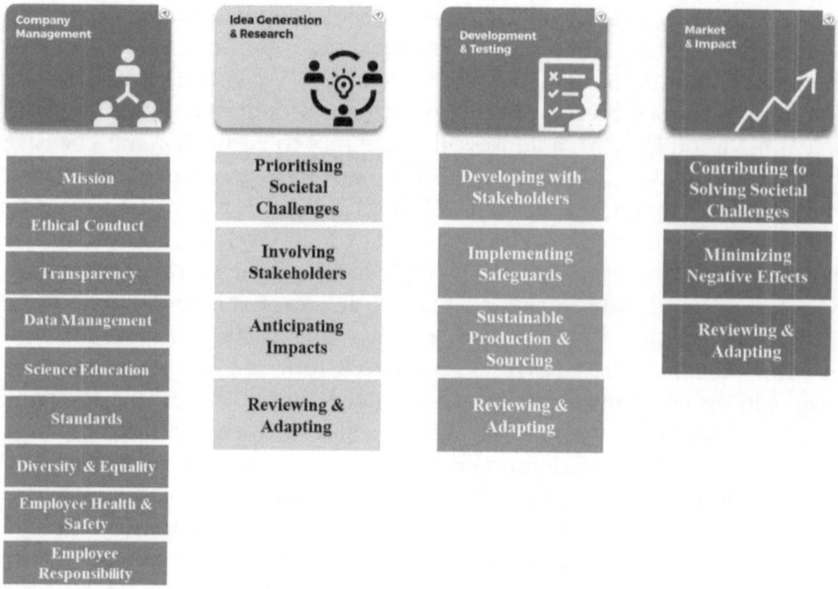

Figure 8.3 Structure and contents of the COMPASS self-check tool.

The four sections of the tool are presented in Figure 8.3 and described below:

1. The *company management* section addresses company practices and procedures that reflect company objectives and values as well as overall strategic orientation as it relates to and affects innovation. It defines the basic rules and conditions for all company activities to take place in, including but not limited to innovation activities. This includes company objectives to create positive societal impact, codes of conduct, commitment to transparency and a certain level of open access to data as well as employee conduct and health-and-safety issues.
2. The *idea generation* section asks about practices that a company has implemented to structure idea generation and selection. This includes questions about if and how the company prioritizes ideas for new products or services that are expected to make a positive contribution to solving societal or environmental challenges, measures to anticipate potential positive and negative impacts of an innovation and circumstances that will lead to adaptation or abortion of further idea development.
3. The *development and testing* section addresses decision-making processes about production, testing and market entry of innovations,

touching upon how these decisions are made and who is involved and consulted in making them. It also deals with the issue of planning for and implementing safeguards against unintended negative impacts of innovations.

4. The *market and impact* section includes questions about practices in the process of launching and evaluating the impact of an innovation. It addresses company strategies for soliciting and integrating feedback and the management of unintended or negative impacts.

The final version of the self-check tool comprises 43 questions and 249 answer options pointing to good practices in responsible innovation. Each question addresses a specific and observable company practice or policy. Answers to the individual questions are scored for verifiable company practices. The questions and the scoring system do not assess the user's awareness, understanding or opinion about specific issues related to responsible innovation in the company, but rather investigate what the company does and does not do (see question examples in Figure 8.4). For example, the COMPASS self-check tool asks about what policies the company has *implemented* for anticipating potential impacts of its innovations, rather than whether the company/user is *aware* of the impacts its innovations may cause. This approach also guides the user to concrete action points or practices, that can help implement and ensure responsible innovation in company processes in the future. In this way, diagnosis and recommendation happen simultaneously.

8.4.2 Actionable results for organizational improvement

There are three main outputs that the tool provides the user with: (1) percentages of total possible points scored in each of the four sections; (2) scores benchmarked against all others who have completed the tool; (3) a summary of questions that the user has marked for follow-up actions for improvement.

The tool is based on a principle of equal weighting of elements of responsible innovation. This is implemented by each sub-section carrying the same weight (same total available points per subsection). In this way, the questionnaire does not prioritize any responsible innovation element over others. All questions within one sub-section can score the same share of points. Two different weighting systems of answer options per question are possible:

1. Each positive response scores an equal fragment of total points available in the question.
2. Any one positive response scores total points available in the question.

By providing an overview of a user's performance across different functions and topics the tool allows for a structured and informed prioritization of

Question example 1:
Who do you involve in the initial idea generation and research stages of new innovations?
❑ *Shareholders.*
❑ *Customers.*
❑ *End-users/beneficiaries.*
❑ *Supply chain partners.*
❑ *Experts.*
❑ *Governmental agencies.*
❑ *Civil society organisations.*
❑ *Consumer bodies, patient associations or other civil society organisations.*
❑ *Potential future customers from general public.*
❑ *Most of the above – we strive for diverse stakeholder involvement.*
❑ *Other, please specify.*
❑ *None of the above.*

Question example 2:
In what roles do you involve stakeholders in your company's innovation processes?
❑ *Being informed of the research and innovation outcomes and/or research processes.*
❑ *Participating in our research through joint-research (e.g. respondents to market research, focus group participants).*
❑ *Providing feedback on innovation processes and the innovation itself.*
❑ *Identifying societal needs and setting the agenda for research and/or innovation.*
❑ *Co-designing with us.*
❑ *Other roles, please specify.*
❑ *We do not consult stakeholders in the initial stages of innovation.*

Question example 3:
How do you ensure effective, productive and fair stakeholder involvement?
❑ *We explain stakeholder roles and the objectives of stakeholder engagement prior to starting the process to ensure these are accepted by participants.*
❑ *We are transparent about the tasks and returns for stakeholders participating in consultations.*
❑ *We share all information relevant to the consultation in an understandable format.*
❑ *We use a third party moderator (or alternatives) to ensure that trust is built up and a broad range of arguments, evidence, and views are exchanged.*
❑ *We ensure that diverse views and perspectives of potential stakeholders are considered.*
❑ *Other measures, please specify.*
❑ *None of the above.*

Figure 8.4 Question examples from the COMPASS self-check tool.

which areas of company functions and responsible innovation should be tackled first. As the user goes through the questionnaire he or she can bookmark specific questions as action points for their company. At the results stage, the user can download a list of bookmarked questions for further action.

8.5 Discussion and conclusions

Based on insights from the discourses on responsible innovation and organizational learning, this chapter has set out to introduce a self-assessment tool tailored to evoke organizational learning for responsible innovation within companies. Using the high diagnostic capability of self-assessment tools, the COMPASS self-check tool strives to enhance organizational learning for responsible innovation by aligning corporate innovation processes and responsible innovation.

The COMPASS self-check tool has laid a foundation for operationalizing responsible innovation practices in industry through a comprehensive and organizational change-oriented self-assessment tool. The questions and answer options (i.e. good practices) that the tool proposes reflect the current state of knowledge and practice in business with regard to responsible innovation.

The usefulness of the tool and its effect on organizational learning and practice will emerge as its usage grows. As more users work through the online tool, the benchmark will become increasingly informative for those who have completed one or more sections and would like to see how their company compares to those of other users. In its current form, the tool strikes a delicate balance between keeping the spirit and ambition of responsible innovation, while providing a relevant and actionable assessment of corporate practices.

To increase the body of knowledge about organizational learning for responsible innovation, future research could investigate changes in company practices or policies that can be causally related to one or more company representatives having used the tool to self-assess the company in terms of responsible innovation. To test the hypothesis that self-assessment can trigger double-loop learning, future research could further investigate if tool users have had the experience of "differences, which make a difference" (Bateson 1972), or, in other words, have been surprised by the tool's question, answer options or scores in the context of how they had perceived responsible innovation or their company's performance according to responsible innovation beforehand.

The conversations that have led to the development of the COMPASS self-check tool will continue beyond its publication and the tool will need to evolve as more aggregated insights into its use and relevance become available. Areas for improvement might, for instance, include the scoring and benchmarking system, its application at a sectoral level to enable within-sector comparisons, as well as additional incentives for regular application within adopting organizations. These data can then be used not only to inform general insights into the state of responsible innovation in companies but also to provide additional incentives to companies to enhance organizational learning for responsible innovation practices.

Notes

1 For instance, the Edelmann Trust Barometer (Edelmann, 2017), an annual survey of more than 33,000 respondents across 28 countries, revealed that 51% of respondents were concerned about the pace of innovation. In this context, about two-thirds of respondents did not believe information shared by the CEOs of companies was credible and expect business to lead through action rather than

words. That is, they are expected to take on responsibility within and beyond the boundaries of their organization that is commensurate with the power they wield over consumers' lives.

2 Seminal works include the model of intellectual development in children by psychologist and philosopher Jean Piaget (1896–1980), the concept of cognitive dissonance (Festinger *et al.* 1956) or the concept of "differences, which make a difference" Bateson 1972).

3 A similar distinction was made by Bateson (1972), who differentiated between first-order learning (or Learning I), which corresponds to Argyris and Schön's (1978) single-loop learning, and second-order learning (or Learning II) based on (self-)reflection, which corresponds to Argyris and Schön's (1978) double-loop learning.

4 Development and testing of the COMPASS self-check tool is described in detail in COMPASS project Deliverable 3.1 "Responsible Innovation Self-Check Tool" and Deliverable 4.3 "Review & Recommendations for Revision of the Responsible Innovation Self-Check tool", which are both available to download at https://innovation-compass.eu/deliverables-2/. Deliverable 4.3 also offers detailed accounts of suggested changes and implementation measures taken in all stages of piloting and testing the tool offline and online.

5 The COMPASS self-check tool is described in detail in COMPASS project Deliverable 3.1 "Responsible Innovation Self-Check Tool" (https://innovation-compass.eu/wp-content/uploads/2019/04/D3.1_Responsible_Innovation_Self_check.pdf).

6 This has been confirmed by company representatives from diverse backgrounds in the tool-testing phase (see Deliverable 4.3 "Review & Recommendations for Revision of the Responsible Innovation Self-Check tool"; https://innovation-compass.eu/wp-content/uploads/2019/04/D4.3_Self-check-tool-testing_final.pdf).

References

Antal, A.B., and Sobczak, A., 2004. Beyond CSR: Organisational learning for global responsibility. *Journal of General Management*, 30 (2), 77–98.

Argyris, C., and Schön, D.A., 1978. *Organizational Learning. A Theory of Action Perspective*. Reading, Mass: Addison-Wesley.

Argyris, C., and Schön, D.A., 1996. *Organizational Learning II. Theory, Method and Practice*. Reading, Mass: Addison-Wesley.

Auer, A., and Jarmai, K., 2018. Implementing responsible research and innovation practices in SMEs. Insights into drivers and barriers from the Austrian medical device sector. *Sustainability*, 10 (1), 17.

Balbaster Benavent, F., Cruz Ros, S., and Moreno-Luzon, M., 2005. A model of quality management self-assessment: An exploratory research. *International Journal of Quality & Reliability Management*, 22 (5), 432–451.

Balbastre, F., *et al.*, 2003. Individual, group, and organizational learning levels and their interactions: An integrative framework. *Management Research: Journal of the Iberoamerican Academy of Management*, 1 (3), 253–267.

Balbastre, F., and Luzón, M.M., 2003. Self-assessment application and learning in organizations: A special reference to the ontological dimension. *Total Quality Management & Business Excellence*, 14 (3), 367–388.

Basu, K., and Palazzo, G., 2008. Corporate social responsibility: A process model of sensemaking. *Academy of Management Review*, 33 (1), 122–136.

Bateson, G., 1972. *Steps to an Ecology of Mind*. New York: Chandler.

Bennet, A., and Shane Tomblin, M., 2006. A learning network framework for modern organizations. *VINE*, 36 (3), 289–303.

Blok, V., Hoffmans, L., and Wubben, E.F.M., 2015. Stakeholder engagement for responsible innovation in the private sector. Critical issues and management practices. *Journal on Chain and Network Science*, 15 (2), 147–164.

Christensen, C.M., Raynor, M.E., and McDonald, R., 2015. What is disruptive innovation? *Harvard Business Review: HBR (*12).

Conti, T., 1997. Optimizing self-assessment. *Total Quality Management*, 8 (2–3), 5–15.

Crossan, M.M., and Apaydin, M., 2010. A multi-dimensional framework of organizational innovation: A systematic review of the literature. *Journal of Management Studies*, 47 (6), 1154–1191.

Dess, G.G., and Picken, J.C., 2000. Changing roles: Leadership in the 21st century. *Organizational Dynamics*, 28 (3), 18–34.

Dreyer, M., *et al.*, 2017. Responsible innovation. A complementary view from industry with proposals for bridging different perspectives. *Sustainability*, 9 (10), 1719.

Edelmann, 2017. *2017 Edelman Trust Barometer™ Global Results* [online]. Available from: www.slideshare.net/EdelmanInsights/2017-edelman-trust-barometer-global-results-71035413

European Commission, 2012. *Responsible Research and Innovation. Europe's Ability to Respond to Societal Challenges*. Luxembourg: Publications Office of the European Union.

Festinger, L., Riecken, H.W., and Schachter, S., 1956. *When Prophecy Fails*. Minneapolis, Minn.: University of Minnesota Press.

Fortis, Z., *et al.*, 2018. Unknown knowns and known unknowns: Framing the role of organizational learning in corporate social responsibility development. *International Journal of Management Reviews*, 20 (2), 277–300.

Gallego-Álvarez, I. , Manuel Prado-Lorenzo, J., and García-Sánchez, I.-M., 2011. Corporate social responsibility and innovation. A resource-based theory. *Management Decision*, 49 (10), 1709–1727.

Godin, B., 2006. The linear model of innovation. *Science, Technology, & Human Values*, 31 (6), 639–667.

Gond, J.-P., and Herrbach, O., 2006. Social reporting as an organisational learning tool? A theoretical framework. *Journal of Business Ethics*, 65 (4), 359–371.

Goodman, J., Korsunova, A., and Halme, M., 2017. Our collaborative future. Activities and roles of stakeholders in sustainability-oriented innovation. *Business Strategy and the Environment*, 26 (6), 731–753.

Hansen, M., and Birkinshaw, J., 2007. The innovation value chain. *Harvard Business Review: HBR* (Reprint R0706J), 2–12.

Hillman, G.P., 1994. Making self-assessment successful. *The TQM Magazine*, 6 (3), 29–31.

Iatridis, K., and Schroeder, D., 2016. *Responsible Research and Innovation in Industry*. Cham: Springer International Publishing.

Jiménez-Jiménez, D., and Sanz-Valle, R., 2011. Innovation, organizational learning, and performance. *Journal of Business Research*, 64 (4), 408–417.

Lau, A.K.W., Yam, R.C.M., and Tang, E.P.Y., 2010. The impact of technological innovation capabilities on innovation performance. *Journal of Science and Technology Policy in China*, 1 (2), 163–186.

Lloria, M.B., and Moreno-Luzon, M.D., 2014. Organizational learning: Proposal of an integrative scale and research instrument. *Journal of Business Research*, 67 (5), 692–697.

Lubberink, R., *et al.*, 2017. A framework for responsible innovation in the business context: Lessons from responsible-, social- and sustainable innovation. In: L. Asveld, *et al.*, eds. *Responsible Innovation 3. A European Agenda?* Cham: Springer International Publishing, 181–207.

Martinuzzi, A., *et al.*, 2018. Responsible research and innovation in industry – challenges, insights and perspectives. *Sustainability*, 10 (3), 1–9.

Mitchell, M., Curtis, A., and Davidson, P., 2012. Can triple bottom line reporting become a cycle for "double loop" learning and radical change? *Accounting, Auditing & Accountability Journal*, 25 (6), 1048–1068.

Mone, M.A., McKinley, W., and Barker, V.L., 1998. Organizational decline and innovation: A contingency framework. *Academy of Management Review*, 23 (1), 115–132.

Nonaka, I., 1994. A dynamic theory of organizational knowledge creation. *Organization Science*, 5 (1), 14–37. Available from: www.jstor.org/stable/2635068.

Nonaka, I., and Takeuchi, H., 1995. *The Knowledge-creating Company: How Japanese Companies Create the Dynamics of Innovation.* Oxford: Oxford University Press.

Nwafor, C., Jarmai, K., Stacherl, B., and Montevecchi, F., 2017. *Benchmark Report & Policy Paper: Integration of the RRI Approach into Collaborative R&D&I and SME Participation in European Funded Collaborative Research in Healthcare, Nanotechnology and ICT* [online]. Available from: https://innovation-compass.eu/wp-content/uploads/2017/09/D1.4-Benchmark-Report_Integration-of-the-RRI-approach-into-collaborative-Research-Development-Innovation-.pdf

Owen, R., Macnaghten, P., and Stilgoe, J., 2012. Responsible research and innovation: From science in society to science for society, with society. *Science and Public Policy*, 39 (6), 751–760.

Pirson, M., Martin, K., and Parmar, B., 2019. Public trust in business and its determinants. *Business & Society*, 58 (1), 132–166.

Richter, U.H., and Arndt, F.F., 2018. Cognitive processes in the CSR decision-making process: A sensemaking perspective. *Journal of Business Ethics*, 148 (3), 587–602.

Schwab, K., ed., 2016. *The Global Competitiveness Report 2016–2017. Insight Report.* Geneva: World Economic Forum.

Sonenshein, S., 2005. Business ethics and internal social criticism. *Business Ethics Quarterly*, 15 (03), 475–498.

Stahl, B., *et al.*, 2017. The Responsible Research and Innovation (RRI) maturity model. Linking theory and practice. *Sustainability*, 9 (6), 1036.

Stata, R., 1994. Organizational learning – The key to management innovation. In: C.E. Schneier, ed. *The Training and Development Sourcebook.* Amherst, Mass.: Human Resource Development Press, 31–42.

Stilgoe, J., Owen, R., and Macnaghten, P., 2013. Developing a framework for responsible innovation. *Research Policy*, 42 (9), 1568–1580.

Tamayo-Torres, I., *et al.*, 2016. Organizational learning and innovation as sources of strategic fit. *Industrial Management & Data Systems*, 116 (8), 1445–1467.

Tarí, J.J., 2008. Self-assessment exercises: A comparison between a private sector organisation and higher education institutions. *International Journal of Production Economics*, 114 (1), 105–118.
van de Poel, I., *et al.*, 2017. Company strategies for Responsible Research and Innovation (RRI). A conceptual model. *Sustainability*, 9 (11), 2045.
Wicki, S., and Hansen, E.G., 2019. Green technology innovation: Anatomy of exploration processes from a learning perspective. *Business Strategy and the Environment*. Available from: 10.1002/bse.2295.
Zietsma, C., *et al.*, 2002. The war of the woods: Facilitators and impediments of organizational learning processes. *British Journal of Management*, 13 (S2), 61-S74.

Best practice II Societal Readiness Thinking Tool by NewHoRRIzon[1]

Tung Tung Chan and Ingeborg Meijer

Description of the good practice example

The primary objective of the Societal Readiness Thinking Tool (SR TT) by Nielsen et al. (2019) is to translate responsible research and innovation (RRI) into practice: how can we help scientists and engineers to increase the societal readiness of their research projects?

The SR TT (https://thinkingtool.eu/tool) provides questions to guide researchers to reflect on the societal alignment of their work at critical stages in the project lifecycle. The tool adopts a stage-gating-inspired approach that divides the development process into discrete stages punctuated by decision gates. Decision gates are characterized by the four phases common to most research projects (Figure II.I):

- Phase 1: research design and formulation
- Phase 2: implementation, data collection and testing
- Phase 3: data analysis and evaluation
- Phase 4: launching and dissemination.

In each phase, a set of questions are presented to stimulate researchers to pause and reflect on the societal impact of their research, in relation to the four RRI dimensions (anticipate, reflect, include, respond) and five RRI key areas (public engagement, science education, gender equality, open access and ethics). To increase usability, 'entry points' are available for selection in each of the four gates. An 'entry point' anticipates why a researcher might be motivated to use the SR TT in the first place.

Each question is accompanied by descriptions of existing projects, methodologies and resources that have proven useful in strengthening the societal appropriateness of research and innovation activities. The purpose of these descriptions is twofold: (1) to guide researchers on how to best address the questions provided in the tool; and (2) to provide real-world examples of how research projects have benefitted from using existing RRI methods and resources in practice.

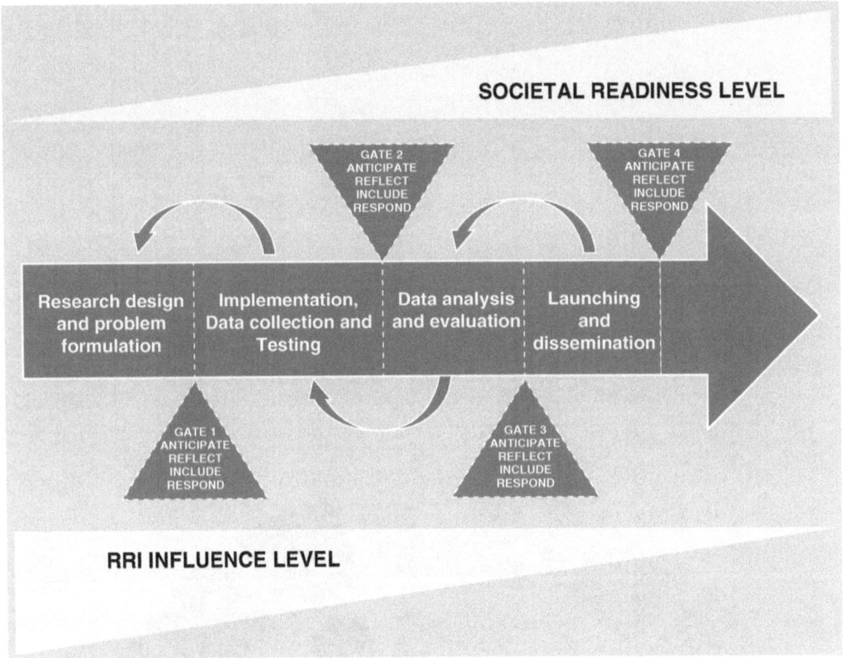

Figure II.I Societal Readiness Thinking Tool.
Note: Some elements in this figure are inspired by Lettice et al. (2017).

Six social labs covering funding schemes across all pillars of Horizon 2020 in the NewHoRRIzon project (Figure II.II) took part in the validation and testing of the tool:

1. Marie Sklodowska Curie Actions
2. Research Infrastructures
3. Leadership in Enabling Industrial Technologies
4. Secure, Clean and Efficient Energy
5. Secure societies
6. European Institute of Innovation and Technology.

Beyond the social labs, researchers, research administrators and managers across various European universities were included in the validation and testing process. Moreover, ad hoc feedback was collected, when the tool was presented orally or as a poster in workshops and conferences from September 2018 to June 2019. The result of the validation and testing led to an improved formulation concerning specific questions and further technical sophistication of the tool. Users also requested to supply information

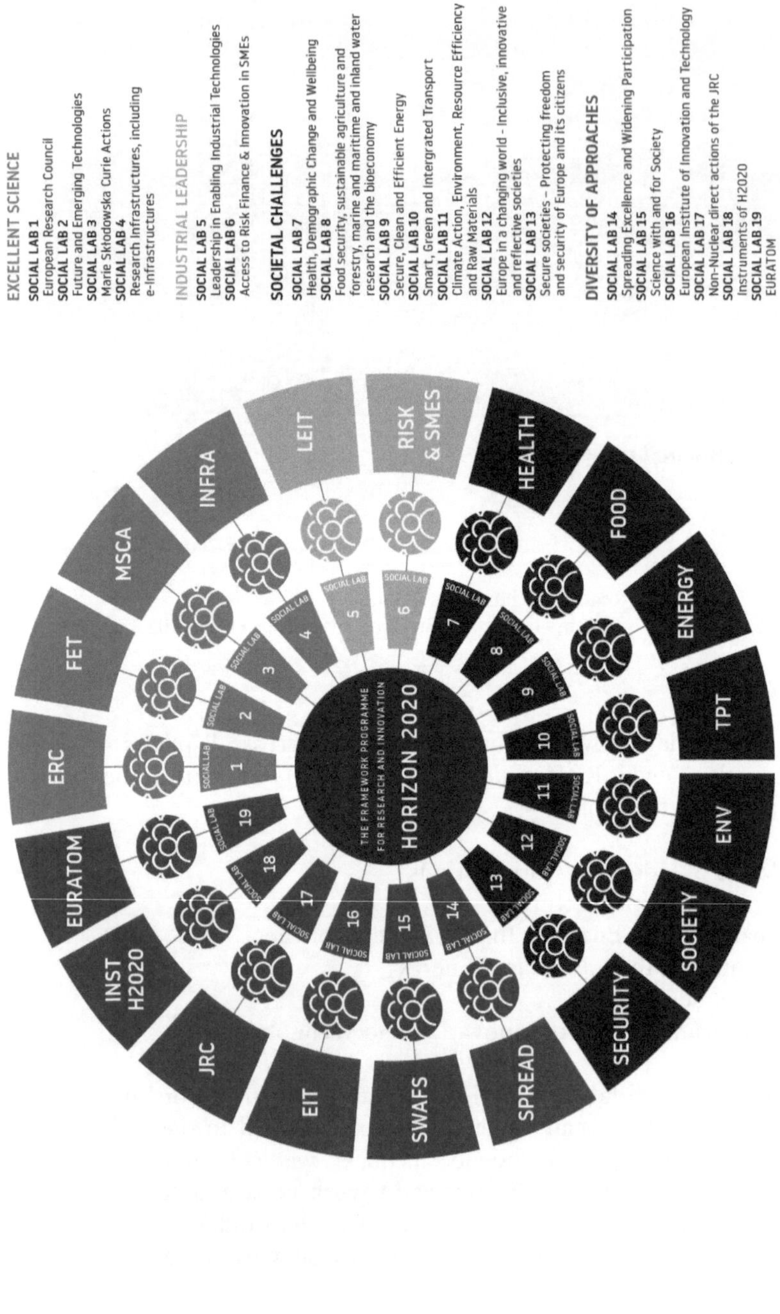

Figure II.11 NewHoRRIzon social labs.

about ways of working with the SR TT, i.e. the practical implementation of the tool in a research project context. In particular, the website should give advice about *who* in a research group should use the tool and *how* a research group should start working with the tool. In order to accommodate these suggestions, the introductory section on the website of the SR TT will be extended. Overall, the SR TT is considered technically adequate and the contents succeed in capturing core elements of RRI.

Aims and goals

Insights derived from literature, co-creation activities and validation with multiple stakeholders within the NewHoRRIzon project have led to the development of the SR TT. The tool is aimed at cultivating researchers' knowledge on RRI, and stimulating integration of RRI in their research project.

Benefit for organisation

Researchers could foresee and anticipate the possible risks and social impacts of their research projects, made apparent by the questions presented in the SR TT. Internal or external evaluators may use the tool to monitor the implementation of RRI in research projects.

Note

1 NewHoRRIzon is a Horizon 2020 project funded by the European Commission, which aims at further integrating Responsible Research and Innovation (RRI) in the research and innovation systems on national and international levels. From May 2017 – April 2021, multiple stakeholders from research, business, policy making, education and civil society are actively recruited to be involved in the project, to better align its processes and outcomes with the values, needs and expectations of society.

References

Lettice, F., Rogers, H., Yaghmaei, E., & Pawar, K. S. (2017). Responsible Research and Innovation Revisited: Aligning Product Development Processes with the Corporate Responsibility Agenda. In A. Brem & E. Viardot (Eds.) *Revolution of Innovation Management* (pp. 247–269). London: Palgrave Macmillan.

Nielsen, M. W., Mejlgaard, N., Alnor, E., Bergmans, J., Bernstein, M., Birkving, A. L., P., Chan, T. T., Griessler, E., Jensen, M. F., van der Klippe, W., Kolby, S., Larsen, S. S., Lorentz, J., Meijer, I., Nicolaisen, P. B., Nieminen, M., & Novitzky, P. (2019). The Societal Readiness Thinking Tool. Available online at: http://thinkingtool.eu/.

9 Reflexive monitoring in action as a methodology for learning and enacting Responsible Research and Innovation

Pim Klaassen, Lisa Verwoerd, Frank Kupper, and Barbara Regeer

9.1 Introduction

In recent years, a considerable amount of conceptual and practical work has been done under the umbrella term Responsible Research and Innovation (RRI). From its onset, RRI has been framed and developed in both policy arenas and academic spheres, leading to a rich variety of interpretations and approaches (Ribeiro et al., 2017). Certain actors emphasize the framing of RRI as a governance framework, whereas others frame it as actively co-shaping research and innovation (R&I) trajectories in interaction with a broad range of actors involved. In general terms, RRI is seen as an approach to align the values and purposes that drive science and innovation with the values, interests and needs of society in order to steer towards outcomes that are societally beneficial, ethically acceptable and environmentally sustainable. The conceptual work sometimes aims at tracing RRI's roots or localizing it in the geography of (like-minded) ideas, movements or disciplinary fields (Blok & Lemmens 2015; Gianni et al. 2018; Iatridis & Schroeder 2016; Oftedal 2014; Rip 2014; Van Lente et al. 2017). Although sometimes such conceptual work is rather critical of RRI – whether *dismissive* (Zwart et al. 2014) or in the sense of investigating the conditions in which RRI might be truly responsible (Grinbaum and Groves 2013) – this arguably stimulated the route RRI travelled from being a rather esoteric subject to becoming almost a household name, at least in fields pertaining to the governance and execution of R&I. In parallel to the conceptual work undertaken on and in RRI, increasing attempts were also made to *practise* RRI. Much of this work builds on older traditions of thought and action, which can be found in disciplines or fields such as (constructive) technology assessment, (applied) ethics, science communication or gender studies, and the conceptual work RRI borrows from, adds to or relates to this. Indeed, in our view one of the merits of RRI is precisely located in the fact that it brings together valuable conceptual and methodological resources that previously were dispersed across various communities.

With this merit, however, there is also an associated risk – namely, that with the ensuing complexity of RRI it becomes difficult to monitor or assess how people, institutions, companies, fields or the R&I system as a whole are influenced by RRI. In this chapter, we argue that reflexive monitoring in action (RMA), an interactive and action-oriented methodology for monitoring complex projects that aim to contribute to sustainable system innovation in the context of 'wicked problems', could well be used to monitor and assess RRI. As such, our approach resonates with the 2015 report by the European Commission's (EC) Expert Group on Policy Indicators for Responsible Innovation, which emphasized the importance of experimental and reflexive learning in assessing RRI (Spaapen et al. 2015). To make our case, we build both on our experiences with formulating and using criteria for assessing RRI gained in the RRI Tools project, funded by the EC, as well as on our experiences with practising RMA during a research project in the context of a Dutch nature policy programme, Natuurpact.

9.2 Background

As we see it, the emergence of RRI as a governance framework for R&I makes immediate sense in the light of two trends. First, R&I play an increasingly important role in society (Acs et al. 2013). That is to say, in today's knowledge economy, a growing number of people are working in R&I (Pyöriä 2005), and the fruits of such work become ever more deeply ingrained in all aspects of our lives. A large part of research efforts concern the production of so-called 'pure' or 'fundamental' knowledge with, at its core, norms of objectivity and independence. It is, however, equally true that R&I are often driven by values of applicability and making a positive societal, environmental or economic impact – working towards improvements for, as it is often put, people, planet and profit. This holds true also for research that is *not* primarily or purposively done with an eye to application or improvement, as this too can unexpectedly find its way to the market – think, for instance, of Alexander Fleming's serendipitous discovery of penicillin (Bosenman 1988).

In parallel to the expansion of the role of scientific knowledge in developing, for instance, technical solutions, health interventions or evidence-based policy-making, so too an increase can be found in the realization that R&I can have unintended and sometimes detrimental consequences (Swierstra & Waelbers 2012). For example, X-rays can be used for diagnostic purposes, but also cause cancer; combustion engines can be used in vehicles to transport people and goods rapidly and cheaply, but at the cost of a negative impact on both the environment and human health; smartphones help connect people to each other and to the virtually endless resources the internet has on offer, but bring with them possible threats to individuals' privacy and risks for mental health (Kawabe et al. 2016).

Arguably, this combination of trends calls for investments to ensure that R&I are imbued with societal values. Insofar as the general public is affected by the (in)direct consequences of R&I, it arguably should have a say in how the public interests should be respected (Dewey 1927). In societies characterized by a plurality of values, the challenge is to incorporate the values of different stakeholders in the development of R&I, so as to work towards ethically acceptable, environmentally sustainable and socially desirable outcomes, products and impacts. This challenge is far from straightforward due to the often fundamental differences between stakeholders in their interests and values and therefore their framings of both problems and solutions (Blok & Lemmens 2015). RRI provides a framework for the governance of R&I that helps facilitate the development of practices in which this is addressed.

Then, what does that framework consist of? This question has received various answers. One important source in the field, the EC, states that RRI is 'an inclusive approach [that] aims to better align both the process and outcomes of [R&I], with the values, needs and expectations of European society'.[1] In the UK, the Engineering and Physical Sciences Research Council (EPSRC) has instituted a framework for responsible innovation built around the process elements of anticipation, reflection, engagement and action (AREA[2]), and along similar lines, Wickson and Carew have proposed the following four core constituents of RRI:

> (1) a focus on addressing significant socio-ecological needs and challenges; (2) a commitment to actively engaging a range of stakeholders for the purpose of substantively better decision-making and mutual learning; (3) a dedicated attempt to anticipate potential problems, assess available alternatives and reflect on underlying values, assumptions and beliefs; and (4) a willingness among all participants to act and adapt according to these ideas.
>
> (Wickson & Carew 2014, p. 255)

In this chapter, we propose a heuristic and analytical model of RRI that shows clear affinity with such understandings, with an additional specification of more fine-grained criteria and a methodology that can be used in applying these criteria in enacting RRI. The combination of criteria and methodology can help bring together the simultaneous practice and assessment of the rich, yet fairly abstract, prevailing understanding of RRI.

9.3 Formulating criteria for RRI – and putting them to good use

With developments towards the conceptualization and enactment of RRI, there is a need to establish whether RRI leads to the assumed outcomes. Looking beyond its good intentions, can the environmental and societal merit of RRI be proven? If so, can these proven outcomes be linked to RRI processes (i.e. can we draw lessons on what determines 'quality' in RRI practice)? In other words, can we validate the value of RRI? To help answer these questions, there have been several attempts to further specify what it entails to engage in RRI (sometimes dedicated to RRI in a specific context), and to assess RRI's impact (Davis & Laas 2014; Heras & Ruiz-Mallén 2017; Kupper et al. 2015; Peter et al. 2018; Wickson & Carew 2014). In the context of the EC-funded project, RRI Tools, two of the authors contributed to the formulation of criteria for RRI.

This attempt commenced with the deceptively simple question: 'What is RRI?' Deceptively simple, because we wanted to avoid pure theory-based conceptualizations, and to steer clear of the implicit suggestion that it would be possible to 'discover' a single fixed and context-independent answer to this question. Thus, even if hypothetically it would be possible to define rules for behaving responsibly in R&I, we would still see these rules in themselves as being of little or no value. In our view, such a 'sterile' theory-driven code-book would be self-contradictory, as it would absolve those applying the model from the very kind of responsibility it seeks to instil. Rather, the burden of deciding what are responsible decisions and actions always remains with those directly or indirectly taking part in R&I processes. These actors themselves need to reflect upon what is responsible in the specific contexts of their work, and this requires careful and systematic reflection on a case-by-case basis. To give expression to this view, the criteria that emerged as our answer to the question 'What is RRI?' were eventually translated into questions inviting reflection and deliberation, rather than assertions.

To arrive at a comprehensive model of RRI and its criteria, we engaged in a process of iterative conceptual modelling (Figure 9.1, and see Klaassen et al. 2017 for a more extensive description). Central to this methodology for concept development are different and disparate forms of expertise, confronted in a series of iterative steps which, in this case, sought to answer our question 'What is RRI?'. Figure 9.1 presents the six different steps of: (1) literature review on RRI and the many disciplinary and conceptual resources RRI is built up upon; (2) expert consultation; (3) stakeholder workshops; (4) cataloguing promising RRI practices; (5) deriving RRI criteria; and (6) in-depth case-study analyses, as well as the iterations to arrive at and subsequently test and adapt our conceptualization of RRI and its associated criteria.

The starting point for any specification of ideals is to ask *why* the ideals are needed. Specifically, why would responsibility need to be introduced in the R&I environment? Or, to paraphrase a famous super-villain: *Why so responsible?* The answers to this question can be linked to two perspectives from which the need for responsibility can be approached. First, as seen by those *inside* the R&I environment, responsibility is an obligation that follows from the (social, financial and moral) power granted to those in a position to carry out or regulate R&I processes. Writing from this perspective, Owen et al. (2013) refer to responsibility as an 'imperative'. Second, from the perspective of those actors traditionally located *outside* R&I, gaining responsibility is a *right*, derived from the breadth of impacts they

Figure 9.1 The iterative conceptualization of Responsible Research and Innovation (RRI).
Source: Klaassen et al., 2017.

(may) experience as result of those processes. This second view asserts that scientists should not be isolated from society for the rather straightforward reason that the impacts of their work are not.

In a European context, these ideas have been integrated in research policy, and resonate with what are known as the seven grand social challenges (Klaassen et al. 2018; Lund Declaration 2009) – although it is now more usual to refer to the sustainable development goals (SDGs) adopted by the United Nations. As we interpret them, the grand challenges or SDGs are not so much research topics, but rather points of intersection between science and society (Cummings et al. 2018). Indeed, these challenges can *also* function as the abstract, but palpable, purposes for RRI.[3]

Taking challenges or the SDGs as long-term purposes, in order to achieve them, we need to think of more concrete products that can function as intermediate stepping stones on our journey. Through our discussions with experts and laypersons alike, we have discovered that, in pursuing responsibility, the associated learning is as important as the results of R&I. We thus make a distinction between R&I products (i.e. the artefacts created through R&I) and learning products (i.e. the skills acquired and practices established in creating R&I products). Responsibility, we maintain, is to be equally sought in both types of products: it is simultaneously a dimension of the products of R&I, and of the hard and soft institutions thereby brought to life. If solving the seven grand challenges can be compared to crossing a river, then R&I products are stepping stones and learning products are the skills needed to discover and jump from one stepping stone to the other. If we have the former and not the latter, we might end up stranded in the middle of the river with little idea of what to do next. In brief, if responsibility is to be 'fostered', the cultural by-products of R&I activities are as important as the activities themselves.

To take an example: if transgenic crops are an R&I-driven solution supporting the second grand challenge (food security, sustainable agriculture and forestry), then the practices institutionalized in the process of researching transgenic crops responsibly are as important as the actual crops. The two might affect technological progress in different timespans, R&I products having a direct and immediate impact while learning products have an indirect and a delayed one – but the two are of equal importance. The questions are 'Do genetically modified organisms (GMOs) constitute a step towards the second grand challenge?', and 'How can research into GMOs be carried out responsibly and how can we instil this responsibility in the future?'

Having delineated the destinations of RRI, we can reason backwards to clarify the processes needed to reach them. The conceptualization of RRI in terms of four process dimensions was arrived at, much in line with much-cited sources on RRI such as Owen et al. (2012) and Stilgoe et al. (2013). Thus, *anticipation and reflection*, *openness and transparency*, *diversity and inclusion* and *responsiveness and adaptive change* were identified as core elements in the process of doing R&I responsibly. These dimensions were then further specified in terms of criteria and sub-criteria, and ultimately one or more inviting questions. To emphasize: we refer to these questions as *inviting questions* to highlight their intended status as points of departure for various forms of reflection and deliberation, rather than as an instrument of accounting. To illustrate what this has reaped, Table 9.1 presents criteria, sub-criteria and inviting questions for the process dimension of anticipation and reflection (and for the other three dimensions, see Kupper et al. 2015). Criteria for RRI can be used beforehand both by researchers and innovators in designing R&I responsibly or by funding organizations for assessing proposals, as well as afterwards for assessing to what degree R&I have been undertaken responsibly, and whether desired outcomes and impacts have been achieved. Such criteria can also be used during the execution of R&I projects or programmes, as a means to monitor and continuously improve them. Arguably, the latter constitutes a way of assessing (attempts at) RRI that is most in line with the earlier description of RRI, especially in relation to RRI's intrinsic openness and connection with experimental learning. A complex new field of its own might appear to open up with this entrance into the realm of evaluation of R&I, but we in fact enter a field in which there is significantly more scholarly experience than in the world of RRI alone.

Depending on the prevailing needs and intentions, evaluators of R&I projects or programmes can choose a method from a wide variety of available (theoretical) approaches. Traditionally, evaluations function as a

Table 9.1 Criteria, sub-criteria and inviting questions for the Responsible Research and Innovation (RRI) process dimension of reflection and anticipation

Criteria	*Specification*	
	Sub-criteria	*Inviting questions*
Analysis of the background, current situation and context of the (planned) research or innovation. (Nordmann, 2014)	Up-to-date information	Has content research been done on relevant background knowledge and up-to-date information?
	Influence other R&I	Has the influence of other innovations/ research on the course of this practice been taken into consideration (e.g. alternative and complementary R&I)?
	Actor analysis	Did an actor analysis take place, identifying on whom the practice might have an impact or who might have an interest in, and might have relevant expertise for, the practice – and identifying how these actors relate to each other?
	Diverging problem definitions	Have efforts been made to address potentially diverging definitions of the problem at stake in the practice?
	Societal role in problem definition and course of practice	Have efforts been made to give a role to societal values, perceptions and interests in defining the problem addressed in the practice and the further course of the practice?
Envisioning of plausible futures (Nordmann, 2014)	Variety of future parameters and impacts	Is there active identification and consideration of immediate, mid-term impacts? Are long-term social, environmental and economic impacts and consequences of the practice – intended and unintended – identified?
	Variety of established methods	Did a well-considered selection and implementation of the methods for anticipation take place (based on previous experience) (e.g. scenario development, real-time technology assessment)?
	Variety of R&I trajectories	Have alternative R&I trajectories been considered (process of R&I)?
Variety of impacts	Ethics	Are ethical aspects and impacts of the practice sufficiently addressed? For example, are research ethics honoured, by protecting the subjects of research, approval from an ethical committee and documented compliance with research ethics and voluntary codes of conduct – in which for example fraud and plagiarism are prohibited (Wickson & Carew 2014)?

Table 9.1 (*Cont.*)

Criteria	*Specification*	
	Sub-criteria	*Inviting questions*
	Legislation	Are legal aspects and impacts of the practice sufficiently addressed? For example, is there documented compliance with highest-level governance requirements Wickson & Carew 2014)?
	Society	Are social aspects and impacts of the practice sufficiently addressed?
	Environment	Are environmental aspects and impacts of the practice sufficiently addressed?
	Grand challenges	Are one or more of the grand challenges set by the European Commission addressed in the practice?
Facilitating deliberation on values, perceptions, needs, interests, choices and definition of the problem at issue in the practice	Integrated reflection and deliberation	Has room for reflection and deliberation on e.g. impacts, alternatives, possibly changing social values, perceptions and needs/ interests and choices made during the practice been built in (Stilgoe et al. 2013)?
	Deliberating values	Do the actors involved regularly engage in a critical analysis of the values, perceptions, needs, interests, choices and definition of the problem at issue underlying their practice?
Addressing roles in RI trajectories	Awareness of differences	Do the actors involved develop an awareness of their own assumptions, values and purposes in relation to the perspectives of others?
	Awareness of responsibilities	Are involved actors aware of and open to reflection on their role responsibilities and accountability (Stilgoe et al. 2013)?

Note: R&I, research and innovation; RI, responsible innovation.

Source: Adapted from Kupper et al. (2015).

type of disciplined inquiry to assess whether a programme or project has contributed to its designated purposes. From an instrumental perspective, this approach primarily serves the needs of project funders in terms of accountability (also understood as upwards accountability; Ebrahim 2005). Especially, for instance, large-scale innovation and research projects that depend on private-sector investments are prone to evaluation focused on accountability based on criteria set by investors. Such criteria tend to spring from a technical–economic perspective, narrowing down the set of interpretations of a project's success to a matter of pre-defined deliverables.

Academic evaluation based on output criteria is also a familiar practice, for instance through quantitative metrics meant to capture the (academic) impact of publications. For example, the Centre for Science and Technology Studies at Leiden University is invested in the development of bibliometrics, and evaluates academic impact through the analysis of citation patterns of different research groups. Through this approach they also attempt to determine the use and value of publications for marketable applications, public policy development or other ways of attaining public value. However, although they offer some insights, the metrics produced have limited use in informing action to improve research practices. This is a known argument against evaluation approaches that focus solely on accountability. Although a legitimate approach, it tends to assume rather linear input–output relations, and the evaluation findings are seldom suitable to feed back into the practice of those actors who implemented the project to help them improve their work. The recognition that *learning* by those actors is required to improve a project's conduct and increase the likelihood of achieving the designated purposes contributed to the emergence of so-called 'learning-oriented' evaluation approaches. In such approaches, evaluations are designed to focus on the understandings, concerns and learning needs of actors and stakeholders involved in or surrounding the project or programme in question.[4] Arguably, this type of perspective also inspired the 'Leiden manifesto for research metrics',[5] which proclaims that the scope of research metrics should be extended beyond output criteria alone, by developing more 'meaningful metrics' in order to inform research performance (Hicks et al. 2015). These two functions of evaluation (i.e. accountability and learning) and their supposed irreconcilability have received considerable attention in the recent scholarly literature in evaluation studies (Adelle & Weiland 2012; Guijt 2010; Kunseler & Vasileiadou 2016; Owens et al. 2004; Regeer et al. 2016).

For, while learning through evaluation is generally professed to be important, evaluation for accountability still predominates. This is especially problematic for long-term R&I projects that search to address complex societal problems that are of a more goal-seeking and emergent nature, and for which the societal impacts and technological advancements weigh heavily but are difficult to predict. Such research projects very rarely take the shape of a linear process of problem formulation, project design and implementation, as they tend to unfold as dynamic and experimental interactions between multiple actors from multiple levels or domains – academic, industrial, societal, policy – during which mutual learning is essential in order to produce the types of responsible outcomes that are sought (i.e. outcomes contributing to the SDGs or grand challenges, that are for instance ethically and socially acceptable and environmentally sustainable) (Regeer et al. 2009). As such projects progress, greater insights and developments in their institutional and societal context may require the reformulation of the project's goals and, subsequently, the research design. Adaptive capacity is required for the project team to be able to anticipate and accommodate such changes. From this perspective, evaluating for accountability based on pre-defined criteria becomes meaningless. Simultaneously, evaluation that focuses solely on learning without considering the developments in institutional and societal contexts might improve specific procedures, but would lose sight of a project's greater ambitions of achieving societal impact. For complex R&I projects, there is a call for evaluation approaches that reconcile the dual purpose of accountability and learning (Lehtonen 2014; Regeer et al. 2016). More specifically, there is a call for evaluation approaches that allow assessment of R&I projects' progress towards their greater ambitions, *while* learning – about goals, strategies, actions and contexts – is promoted (Klerkx et al. 2010). By creating space for learning processes, the project team and other stakeholders may build their adaptive capacities and abilities to inform adaptive change, in light of new insights and a changing environment. The evaluation approach that is adopted should be as flexible as the project itself.

From different fields – including development studies, system innovation, transdisciplinary research, transition management and RRI – scholars have argued that the answer lies in evaluation approaches geared towards enhancing *reflexivity* (Arkesteijn et al. 2015; Botha et al. 2017). Reflexivity is a capacity that allows projects' participants to act in greater accordance with, and are responsive to, their institutional and societal context, in line with their understanding(s) of a project's success (Elzen et al. 2017). Indeed, reflexive approaches seek to encourage participants' adaptive capacities. Reflexive evaluation occurs during a project's course and is ideally integral to it, as part of iterative cycles of defining, implementing and adjusting its design in order to move towards responsible research outcomes (Regeer et al. 2009). As such, reflexive evaluation methods simultaneously function as promoters for and assessors of learning processes to achieve environmentally sustainable and socially acceptable goals. Table 9.2 shows the three conceptualizations of evaluation approaches and their characteristics, which we have discussed here.

In this chapter, we further focus on one specific reflexive approach that has proliferated in various domains and explore its value for RRI projects: RMA. RMA is an interactive, action-oriented monitoring and evaluation method, originally developed to support projects with ambitions to make system innovations and which require major institutional and social change (Van Mierlo et al. 2010). As for most reflexive approaches, RMA is intended to increase a project's reflexivity – its ability to affect and interact with the context within which it operates – by encouraging its participants' collective learning processes, through which institutional and societal barriers to system innovation are identified and overcome. As Arkesteijn et al. (2015) explain, RMA builds on the premise that, while the contribution of a single system innovation project to the overarching, complex system innovation processes cannot be assessed, it is possible to understand and work (and

Table 9.2 Three ideal–typical conceptualizations of evaluation approaches

	Evaluation approaches for accountability	*Learning-oriented approaches*	*Reflexive evaluation approaches*
Evaluation purpose	Evaluation as mechanism for accountability assurance, assessing a project's performance	Evaluation as mechanism to enhance learning processes for improved (research) practice (procedures, processes)	Evaluation as mechanism to enhance reflexivity through monitoring of and reflection on goals, strategies, actions, institutional and social contexts
Characteristics of evaluand	Simple project: mostly linear causal chain of inputs, outputs and outcomes	Complicated project: multi-actor interactions with complicated multi-directional causal mechanisms	Complex project: multi-actor, multi-level governance constellations with unclear/unknowable and uncertain causal mechanisms
Evaluation criteria	Pre-defined deliverables (technical–economic; (cost) efficiency, effectiveness)	Learning needs of project team in light of their own practice	Learning needs in light of jointly determined goals and principles (e.g. ethics, inclusion, responsiveness)
Evaluation design	Technical–analytical evaluation design by evaluators	Deliberative evaluation design in concert between evaluators and project team	Deliberative–analytical, emergent and responsive evaluation design in concert between evaluators, project team and other relevant stakeholders
Role of evaluator	Distant observer, judge	Facilitator of learning processes	Facilitator, mediator, observer: 'critical friend'
Time aspect	Evaluation takes place prior or after the project	Evaluation occurs parallel to the project	Evaluation is an integral part of the project

Source: Adapted from Kunseler & Vasileiadou (2016).

document!) towards a project's design and outcomes in terms of relevance to long-term ambitions of system innovation. Key to the methodology is recurrent reflection on the institutional and societal context of a project in relation to its long-term ambitions, and its concrete project design and effects (i.e. on the theory of change on which a project's design is founded). We argue that this holds merit for the evaluation of RRI. While a project's societal impact cannot be foreseen or anticipated in full, what *is* possible is to guide action along the RRI criteria framework, as described above. RMA methodology may help to enhance the reflexivity of a project team and build its capacity to enact RRI criteria. As a result, one has worked towards responsible R&I outcomes, while simultaneously the process towards these becomes more responsible.

The actual monitoring activities of RMA are integral to the project, and are usually guided by an appointed reflexive monitor. The monitor observes moments of interaction (like team meetings) and focuses on whether learning and actions towards the intended purposes take place, assisting such processes by making use of a range of reflexive monitoring tool activities (see Van Mierlo et al. 2010). As such, the reflexive monitor is not merely an observer, but rather a facilitator or 'critical friend'; close enough to fully understand the issues encountered, but with sufficient distance to legitimately and critically reflect on the process. What it is that needs to be learned, or by whom, is not a given in this method, but rather something that itself needs to be recurrently assessed and reviewed. RMA does not provide a blueprint or a strictly structured action plan, but should rather be seen as a flexible methodology that can be customized to support any endeavour that aspires to contribute to sustainable development and societal change.

9.4 Illustrating RMA's value as an RRI way of assessing RRI

We will now illustrate the added value of integrating the RMA methodology in research and (system) innovation projects to simultaneously promote and assess RRI. The case we present is the first period (2014–2017) of the Natuurpact project. This large-scale transdisciplinary research project took place in the context of a Dutch nature policy programme (PBL & WUR 2017). In 2013, the Dutch government, together with governments of the 12 Dutch provinces[6] and a range of social organizations, signed the so-called *Natuurpact agreement*. In this pact, all parties agreed on a joint approach to achieve national, long-term ambitions to improve the conservation and resilience of Dutch nature by 2027. The years prior to the pact had been characterized by the reduction of biodiversity, low levels of societal engagement with nature and disputes over the value and function of nature, specifically between the agricultural and economic domain on the one hand and nature conservation advocacy groups on the other. To achieve the pact's ambitions, it was argued, innovative policy strategies were required and, in support of this, the parties decided upon the Natuurpact project, to study and simultaneously support the innovation of policy strategies responsibly. The project was intended to be inclusive and responsive to all relevant policy actors and geared towards mutual learning through deliberation and experimentation, to overcome the value

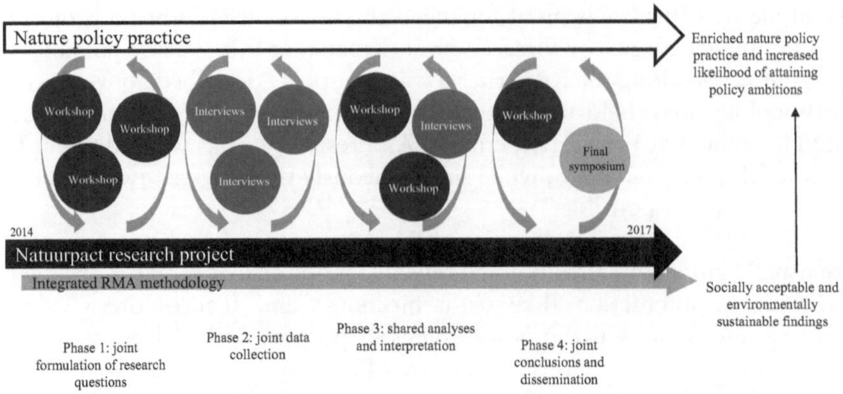

Figure 9.2 Schematic overview of the Natuurpact research project.

disputes and encourage social engagement with nature. The project ran in parallel to regional policy processes of development and implementation, in order to directly inform policy practice with socially robust findings from a multitude of scientific disciplines pertinent to environmental policy. It was expected that as outcomes of this approach there would be societally acceptable and environmentally sustainable findings that could inform policy practice, thereby enhancing the likelihood of attaining the ambitions recorded in the Natuurpact.

The Natuurpact project was led by a multi-institutional and interdisciplinary project team, which we will refer to as 'the researchers'. The researchers collaborated with relevant policy actors during roughly four different phases of the research, such as the joint formulation of research questions, joint data collection, shared analyses and interpretation, and drawing joint conclusions and dissemination (Figure 9.2). This, for instance, occurred during diverse research activities in each phase, such as workshops and sessions, but also through bilateral interactions (such as interviews) between regional policy actors and members of the research team. The first period of the Natuurpact project was concluded in 2017. Two of this chapter's authors were part of the research team in the role of 'reflexive monitors'; their task was to simultaneously promote and assess the reflexivity of the Natuurpact project while it was conducted, to enhance the enactment of its RRI ideal throughout the project.

Concretely, this meant they took part in project team meetings and simultaneously monitored research progress during other research activities. During the meetings, they would ask inviting questions to reflect on RRI criteria and how they might be enacted during the Natuurpact project. For example, when planning the first workshops at the start of the project, they facilitated the researchers to make a stakeholder analysis for the project.

While doing so, they encouraged deliberation on stakeholder diversity and inclusion: who to include, why, how to encourage participation, and what type of participation would be pertinent? An important part of their role as reflexive monitors was to help the researchers make explicit the challenges that they perceived with stakeholder inclusion, and formulate actions to overcome these. For instance, the researchers anticipated minor conflicts between some of the stakeholder groups and were concerned this would obstruct an open and constructive dialogue during the workshop. The project leader retrospectively reflected:

> We had a lot of discussion on who to invite during which workshops. We reasoned that the conflict between regional and national government was already so significant that also bringing in societal organizations too early in the project would only work against our purpose of deliberation.

In response, it was decided to first hold a few smaller parallel workshops with individual stakeholder groups (e.g. regional policy actors) before a larger heterogeneous session attended by multiple groups. This, it was reasoned, would enable the different actors to become accustomed to the idea of collaborating on a research project, spurring their willingness to work together.

In a similar fashion, as reflexive monitors they invited reflection on how to ensure criteria such as openness and transparency towards the participating actors would be achieved throughout the project. For example, during the third phase of data analysis and interpretation, the researchers found being open and transparent about their research process quite unnerving. They felt vulnerable to the critiques of the participation policy actors: would their expertise be questioned? Making this discomfort explicit made it possible to expose the underlying concern, which allowed for action to be formulated to address it. In discussing this matter of openness, a member of the research team explained:

> We realized, I think, we were asked to do the same thing we asked of the participating policy actors: to open up our ways of working to the eyes of others to benefit mutual understanding and sense-making. That is scary. But if we asked it from them, we had to also do it ourselves.

This insight led the researchers to decide to open up the 'black box' of their analyses and explain their work in a way that allowed for mutual understanding, including the use of visualizations and steering clear of using research jargon. In essence, the work of reflexive monitors for promoting core research principles or elements – such as diversity and inclusion, openness and transparency, and responsiveness and adaptive change – encompasses facilitating recurring and systematic reflection on these elements.

As part of promoting RRI elements, they simultaneously monitored how the participating actors appreciated the responsible approach adopted by the Natuurpact project team during research activities, to oversee whether the actions taken indeed led to the intended outcomes of socially acceptable and environmentally sustainable findings, and knowledge-enriched policy practice. In addition, they conducted ten in-depth interviews with selected participants (from national and regional government, and societal organizations) and a focus group session with another 12 (primarily representatives from regional governments) after the project was finalized to assess the different ways the Natuurpact project had been of value to their policy practice.[7] The findings demonstrated that (participating with) the Natuurpact project manifested five different types of policy value (Verwoerd et al. 2017), which were much in line with other conceptualizations of social effects of transdisciplinary research projects (e.g. Walter et al. 2007; Wiek et al. 2014). Importantly, these value types were found to occur both at the regional level and across tiers of government. Table 9.3 presents the value types, including some examples from our empirical experience. To start with, the research was found to have *network value* to most of its participants, including network expansion, the strengthening of existing relationships, and building trust between stakeholder groups. Second, it was established that the project effected *affective value*, comprised of a sense of relief and reassurance to see other policy actors encountering similar challenges, and a sense of togetherness and belonging to a community. Third, the project proved to have significant *cognitive value*, as it generated new knowledge and increased understanding of the workings of nature policy strategies and contributed to conceptual coherence in terms of a shared language and frame of reference among the participants. Fourth, the project had *instrumental value*, which was visible in the direct uptake of new insights into policy development and implementation. Finally, the Natuurpact project was shown to have *strategic value*, as it contributed to policy agenda-setting, societal support for nature policy and legitimizing certain policy decisions.

In their role as reflexive monitors during the Natuurpact project they found that inviting deliberation and reflection on responsible research criteria encouraged the reflexivity of the project team. More specifically, their 'sensitivity' (Sarkki et al. 2013) was found to increase their attention towards the diverse needs and perspectives of stakeholders involved with the research, and the capacity to act on this knowledge (Verwoerd et al. 2017). Furthering this concept of researcher sensitivity, throughout the project the researchers built and drew from what we have termed 'reflexive capacities' to ensure the research sufficiently enacted RRI core elements. For example, to enact stakeholder diversity and inclusivity, the researchers built the capacity of *developing shared ownership* of the participants over the research. This included their growing understanding of complex power dynamics between stakeholder groups and the building of robust and trustworthy relationships. To enact openness and transparency, the researchers built the reflexive capacity of interacting with participants through a *mode of humility*. This meant they organized a safe space for deliberations: the

Table 9.3 Overview of five value types manifest at different government levels, with illustrative quotes

Value type	*Government level at which value type was manifest*	*Quotes from interviews with regional policy actors who participated in the Natuurpact research*
Network value	Regional level	Network expansion: *'It was during this workshop on agriculture when I learned Campina* [Dutch dairy cooperation] *was also employing nature-inclusive agricultural methods. Then I thought, well, perhaps we should also get in touch with them'*
	Across government tiers	Strengthening existing professional relationships: *'You now also have a face and a name, you know who you are dealing with or who to get in touch with* [within national government]'
Affective value	Regional level	Relief and reassurance: *'As a result of all the workshop sessions and also the final report I think "well, we are not doing too bad as a province". Earlier I was afraid we had missed opportunities, but now I feel quite relieved'*
		Sense of togetherness: '[Participating with the Natuurpact project] *creates a sense of "we're in this together". It makes you aware that you share a responsibility'*
	Across government tiers	Building of trust across government tiers: *'Last year I also spoke with* [representative of national government]. *And you notice that – on the base of increased trust – there's more openness in discussion compared to five years ago'*
Cognitive value	Regional level	New knowledge and increased understanding of the workings of and perspectives on nature policy strategies: *'Private parties commented on aspects they run into with nature policy that I had never considered. So* [the Natuurpact project] *provided a different lens than just the administrative one'*
	Across government tiers	Conceptual coherence, shared language, shared ambitions: *'The benefit of working together during this project is that you learn to speak the same language and can more easily connect'*
Instrumental value	Regional level	Informing of regional policy formulation: *'In parallel to the Natuurpact research, we were writing our "Nature Vision"* [regional nature policy]. [...] *In our final version we have integrated the recommendations of the research, it was a valuable addition to our policy'*

(*continued*)

Table 9.3 (*Cont.*)

Value type	*Government level at which value type was manifest*	*Quotes from interviews with regional policy actors who participated in the Natuurpact research*
	Across government tiers	Policy alignment/coordination: *'The findings show where national and regional policy enforce or work against each other'*
Strategic value	Regional level	Regional policy agenda-setting, societal support for regional nature policy, legitimization of regional policy decisions: *'In talking with societal partners it is highly useful to have a report that backs up our decisions. It justifies some of our work'*
	Across government tiers	National policy agenda-setting, legitimization of decentralization: *'The research pinpoints areas in nature policy that need a lot of work. The report has created collective urgency to focus on these'*

researchers positioned themselves on equal footing with the participating policy actors and, from there, invited and gave constructive feedback on nature policy research and practice, respectively. This benefited knowledge integration.

Finally, to enact responsiveness and adaptive change, the researchers built the capacity of *purposeful responsiveness to emergent research needs.* This meant that, while the researchers sought to be responsive to all policy research needs of the participating policy actors, they also kept in mind the feasibility of these research needs as regards the project's budgetary limitations. In sum, the researchers' ability to conduct more responsible research, and to operationalize and enact responsible research criteria, grew. Here, the role of reflexive monitors should be modestly assessed: we cannot claim the efforts of the monitors were entirely responsible for the development of these reflexive capacities. The researchers have likely already had these abilities – as reflexive monitors our main task was to organize reflection and deliberation to help them advance. Indeed, the assumption was that, as the researchers' reflexive skills grew, the reflexive monitors would eventually become dispensable to the Natuurpact research project.

Although this case demonstrates how RMA encouraged researchers' reflexive capacities and contributed to a variety of responsible, societal effects, this is not to say that everything went as planned. For instance, there were times when the reflexive monitors were unsuccessful in ensuring time for reflection or encouraging a 'reflective mindset' of the researchers, especially when the project's workload intensified towards the end. Despite efforts to make their work integral to theirs and to encourage ownership, the authors' presence as monitors of the responsibility of the research process and outcomes at times rather seemed to encourage externalization and absolving of responsibility – precisely the opposite of what was intended. To address this issue, for the following Natuurpact research period (2018–2020) it was decided to formally record allocated time and resources to process monitoring within the project proposal. This has led to a formal sub-project in which researchers, together with reflexive monitors, have been appointed with the task of monitoring the responsibility of their own project. Although the results of this novel approach are yet to be assessed, it would be interesting to see if and what generalizable lessons can be drawn from this for RRI monitoring practices.

9.5 Discussion and conclusion

Recent years have seen much conceptual and practical work geared towards opening up research to society and aimed at redirecting science and innovation towards offering societally beneficial, ethically acceptable and environmentally sustainable outcomes. Under the label of RRI, an increasingly comprehensive framework is emerging to guide the governance and execution of responsible R&I. In this chapter we have shared our understanding of what such a framework should look like. Comprised of process dimensions that can be operationalized in the form criteria, sub-criteria and inviting questions, the framework is rooted in a philosophy that simultaneously underlines the importance of producing responsible R&I outcomes, as well as of a responsible research process towards these. In our view, the framework is most useful when adopted by actors who take part in R&I processes to carefully and systematically reflect on what *responsible* means in their given situation. In doing so, we emphasize, researchers and innovators learn to act in a more responsible manner. We have illustrated the potential role that a specific methodology for reflexive monitoring, RMA, can play to promote such learning. At the heart of this is RMA methodology's encouragement of the responsibility of actors involved with R&I, through seeking to increase these actors' reflexivity.

Most experience with RMA so far has primarily been geared towards supporting action-oriented, transdisciplinary projects that aspire to contribute to system innovations. In these contexts, RMA has become a proven methodology for reflexive evaluation that promotes learning and reflexivity through precisely those values that are also central to the process dimensions of RRI: reflection, anticipation, openness and responsiveness. It will be interesting to see whether RMA can make a similarly significant contribution to learning for responsibility in the context of other types of R&I. The outlooks for this are bright, given the close alignment of RMA with general features of RRI and the potential to use it simultaneously to assess the degree to which R&I projects or programmes can be seen as RRI, to contribute to learning in or on RRI, and, hence, to promote RRI in these projects or programmes.

Notes

1 https://ec.europa.eu/programmes/horizon2020/en/h2020-section/science-and-society
2 See https://epsrc.ukri.org/research/framework/area/.
3 Perhaps needless to add: such purposes will always remain up for discussion, and their concrete interpretation should be left to those taking responsibility for pursuing one purpose or the other.
4 This type of evaluation can be understood as part of the 'fourth-generation evaluation' paradigm proposed by Guba and Lincoln (1989).
5 See www.leidenmanifesto.org.
6 From now on, we will refer to the governmental level of the provinces as the regional level.
7 To this end, we involved colleagues who were not previously involved with the Natuurpact project to mitigate the risk of biased insights due to our investment in the project.

References

Acs, Z. J., de Groot, H. L., & Nijkamp, P. (Eds.). (2013). *The Emergence of the Knowledge Economy: A Regional Perspective*. Heidelberg, Germany: Springer Science & Business Media.

Adelle, C., & Weiland, S. (2012). Policy assessment: The state of the art. *Impact Assessment and Project Appraisal*, *30*(1), 25–33.

Arkesteijn, M., van Mierlo, B., & Leeuwis, C. (2015). The need for reflexive evaluation approaches in development cooperation. *Evaluation*, *21*(1), 99–115.

Blok, V., & Lemmens, P. (2015). The emerging concept of responsible innovation. Three reasons why it is questionable and calls for a radical transformation of the concept of innovation. In B. J. Koops, I. Oosterlaken, H. Romijn, T. Swierstra, & J. van den Hoven (Eds.), *Responsible Innovation 2* (pp. 19–35). Dordrecht, the Netherlands: Springer International Publishing.

Botha, N., Coutts, J., Turner, J. A., White, T., & Williams, T. (2017). Evaluating for learning and accountability in system innovation: Incorporating reflexivity in a logical framework. *Outlook on Agriculture*, *46*(2), 154–160.

Bosenman, M. F. (1988). Serendipity and scientific discovery. *The Journal of Creative Behavior*, *22*(2), 132–138.

Cummings, S., Regeer, B., de Haan, L., Zweekhorst, M., & Bunders, J. (2018). Critical discourse analysis of perspectives on knowledge and the knowledge society within the Sustainable Development Goals. *Development Policy Review*, *36*(6), 727–742.

Davis, M., & Laas, K. (2014). “Broader impacts” or “responsible research and innovation”? A comparison of two criteria for funding research in science and engineering. *Science and Engineering Ethics*, *20*(4), 963–983.

Dewey, J. (2012) [1927]. *The Public and its Problems: An Essay in Political Inquiry*. Pennsylvania: Penn State Press.

Ebrahim, A. (2005). Accountability myopia: Losing sight of organizational learning. *Nonprofit and Voluntary Sector Quarterly*, *34*(1), 56–87.

Elzen, B., Augustyn, A., Barbier, M., & van Mierlo, B. (2017). Agroecological transitions: changes and breakthroughs in the making. In *AgroEcological Transitions* (pp. 9–16). Wageningen University & Research. DOI: http://dx.doi.org/10.18174/407609

Gianni, R., Pearson, J., & Reber, B. (Eds.). (2018). *Responsible Research and Innovation: From Concepts to Practices*. Abingdon, UK: Routledge.

Grinbaum, A., & Groves, C. (2013). What is “responsible” about responsible innovation? Understanding the ethical issues. In R. Owen, J. Bessant, & M. Heintz (Eds.), *Responsible Innovation: Managing the Responsible Emergence of Science and Innovation in Society* (pp. 119–142). Chichester, UK: John Wiley.

Guba, E. G., & Lincoln, Y. S. (1989). *Fourth Generation Evaluation*. Newbury Park, CA: Sage Publications.

Guijt, I. (2010). Exploding the myth of incompatibility between accountability and learning. In J. Ubels, N. A. Acquaye-Baddoo, & A. Fowler (Eds.), *Capacity Development in Practice* (pp. 277–291). Abingdon, UK: Earthscan.

Heras, M., & Ruiz-Mallén, I. (2017). Responsible research and innovation indicators for science education assessment: How to measure the impact? *International Journal of Science Education*, *39*(18), 2482–2507.

Iatridis, K., & Schroeder, D. (2016). *Responsible Research and Innovation in Industry*. Doi: 10.1007/978-3-319-21693-5

Hicks, D., Wouters, P., Waltman, L., Rijcke, S. D., & Rafols, I. (2015). Bibliometrics: The Leiden Manifesto for research metrics. *Nature*, *520*(7548), 429.

Kawabe, K., Horiuchi, F., Ochi, M., Oka, Y., & Ueno, S. I. (2016). Internet addiction: Prevalence and relation with mental states in adolescents. *Psychiatry and Clinical Neurosciences*, *70*(9), 405–412.

Klaassen, P., Kupper, F., Vermeulen, S., Rijnen, M., Popa, E., & Broerse, J. (2017). The conceptualization of RRI: An iterative approach. In L. Asveld, R. van Dam-Mieras, T. Swierstra, S. Lavrijssen, K. Linse, & J. van den Hoven (Eds.), *Responsible Innovation 3* (pp. 69–92). Cham: Springer.

Klaassen, P., Rijnen, M., Vermeulen, S., & Kupper, F. (2018). Technocracy versus experimental learning in RRI: On making the most of RRI's interpretative flexibility. In R. Gianni, J. Pearson, & B. Reber (Eds.), *Responsible Research and Innovation* (pp. 91–112). Abingdon, UK: Routledge.

Klerkx, L., Aarts, N., & Leeuwis, C. (2010). Adaptive management in agricultural innovation systems: The interactions between innovation networks and their environment. *Agricultural Systems, 103*(6), 390-400.

Kunseler, E. M., & Vasileiadou, E. (2016). Practising environmental policy evaluation under co-existing evaluation imaginaries. *Evaluation*, 22(4), 451–469.

Kupper, J. F. H., Klaassen, P., Rijnen, M. C. J. A., Vermeulen, S., & Broerse, J. E. W. (2015). Report on the quality criteria of Good Practice Standards in RRI. Retrieved from: www. rri-tools. eu/documents/10184/107098 D, 1.

Lehtonen, M. (2014). Evaluating megaprojects: From the 'iron triangle' to network mapping. *Evaluation*, *20*(3), 278–295.

Lund Declaration. (2009). Conference: New Worlds – New Solutions. Research and Innovation as a Basis for Developing Europe in a Global Context. Lund, Sweden, 7–8 July 2009.

Nordmann, A. (2014). Responsible innovation, the art and craft of anticipation. *Journal of Responsible Innovation*, *1*(1), 87–98.

Oftedal, G. (2014). The role of philosophy of science in Responsible Research and Innovation (RRI): The case of nanomedicine. *Life Sciences, Society and Policy*, *10*(1), 5.

Owen, R., Macnaghten, P., & Stilgoe, J. (2012). Responsible research and innovation: From science in society to science for society, with society. *Science and Public Policy*, *39*(6), 751–760.

Owen, R., Stilgoe, J., Macnaghten, P., Gorman, M., Fisher, E., & Guston, D. (2013). A framework for responsible innovation. In R. Owen, J. Bessant, & M. Heintz (Eds.), *Responsible Innovation: Managing the Responsible Emergence of Science and Innovation in Society* (pp. 27–50). Chichester, UK: John Wiley.

Owens, S., Rayner, T., & Bina, O. (2004). New agendas for appraisal: Reflections on theory, practice, and research. *Environment and Planning A*, *36*(11), 1943–1959.

PBL & WUR (2017). *Lerende evaluatie van het Natuurpact. Naar nieuwe verbindingen tussen natuur, beleid en samenleving*. The Hague: PBL.

Peter, V., Maier, F., Mejlgaard, N., Bloch, C. W., Madsen, E. B., Griessler, E., Wuketich, M., Meijer, I., Woolley, R., Lindner, R., & Bührer, S. (2018). Monitoring the Evolution and Benefits of Responsible Research and Innovation in Europe: Summarising Insights from the MoRRI Project. Brussels: European Commission DG Research and Innovation.

Pyöriä, P. (2005). The concept of knowledge work revisited. *Journal of Knowledge Management*, *9*(3), 116–127.

Regeer, B. J., Hoes, A. C., van Amstel-van Saane, M., Caron-Flinterman, F. F., & Bunders, J. F. (2009). Six guiding principles for evaluating mode-2 strategies for sustainable development. *American Journal of Evaluation*, *30*(4), 515–537.

Regeer, B. J., de Wildt-Liesveld, R., van Mierlo, B., & Bunders, J. F. (2016). Exploring ways to reconcile accountability and learning in the evaluation of niche experiments. *Evaluation*, *22*(1), 6–28.

Ribeiro, B. E., Smith, R. D., & Millar, K. (2017). A mobilising concept? Unpacking academic representations of responsible research and innovation. *Science and Engineering Ethics*, *23*(1), 81–103.

Rip, A. (2014). The past and future of RRI. *Life Sciences, Society and Policy*, *10*(1), 17.

Sarkki, S., Heikkinen, H. I., & Karjalainen, T. P. (2013). Sensitivity in transdisciplinary projects: A case of reindeer management in Finland. *Land Use Policy*, *34*, 183–192.

Spaapen, J., Strand, R., Bauer, M. W., Hogan, E., Revuelta, G., Stagl, S., Paula, L., & Guimaraes Pereira, A. (2015). *Indicators for Promoting and Monitoring Responsible Research and Innovation – Report from the Expert Group on Policy Indicators for Responsible Research and Innovation*. Brussels: Directorate-General for Research & Innovation Science with and for Society.

Stilgoe, J., Owen, R., & Macnaghten, P. (2013). Developing a framework for responsible innovation. *Research Policy*, *42*(9), 1568–1580.

Swierstra, T., & Waelbers, K. (2012). Designing a good life: A matrix for the technological mediation of morality. *Science and Engineering Ethics*, *18*(1), 157–172.

Van Lente, H., Swierstra, T., & Joly, P. B. (2017). Responsible innovation as a critique of technology assessment. *Journal of Responsible Innovation*, *4*(2), 254–261.

Van Mierlo, B. C., Regeer, B., van Amstel, M., Arkesteijn, M. C. M., Beekman, V., Bunders, J. F. G., ... & Leeuwis, C. (2010). *Reflexive Monitoring in Action. A Guide for Monitoring System Innovation Projects*. Wageningen: Communication and Innovation Studies, WUR; Amsterdam: Athena Institute, VU.

Verwoerd, L., Van der Wildt-Liesveld, R., & Regeer, B. J. (2017). *The Value of Reflexive Evaluation: A Review of the Natuurpact Evaluation (2014-2017)*. Amsterdam: Athena Institute.

Walter, A. I., Helgenberger, S., Wiek, A., & Scholz, R. W. (2007). Measuring societal effects of transdisciplinary research projects: Design and application of an evaluation method. *Evaluation and Program Planning, 30*(4), 325-338.

Wickson, F., & Carew, A. L. (2014). Quality criteria and indicators for responsible research and innovation: Learning from transdisciplinarity. *Journal of Responsible Innovation, 1*(3), 254-273.

Wiek, A., Talwar, S., O'Shea, M., & Robinson, J. (2014). Toward a methodological scheme for capturing societal effects of participatory sustainability research. *Research Evaluation*, *23*(2), 117–132.

Zwart, H., Landeweerd, L., & van Rooij, A. (2014). Adapt or perish? Assessing the recent shift in the European research funding arena from 'ELSA'to 'RRI'. *Life Sciences, Society and Policy*, *10*(1), 11.

Best practice III Data-driven materiality analysis

Donato Calace and Adriana Farenga

A materiality assessment is a systematic analysis of societal needs and expectations, enabling the identification by a company of those issues that are crucial for both internal and external stakeholders, as well as gaps between business and society priorities. In this sense, a materiality analysis is a strategic compass informing Responsible Research and Innovation (RRI) assessments, ensuring that there is clear evidence that RRI efforts are oriented towards relevant and impactful (material) issues.

As of today, key stakeholders, regulators and investors are placing an increasing weight on the topic of materiality. This increased attention is causing a shift in practices around materiality, as a number of key debates and policy initiatives that are currently under way show.

One key example comes from the World Federation of Exchanges (WFE). In its *ESG Guidance and Metrics* (June 2018), the WFE includes a section on materiality, stating that investors wish to understand how companies determine which issues are material (process, timeframe, relevant legal framework, etc.).

Another major development comes from the European Union Commission, which is currently reviewing the Non-Financial Reporting Directive. As part of this update, the Commission has introduced a new definition of materiality (the *double materiality*) while asking for more disclosures on the materiality determination process, with particular reference to the methodology chosen.

Finally, in September 2019 the US Financial Services Committee voted to pass the ESG Disclosure Simplification Act, which will require public companies to disclose information on their environmental, social and corporate governance (ESG) practices, including a description of any process used to determine the impact of ESG metrics on the company's long-term business strategy. In this way, the Act recognizes officially that ESG metrics are material for the purpose of disclosure, so it is crucial to measure them in the most accurate way possible.

In light of all these policy developments, one question that organizations need to consider is: are the current materiality assessment processes adequate? What characteristics should a robust materiality analysis feature?

Modern materiality needs to be *flexible, time-variant*, and *context-dependent*. Consequently, the only defence against subjective, manipulable and self-serving materiality is to ensure that the materiality assessment is accorded with a robust due process – and regularly updated, as emerging issues can rapidly change what is material.

Identifying and monitoring material risk and opportunities are key to business strategy. An inadequate or uninformed assessment approach exposes entities to reputational and legal risks, and this risk is extended along the whole value chain. However, the standards with which companies are required to comply are increasing exponentially, and while frameworks and regulations are asking for a robust process, they do it without describing how such a process should be structured.

The lack of a definition of what a robust materiality process is leads to an even deeper limitation of current materiality assessments, being that the purpose of the analysis is predominantly restricted to reporting only. Currently, sustainability and communications teams are in charge of the materiality and reporting processes – with little or no involvement from other key departments, such as strategy, governance and risk management (Calace, in press). As indicated by Eccles and Krzus (2014), in implementing materiality analyses, companies are expected to choose the issues that are most material in terms of their sustainability implications and then act accordingly. If an issue is identified as material, it should be included and explained broadly in the sustainability report and in the company's allocation of resources and efforts (Khan et al., 2016). In other words, a materiality assessment should go beyond the sole purpose of identifying issues that should be included in the non-financial report.

Scholars and experts are calling for the evolution of quantitative methods in materiality decision making (Calabrese et al., 2015; Edgley et al., 2015), currently relying mostly on anecdotal evidence from stakeholder workshops and roundtables, or from surveys. This broadening of the purpose becomes particularly important in light of the recent development of materiality analysis as the input for risk management processes (World Business Council for Sustainable Development (WBCSD) and Committee of Sponsoring Organizations of the Treadway Commission (COSO), 2018).

Such a debate around the materiality process leads by default to the question of how often companies should perform a materiality assessment.

All evidence leads to the fact that companies should carry out a continuous data-driven materiality analysis, no longer confined in the reporting area but thoroughly embedded in risk management processes. Even more: under every aspect, materiality is in its own essence a risk management process. The report is a simple outcome of this process. The sooner an organization understands this, the sooner it will be able to future-proof its strategy and stay ahead of the competition.

In light of these considerations, the contribution of a data-driven and continuous materiality determination process to RRI appears crucial: without a credible materiality assessment as foundation, it would be difficult to assess to what extent an RRI initiative is an ancillary pet project or truly integrated in the business' core strategy.

Aims and goals

There are vast amounts of unstructured data publicly available representing what matters to stakeholders, and quantifying is crucial to achieve a robust materiality. Datamaran's approach leverages Natural Language Processing (NLP) to quantify narrative disclosures in annual financial reports, sustainability reports, Securities and Exchange Commission filings, mandatory and voluntary regulatory documents, social media and online news.

Benefit for organization

Using Datamaran's approach companies can stay in charge of their materiality and risk processes, identifying trends that are concealed in unstructured data (Datamaran, 2019). The insights gained will enable users to build a credible and auditable process to identify and monitor material issues, enabling them to focus on the implications of the issues identified and strategize leveraging a data-driven approach.

References

Calabrese A., Costa R., Levialdi N. and Menichini T. (2016). A fuzzy Analytic Hierarchy Process method to support materiality assessment in sustainability reporting, *Journal of Cleaner Production*, doi: 10.1016/j.jclepro.2015.12.005

Calace, D. (in press) Materiality: From Accounting to Sustainability and the SDGs. In: Leal Filho W., Azul, A. M., Brandli, L., Özuyar, P., Wall, T. (eds) *Responsible Consumption and Production. Encyclopedia of the UN Sustainable Development Goals*. Cham: Springer.

Datamaran (2019). Materiality Definition: The Ultimate Guide. Retrieved from www.datamaran.com/materiality-definition/

Eccles, R. G., and Krzus, M. P. (2014). *The Integrated Reporting Movement: Meaning, Momentum, Motives, and Materiality*. Hoboken, NJ: John Wiley.

Edgley, C., Jones, M. J. and Atkins, J. (2015). The Adoption of the Materiality Concept in Social and Environmental Reporting Assurance: A Field Study Approach. *The British Accounting Review*, 47(1): 1–18.

Khan, M., Serafeim, G. and Yoon, A. (2016). Corporate Sustainability: First Evidence on Materiality. *The Accounting Review*, 91(6), 1697–1724.

World Business Council for Sustainable Development (WBCSD) and Committee of Sponsoring Organizations of the Treadway Commission (COSO) (2018). Enterprise Risk Management: Applying Enterprise Risk Management to Environmental, Social, and Governance-related Risks.

World Federation of Exchanges (WFE) (2018). ESG Guidance and Metrics. The WFE Research Team. Retrieved from www.world-exchanges.org/our-work/articles/wfe-esg-revised-metrics-june-2018

10 A future-oriented evaluation and development model for Responsible Research and Innovation

Mika Nieminen and Veikko Ikonen

10.1 Introduction

Recently there has been considerable discussion about Responsible Research and Innovation (RRI) (e.g. von Schomberg, 2011; Owen, Bessant, & Heintz, 2013; Gianni, 2016; Yaghmaei, 2018) as a continuation of the various debates on risk anticipation and the ethics of technological development (e.g. technology assessment and ethical, legal and social aspects studies). Consequently, a number of attempts have been made to operationalize the concept of RRI, for instance in the design of research projects and in the implementation of innovations. Most notably, in a number of European Union (EU) projects specific "toolboxes" for applying RRI have been developed (e.g. RRI tools[1]), ideas of RRI governance have been studied (e.g. EGAIS[2] and ResAgora[3]), and evaluation frameworks for RRI have been created (e.g. MoRRI[4] and Satori[5]).

However, such RRI evaluation frameworks typically have a relatively general focus on, for instance, ethics assessment practices, with less attention paid to contextual embeddedness and organizational factors. In addition, it seems that there is hardly any explicit connection to evaluation theory. In the following discussion, we aim to contribute to RRI evaluation by paying specific attention to these elements. The discussion functions as a starting point for our developed RRI evaluation model, providing it with a more solid and contextualized basis.

Our general starting point is that the development of responsibility in research and innovation depends on the complex interactions between various actors and on organizational dimensions and their dynamic co-development with the organization's external environment. Due to this complexity, organizational change is not a fully manageable and controllable process in the organization, contrary to still relatively commonly held beliefs (e.g. Tsoukas & Chia, 2002). This view can also be called systemic, including the idea that various dimensions of an organization, such as political, cultural and structural processes, are intimately intertwined and affect each other directly and indirectly via various feedback loops, forming a complex system (e.g. Cao, Clarke & Lehaney, 2003). Concurrently, societies have also become profoundly complex and interconnected due to changing institutional structures, globalization, rapidly developing technologies and the intertwining of human-made and natural systems (e.g. Urry, 2005; Helbing, 2013).

Recently, this complexity and the systemic nature of our organizations and societies have gained the attention of a growing number of evaluation experts and scholars (e.g. Forss, Marra, & Schwartz, 2011; Williams & Hummelbrunner, 2011; Mowles, 2014; Nieminen & Hyytinen, 2015). The central idea is that input–output–outcome linear thinking, which is manifest in many traditional approaches to evaluation (e.g. Kellogg Foundation, 2004; Wholey, Hatry, & Newcomer, 2010), does not match well with the complexity of processes and the multiple relationships between actors. Therefore, many system thinking-based approaches emphasize participative, developmental and process-based evaluation approaches that better capture this complexity (e.g. Patton, 2011; Williams & Hummelbrunner, 2011), the core idea being the development of something concrete (instead of summative, outcome-oriented approaches) and active stakeholder and citizen engagement in the process. At the same time, attention has been paid to the necessity of including anticipatory elements in evaluation studies: fast societal development may make ex post evaluations less useful, as recommendations based on past operations may turn out to be outdated, especially in rapidly developing societal and industrial sectors (e.g. Nieminen & Hyytinen, 2015; Hyytinen, 2017). Participatory approaches, anticipation of future impacts and reflexivity are also the basic dimensions of the concept of responsible innovation (Owen et al., 2013; Stilgoe, Owen, & Macnaghten, 2013), although the emphasis has been on the assessment of social acceptability and desirability in the context of potential risks.

Using these approaches as a starting point, in this chapter we describe a concrete evaluation approach for integrating RRI into the development principles and procedures of innovation activity. The model was developed in 2018 for an urban development project called 6Aika[6] (a joint development and strategy project of Finland's six largest cities) with the specific aim of developing innovation and business platforms. The developers of the 6Aika project wanted to integrate RRI thinking into the platform development by extending the range of societal impacts beyond economic impacts towards wider societal, cultural and environmental impacts. The model was developed in close collaboration with the developers of an evaluation framework for the open innovation platforms, and further discussed in an international workshop of innovation platform developers participating in the MARIE (MAinstreaming Responsible Innovation in European S3) project. The model will be utilized in future Six City projects and in the development of open innovation platforms. The aim of this chapter is thus to introduce a practical, yet theoretically argued, ex ante evaluation approach to embed responsibility thinking into innovation platforms.

The chapter is structured as follows. In Subsection 10.2.1, we introduce the basic theoretical assumptions behind our approach. In Subsection 10.2.2 we then briefly review the evaluation approaches that take systemic complexity as their starting point and compare these views with the presented RRI evaluation approaches. Based on this, we argue for a need for a systemic and future-oriented evaluation approach for developing and evaluating responsibility-related practices in organizations. In Section 10.3 we introduce our evaluation model, and in the final section we summarize our arguments.

10.2 Theoretical background

10.2.1 Complexity and heterogeneity

A major driver of increasing complexity during the last few decades has been globalization. Although economic globalization began as far back in history as the Middle Ages, recent developments in ICT have made global networks more intensive, intensified economic globalization and coupled people, their organizations and economic structures more tightly with each other across the world (e.g. Castells, 1996; Friedman, 2005). Simultaneously, technological developments have taken place at an increased pace (e.g. Kurzweil, 1999), created huge and complex technological systems (Hughes, 1983) and have become profoundly intertwined with social, environmental and economic developments (e.g. Freeman & Louca, 2002). Business is taking place in global "value chains" and "ecosystems", to use two concepts that describe the complex interdependencies of various business-related actors (e.g. Moore, 1996; Autio & Thomas, 2013). It has also been shown that changes in one part of the global system can crucially affect the performance of other systems and even produce crucial systemic risks. For example, as global energy and food systems, communication systems and production and natural systems become increasingly dependent on each other, the vulnerability of the whole is also increased (Helbing, 2013).

Wide systemic interconnectedness increases complexity, which in turn increases volatility and unanticipated changes across systems as different parts are indirectly coupled with each other (Holland, 1995; Lawrence, 2013). This further highlights the challenge of ambiguity: there may be various legitimate starting points and perspectives for interpreting and evaluating on-going developments, from techno-economic concerns to interpersonal relationships and social and ethical concerns (cf. Rittel & Webber, 1973). Due to the multi-dimensionality of socio-economic and environmental challenges, it is difficult to make decisions based on only certain views or interests. Solving such challenges, whether on a global or national scale (e.g. climate change, migration or unemployment) or on an organizational level (e.g. strategic decisions on product and service development and provision) requires the engagement of various actors in the problem-solving process. It is also possible that there is no definite solution to these challenges – as situations and actors change we therefore need to constantly create new, specific solutions that fit the situation (Eoyang & Holladay, 2013).

The development and implementation of successful innovations have thus become an increasingly complex endeavour. Innovations are a result of the interaction of various actors and contextual factors in an innovation environment, including businesses, knowledge-producing organizations and the wider institutional context in society (e.g. Damanpour, 1996; Geels & Schot, 2007; Borras & Edler, 2014). These actors and their various relationships (e.g. collaboration, transactions, competition, steering) form a system that influences the possibilities for innovation activity (e.g. Smith, 2000; Edquist, 2005). While the concept of "innovation system" emphasizes the heterogeneity of actors and the systemic constellations they may form, recently the concept of innovation ecosystem has also emphasized the mutual dependencies of various innovation- and business-related actors from companies to universities and research institutes (e.g. Moore, 1996; Autio & Thomas, 2013).

Interconnectedness also becomes visible in so-called systemic innovations. A systemic innovation is an innovation that is dependent on changes in other technologies or social subsystems (such as markets or regulation) (Teece, 1988). Innovation may also emerge through a combination of various technologies, services or social innovations, which becomes possible only when various actors and system dimensions are in interaction with each other. We can take the mobile phone as an example. Their success has been dependent on the development and combination of various technologies (e.g. signal processing, battery technology) and services (e.g. applications, music and news services, social media) which together have made possible radical and systemic changes in communication and social activity. Such innovations are "systemic" by nature in the sense that their implementation requires changes in the existing technological system and markets as well as in the institutional context (e.g. regulation) and in organizations (e.g. how work is organized when the new technology is implemented) (e.g. Geels & Schot, 2007; Hyytinen 2017). Therefore, the implementation of an innovation may require diverse communication, shared target setting and joint activities. This also implies the idea of open innovation (Chesbrough, Vanhaverbeke, & West, 2006; cf. von Hippel, 1988) and co-creation to increase the social effectiveness and business impact of innovation.

Co-creation relies especially on the collective creativity of all stakeholders and end-users in innovation processes (Prahalad & Ramaswamy, 2000; Sanders & Stappers, 2008). User participation is especially useful in the early stages of innovation due to the high level of uncertainty and low formalization (Alam, 2006). Whereas in traditional human-centred design only a small number of users are involved in design activities, co-creation suggests wider stakeholder and citizen inclusion, emphasizing the creative nature of participation, seeking new solutions, concepts and ideas, as well as equal and open interaction between different actors (Ainasoja et al., 2012; Kaasinen et al., 2010).

The need to open up innovation processes to various stakeholders and citizens is further emphasized by the normative and relative nature of social impacts. For instance, the social, environmental and economic impacts of an innovation (e.g. economic and environmental impacts of mining or transportation-related innovations) may be in contradiction and need to be negotiated. Responding to them necessitates opening up innovation processes and connecting them to solving problems that people consider important (e.g. Schot & Steinmuller, 2018). Complexity also emphasizes the need for careful anticipation, as there may be systemic impacts that are otherwise difficult to foresee. Engagement of stakeholders increases, in turn, possibilities for reflection on potential future impacts from various perspectives (cf. Owen et al., 2013).

10.2.2 Towards systems thinking in evaluation

As discussed in the previous section, our societies and organizations have become increasingly complex and rapidly changing (e.g. Urry, 2005) and traditional evaluation approaches no longer function well for two main reasons. Firstly, they tend to over-simplify the complexity of impact paths. Typically, a range of factors and actors contribute directly and indirectly to observed results or impacts, yet this has been taken insufficiently into account. Secondly, many evaluations are backward-looking (ex post evaluation), whereas social and economic changes take place rapidly and instead necessitate forward-looking and anticipatory approaches (Nieminen & Hyytinen, 2015; Hyytinen, 2017).

Traditional evaluations are also predominantly based on rather straightforward linear, causal thinking. It is assumed that impacts are produced in a linear manner following the chain inputs–process–outputs–impacts. Evaluations normally also include contextual factors that may affect the delivery of the planned outcomes and impacts. The activity then needs to be adapted to take into account these contextual factors (e.g. Dyehouse, Bennett, Harbor, Childress, & Dark, 2009; Gertler, Martinez, Premand, Rawlings, & Vermeersch, 2011). Mainstream evaluations seem to roughly follow this model in one way or another (Patton, 2011). There are also various sophisticated variations on this theme including, for instance, "contribution analysis". This puts forward the idea that the outcomes of an activity may be attributed also to factors other than the activity being evaluated (which is usually some kind of intervention, programme, etc.). The basic idea is to ask whether "in light of the multiple factors influencing a result, has the intervention made a noticeable contribution to an observed result and in what way" (Mayne, 2012, p. 273). Contribution analysis usually also postulates a theory of change that includes the assumed causal linkages yielding the sought-after targets.

Currently, there are various attempts to evaluate the advancement of RRI. In the following, some recent EU-funded RRI projects representing the state of the art in the field are used as examples.

Some projects explicitly follow the linear model, albeit recognizing its problems. For instance, the researchers of an EU-funded project (ended in 2018) Monitoring the Evolution and Benefits of Responsible Research and Innovation (MoRRI) explicitly state in the final report that "the indicators developed to monitor the emerging patterns of RRI at Member State level (Section 2) were based on a relatively conventional intervention logic: inputs → outputs → outcomes → impacts" (MoRRI, 2018, p. 11). While this concerns macro-level impact monitoring, the researchers also collected survey and case-based data on the impacts of RRI and moved beyond straightforward linear thinking towards contribution analysis. They also recognize the problems of attributing observed impacts to RRI:

> The problem of attribution of very general effects to RRI processes and outcomes is a significant challenge for developing a monitoring system for RRI benefits. Lengthy timelags can occur between observed changes in the R&I [research and innovation] system, which might be monitored through indicators of RRI outcomes, and flow-on or emergent benefits to society at large. ... This is [the] measurement theory challenge in the field of indicator development generally.
>
> (ibid., 22)

While MoRRI can be seen as representing the so-called summative evaluation approaches, another recent EU-funded RRI project (ended in 2017), Stakeholders Acting Together on the Ethical Impact Assessment of Research and Innovation (SATORI), represents a different approach to evaluation by trying to create a formative procedure or framework for ethical assessment. In essence, the framework is an ex ante evaluation approach that also utilizes foresight studies to describe the ethical impacts of research and innovation (Reijers, Brey, Jansen, Rodrigues, Koivisto, & Tuominen, 2018). The proposed framework also suggests wide engagement of stakeholders in the evaluation. The actual framework consists of six consecutive steps, each of which includes various actions, and each starting with an analysis of the need for ethical assessment and ending with a review process aimed at giving feedback to iteratively develop the process (ibid.). The SATORI approach aims at developing standardized practices in organizations performing and funding research and, in doing so, assumes rather direct causal linkages between improved ethical performance in an organization and its social impacts.

In the EU project Responsible Industry (ended in 2017), an evaluation model was created to foster RRI practices in firms. The so-called RRI maturity model was developed in collaboration with five companies with the aim of helping firms to understand and reflect on their current practices and benchmark practices in the industry.[7] In essence, the maturity model identifies categories and components of RRI that could serve as building blocks for further development of RRI. The RRI maturity model:

> is a tool that generally overviews and provides an empirically tested basis for drawing practical conclusions that can inform organizational practice. Its theoretical contribution to the discussion of RRI is thus twofold: It firstly allows for a more comprehensive view of RRI that combines existing accounts and it secondly allows moving beyond the one-dimensional view of RRI to a multi-dimensional account that paves the way for practical interventions.
>
> (Stahl, Obach, Yaghmaei, Ikonen, Chatfield, & Brem, 2017)

Here "more comprehensive view" is understood to refer to the fact that, as RRI was originally developed for the EU research and innovation context, changing the context to company research and innovation calls for a different terminology, approach and deployment. For example, instead of open science, the concept of open innovation can be more relevant to companies while at the same time they may have a need for intellectual property rights and knowledge protection.

This kind of attempt to develop practices by using an inclusive and collaborative evaluation approach comes close to the so-called developmental evaluation approach. Developmental evaluation aims at developing an organizational practice or an innovation. One of the developers of the approach, Michael Patton (2011), considers it as a response especially to the challenges emerging from societal complexity. Thus, according to Patton (ibid. 7), "Developmental evaluation … is an approach to evaluation especially appropriate for situations of high uncertainty where what may and does emerge is relatively unpredictable and uncontrollable". Developmental evaluation does not rely on any specific methodology, but can use various methods to support development. While developmental evaluation does also recognize causality, for it, causality is the "patterns" embedded in "complex systems dynamics and interdependencies" (ibid., p. 24).

Patton's approach reflects the growing dissatisfaction with linear approaches during recent years and increasing interest towards systems thinking in evaluation – even though the linear approaches still form the mainstream (e.g. Cabrera, Colosi, & Lobdell, 2008; Dyehouse et al., 2009). What is common to these various approaches is that they are trying to pay attention to the interrelationships among actors, between the whole and its parts, and between the action and its context (Cabrera et al., 2008; Nieminen & Hyytinen, 2015). The roots of this system thinking can be found in research fields like general systems theory, system dynamics, and complexity science. In practice, the system approach has been applied as an analytical perspective and as a multi-method approach (Williams & Imam, 2006; Dyehouse et al., 2009; Nieminen & Hyytinen, 2015).

Hyytinen (2017) has suggested an evaluation approach based on three perspectives: "futures view, systems view and multi-actor view". A futures view is linked to the observation that, while our environment is increasingly complex, it is also changing rapidly. Recommendations based on traditional ex post evaluations can easily prove out of date. Moreover, innovation is always future-oriented, as the impacts of innovations take place in the future. A systems view emphasizes the need to take into account complex direct and indirect interlinkages between different factors and actors in the activities producing social impacts, and emphasizes that this should be reflected in the approach to and methodologies for evaluation. A multi-actor view pays specific attention to various perspectives, interests and motivational factors "that make innovations happen, spread and gain foothold". As a methodological solution Hyytinen suggests a combination of evaluation, foresight, system dynamic modelling and societal embedding. In this ensemble "evaluation provides information on the past and current state of the system, its structures, operations and historical development" and foresight "yields information about future transformations in the system and endorses the setting of long-term targets". Furthermore, system dynamic modelling provides "analysis of interdependencies and feedback loops among the actors and their environment", and societal embedding "an approach to initiate the dialogue among various stakeholders and key actors who set conditions for social development and for diffusion of social or technological innovations" (Nieminen & Hyytinen, 2015, pp. 453–454).

Based on the ideas of Hyytinen (2017) and of Nieminen and Hyytinen (2015), in the following we suggest a practical organizational approach for future-oriented and multi-perspective evaluation of responsibility. The general aim of the model is to be "developmental" in the sense discussed above, not summative (i.e. the focus is not on outcomes but on processes and the development of actions).

The model was originally developed for the 6Aika innovation platforms.[8] 6Aika is a joint strategy for sustainable urban development of the six largest cities in Finland that aims to tackle the challenges of urbanization and to create new business, know-how and jobs in Finland. This Six City strategy has three large-scale spearhead projects: Open Data and Interfaces, Open Participation and Customership and Open Innovation Platforms. The projects embody the smart city model defined in the strategy: customer-cantered co-creation, opening and utilizing data and developing services in real urban environments. The cities have opened and are opening up real and virtual spaces and turning them into innovation platforms: schools, shopping centres, hospitals, nursing homes, streets and neighbourhoods as innovation platforms and co-creation facilities where new products and services are developed and tested. The aim of the model that we planned for

6Aika was to include responsibility-related elements as important design and evaluation criteria for platform design.

10.3 Future-oriented evaluation and development model for Responsible Research and Innovation

Although various slightly different definitions of RRI have been proposed (e.g. Sutcliffe, 2011; von Schomberg, 2011; Owen et al., 2013), they each share a number of common RRI characteristics, such as a focus on social challenges, engagement of stakeholders, opening up of research and innovation to society and risk avoidance (Gurzawska, Mäkinen, & Brey, 2017; Smallman, 2018). For instance, Owen and his colleagues (2013, p. 38) suggest four basic dimensions of responsible innovation which include the above-mentioned elements. *Anticipation* is a necessary part of the analysis of the social, economic and environmental impacts of innovation activity in a rapidly changing society. *Reflexivity* is an essential part of every actor's operations. Each actor should consider their own underlying motivations and purposes for participating in the innovation activity and ponder them openly together with the other actors. *Inclusiveness* brings into the common discussion various stakeholder and citizen interests, values and perspectives. Finally, the process needs to be *responsive*, leading to learning and changing of target-setting and operative practices. These basic principles of responsible innovation also fit well with our understanding of the increasingly complex and systemic social context, in which research and innovation activities are embedded. Here, we use the definition suggested by Stilgoe and colleagues (2013, p. 1570), according to which RRI aims at "taking care of the future through collective stewardship of science and innovation in the present". The definition is broad enough to be used in various kinds of innovation-related activities and captures well the general (normative) idea of RRI.

Figure 10.1 depicts the basic ideas of our model. As explained earlier, the model incorporates the ideas of a future-oriented, systemic and multi-actor approach, as well as developmental evaluation. The principles of responsibility are intertwined in the step-wise process from the very beginning and define the targets of successive actions. The dominating character of the process is co-creation, which essentially defines the values and operational goals of the innovation activity. At best, the process forms a loop of continuous learning in which participating organizations and persons learn from each other during the evaluation process and reflexively change their activity. Evaluation is not used only for the anticipation of potential impacts, but also for following the results and impacts of the development process. The evaluation of responsibility should be an integral part of "business-as-usual" everyday planning and management in an organization (cf. Reijers at al., 2018) and our model may help by suggesting practical steps towards achieving this. The evaluation and learning loop should ideally be embedded also in the innovation process, which is depicted in the figure as different phases by adapting Kline and Rosenberg's (1986) classical analysis of the innovation process in firms.

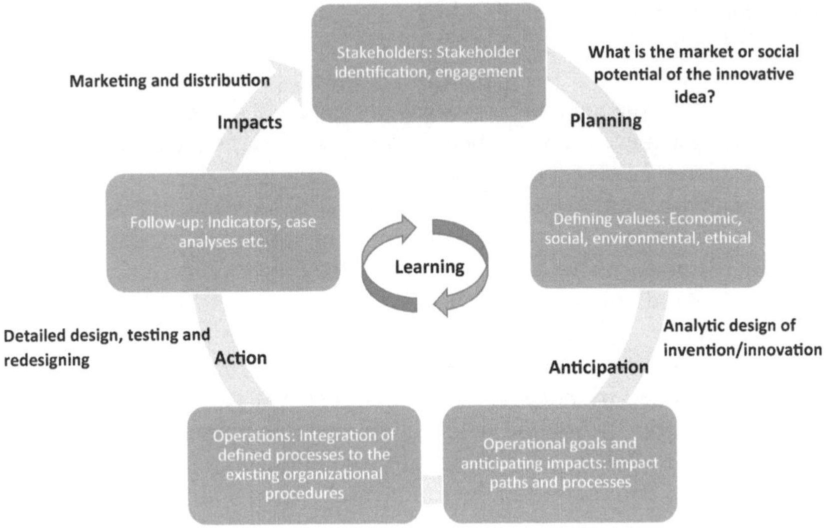

Figure 10.1 Future-oriented Responsible Research and Innovation (RRI) evaluation.

Besides defining the joint targets and shared values for the innovation activity, the goal of the process is also that the participating organizations and people change their own behaviour so that the joint targets are possible to achieve. Broad societal goals or socio-technical innovations are especially systemic by nature and various societal dimensions, organizations and people affect the realization of these targets (e.g. Geels & Schot, 2007). Achieving these goals or implementing innovations necessitates the creation of shared visions, collaboration between and networks of many organizations and people, learning and operational adjustments (e.g. Rotmans & Loorbach, 2009). For instance, utilizing ICT in health care is not only about implementing new technological instruments, but necessitates wide changes in the health care organization, including, for example, work practices, management and customer relationships. In addition, changes are needed in public policy as well as in customers' attitudes towards new services (e.g. Kivisaari, Väyrynen, & Saranummi, 2004). This emphasizes the developmental evaluation approach with its idea of evaluation as an inclusive process in which participants co-develop new socio-technical innovations and ways to implement them.

10.3.1 Stakeholder identification and engagement

In a complex social environment it is important that all relevant stakeholder groups are involved from the very beginning of the process. These participate in defining the targets of innovation activity and related elements of responsibility and values as well as in anticipating the impacts and potential risks. The process connects the values of stakeholder groups to the innovation process, brings out the positive and negative aspects of the innovation activity and increases the acceptability and desirability of innovations (cf. Leikas, 2009; Saariluoma, Cañas, & Leikas, 2016). The stakeholder groups should also be involved in ex post or mid-term evaluation, in which it is assessed whether the defined targets have been reached and, if not, why they have not been achieved. Engagement in these discussions should be a mutual learning experience for all of the participants.

Deciding who is and who is not a stakeholder is, however, not an easy task, which is reflected, for example, in the different stakeholder definitions and analytical frameworks for stakeholder analysis. For instance, a definition by Freeman (1984, p. 46) states that a stakeholder is "any group or individual who can affect or is affected by the achievement of the organization's objectives". This definition is relatively broad compared, for instance, with the definition suggested by Eden and Ackermann (1998, p. 117), who consider stakeholders to be "people or small groups with the power to respond to, negotiate with, and change the strategic future of the organization".[9] For the purposes of RRI and social and environmental responsibility, even the definition by Freeman may be too narrow. When considering the future impacts and consequences of an innovation, we should also take into account those *potentially* affected by the organization or the actions of groups of organizations. These additions would take into account the need for systematic anticipation of the long-term and less obvious impacts of the innovation, as well as the fact that the innovation does not take place in isolation. Organizations are often dependent on other organizations in their innovation activity (e.g. Adner, 2012), which means that organizations involved in collaborative innovation activity should view RRI also as a collective action with other involved firms. Impacts of the innovation may be also global, which makes the definition of "stakeholder" even more challenging. This fuzziness of the concept also emphasizes the fact that from the system theoretical point of view "stakeholder" may be a too restrictive concept, as it emphasizes a central, definable organization and its operational environment. In a systemic and complex environment even the concept of organization may become blurred, and the relationships with the environment and its various actors are difficult to define. While this question remains unanswered here, we take a practical orientation to this dilemma: a systematic approach to thinking about "stakeholders" is better than ad hoc attempts to identify potential groups of people or actors.

Systematic stakeholder-mapping techniques and analysis that support the identification of stakeholders, analysis of their interests and engagement in the process help. There are various variants of stakeholder analysis (e.g. Mitchell, Agle, & Wood, 1997; Mayers & Vermuelen, 2005) and these can be used for different purposes, from analysing stakeholder power resources to inequalities. Reed et al. (2009) have suggested that there are three main variants of stakeholder analysis, including the descriptive (empirically describes stakeholders and their characteristics), normative (aiming to engage stakeholders to reach a shared understanding and legitimization of action) and instrumental approach (aiming to understand how organizations can identify and manage stakeholders to reach their targets). These approaches, in turn, may use various methods to identify stakeholders, categorize them and investigate stakeholder relations. We are unable to discuss these methods here in detail due to length restrictions, but in the following we provide some examples of useful methods. For identifying stakeholders: semi-structured interviews and snowball sampling (initially identified stakeholders identify new stakeholder categories and contacts; these are then asked to identify again more stakeholders and so forth); for stakeholder categorization: interest–influence matrices (stakeholders are placed in a matrix according to their interest and influence; this can be done by stakeholders themselves, e.g. in a workshop) and stakeholder-led stakeholder categorizations (as in the previous method); and for stakeholder relationships analysis: actor–linkage matrices (analysis of linkages in a two-dimensional matrix) and social network analysis (structured analysis of relationships) (ibid.).

In the context of RRI, stakeholder analysis should function as part of a systematic attempt towards more open innovation activity, sharing of information and increasing transparency of the innovation and various actors' roles in the process. Systematic and successful stakeholder analysis may help to identify the relevant actors, ponder their potential roles and integrate them as part of the innovation process or development platform. Unsuccessful identification and engagement of stakeholders may cause challenges in the implementation of innovation (its acceptance and desirability). An example might be the inclusion of citizens in the platform development. The perhaps seemingly simple question of stakeholder identification can turn out to be rather complex when considering potential social groups (e.g. age, gender, religious and ethnic identity), their internal diversity (including e.g. different political opinions) and suitable organizations, associations or persons to represent these groups. In such cases, stakeholder analysis can help to identify relevant and important actors and bring to the fore the diversity of values of different stakeholder groups.

10.3.2 Defining values, operational goals and impacts

As we know, innovation processes are always contextual (e.g. Ortt & van der Duin, 2008), which refers to the fact that implementation challenges and relevant questions vary depending on, for example, the organization, societal sector or industrial branch, the technology or service and the use of the innovation. Following from this, social and ethical questions that are important to biotechnology may differ from those that are important to ICT (cf. Luppicini, 2010). This contextuality emphasizes the importance of the engagement of relevant and concerned parties and their values in the innovation process. Values define the social acceptance and desirability of the innovation. In this context, values can be defined "as something that a person or a group of people consider important in life" (Friedman, Kahn, & Borning, 2006, p. 349, as cited in Simon, 2016, p. 221). This kind of value-sensitivity should be an integral part of the responsible design of technologies, as it enhances successful and sustainable uptake of innovation – even though it means both critical analysis of the values inscribed in the technology development as well as intentional integration of the desired values into the technology development (Simon, 2016).

Desirability aspects may be especially challenging for innovation actors who are accustomed to operating based on acceptability, which is often understood narrowly to cover, for example, technological usability and improvement aspects, instead of asking whether the innovation or technology is actually needed and serves real existing needs in people's lives (Leikas, 2009; Saariluoma et al., 2016).

This value sensitivity can be easily criticized for being overly relativistic. Because values are anchored in people´s life experiences, they are all equally valuable and negotiable. However, this does not hold true for generally accepted values and norms, such as human rights and existing legislation. There are certain values that can be considered generally important for society even though contextuality and desirability may relativize their value and importance. An example of integrating these two perspectives is provided by a recent EU-funded project "Responsible Industry" (2017), the values and core questions of which were defined with a view to advancing the implementation of RRI. The suggested general values included individuals' rights and liberties, personal safety and health, autonomy, quality of life, integration and dignity, and justice. These general values were then integrated into the innovation process by reflecting them on to the more relative values of societal and ethical impacts.

Values are also tightly anchored to the future promises and visions provided by innovations and new technologies. Innovations always have future consequences for us, which should raise the question of whether we value those consequences, whether they are in accordance with our values. It is equally important to anticipate the potential risks and indirect and unintended impacts of innovation activity. These potential negative impacts also have to be compared with and weighed against the positive impacts. How big may acceptable negative impacts be compared to positive ones? As there is no clear-cut answer to this question, it is sensible to discuss it with stakeholders and citizens. Discussion helps to piece together various situational values, ethical questions and responsibilities. Discussion may also help to find new, more acceptable solutions and create compromises and shared visions of desirable or acceptable futures.

These goals can be supported by systematic foresight methods (e.g. Georghiou, Cassingena Harper, Keenan, Miles, & Popper, 2008). At its best, the foresight process integrates different values and interests into the innovation process by asking: what kind of innovation do we want and accept? In addition, it is possible to create alternative scenarios to describe possible futures, and define impact and vision paths or roadmaps to assist goal attainment. A roadmap or an impact path can also include various concrete mid-term targets and operations that help to achieve the shared vision (Auvinen, Ruutu, Tuominen, Ahlqvist, & Oksanen, 2014). In addition, various well-known approaches are available for assessing the impacts of technology development that support the objectives of making various values visible, producing needed information and reaching consensus. These include, for instance, technology assessment and its various variants (Schot & Rip, 1997), combination of risk analysis and technology assessment (van Wezel et al., 2018), ethical, legal and social aspects assessment in life sciences and various impact assessment approaches from environmental to health impact assessments (João, Vanclay, & den Broeder, 2011).

10.3.3 Integration into the organization(s)

RRI should be integrated into the operations of the implementing organization(s) by asking "How can RRI be integrated throughout the value chain and design process?" and "Can foresight and ethical and social impact analysis be performed in the context of the existing organizational procedures and practices?"

This step includes at least two interconnected aspects. Firstly, how to translate the co-created and negotiated visions, values and action steps into concrete operations and, secondly, how to integrate responsible ways of acting into the daily routines of the organization.

It has been suggested that RRI evaluation and anticipation should be integrated into existing organizational practices and action models, such as into the existing corporate social responsibility (CSR) functions (Reijers et al., 2018). In this case, the more traditional follow-up of environmental and social impacts of products would be extended with the development phase of products and services.

A challenge here is that in many cases firms do not focus on the innovation process, but rather on product acceptability and impacts (van de Poel et al., 2017). Another challenge may be that CSR is used by companies more as a branding instrument than as an integrated strategic function. Therefore, it has been suggested that a company-specific RRI strategy is needed, which can then be translated into RRI activities and RRI key performance indicators, supporting the uptake of RRI at the whole-company level (ibid.). In addition, RRI could be integrated (as suggested by proponents of life-based design, e.g. Saariluoma et al., 2016) into the core of the technological design and innovation process. An optimal solution would be for both approaches to be used concurrently.

However, introducing RRI to firms is not an easy task for various reasons. Companies may find RRI discourse difficult or unsuited to them, or may feel that they are already implementing responsibility thinking in various forms (e.g. user-driven innovation, risk assessment and CSR) (ibid.). Gurzawska and colleagues (2017) have put forward three different categories of incentives that may support the implementation of RRI: external and internal stakeholder incentives, instrumental and non-instrumental incentives, and direct and indirect incentives. Examples include corporate reputation, critical consumerism and RRI certification. In addition, they suggest that employer engagement and good governance may support the implementation of RRI. They also consider RRI-related incentives to be contextual. Factors such as size of company, branch of industry and type of business ecosystem may affect the implementation of RRI. While Gurzawska and his colleagues have studied the general dynamics and environmental stimulants that may support the implementation of RRI, they leave it largely open as to who should implement these incentives and how – whether this should be done by public authorities and/or firms and their associations voluntarily, or whether it is better to just wait for incentives, such as critical consumerism, to emerge. Besides external incentives, much depends on the firms' own self-understanding and whether they see RRI as useful. Wider transformation may require the firm to redefine its mission to include, besides economic value creation, also the advancement of social and environmental values (Porter & Kramer, 2011). In so doing, responsibility thinking would be integrated into the firm's core values and operations.

10.3.4 Follow-up

Analysis of the actions of organizations and their social impacts has become increasingly important in our complex and constantly changing societies. As the operational environment is multi-dimensional and includes a temporal dimension with past, current and future developments, we have suggested a multi-method approach based on systematic integration of evaluation and foresight methods to provide a holistic analysis (Nieminen & Hyytinen, 2015; Hyytinen, 2017).

This holistic approach should include a systemic view of RRI. For instance, an innovation aimed at guaranteeing the personal safety and health of the technology users is likely to be a multi-dimensional phenomenon, making it difficult to isolate the influence of the specific innovation or its attributes. Due to the non-linear, systemic nature of the impacts, the evaluation questions should also be as open as possible. The typical evaluation approach, which is to start with formal targets (such as the above-mentioned safety and health) and then focus throughout the evaluation on the question of whether or not these targets have been achieved, can be blind to even significant unintended outcomes, impacts and side effects. Due to the systemic nature of impacts, the monitoring and follow-up of actions should be based on different data sources (quantitative and qualitative), pay attention to alternative impact paths and unintended consequences and include a multi-actor view with various perspectives reflecting different societal values.

Based on the ideas of developmental evaluation (Patton, 2011), the RRI process should be iterative and lead to learning from experiences and re-orientation of activities. Monitoring should be used as a continuous means of steering the RRI activities, with the aim of making them more effective and functional. While we emphasize the significance of developmental evaluation and collective learning, indicators and ex post case studies are needed to make visible what has been achieved. They should be analysed and interpreted in collaboration with the stakeholders and ecosystem actors in order to transform these "snapshots" into collective learning experiences. The achievements, problems and weaknesses should be analysed from various perspectives by engaging the actors in the process. This is important, as assessment of outcomes always includes a number of value- and interest-related questions.

10.3.5 Critical questions of the approach

In order to further develop and validate the model, the premises of the model were discussed in a separately organized international workshop of regional developers of RRI participating in the meeting of the EU-funded MARIE consortium (2017–2021[10]) in Tampere, Finland on 29 May 2018. In total, 20 people from eight European regions participated in the workshop. The group consisted of people with diverse backgrounds dealing with regional developmental activities and supporting the implementation of RRI in their own regions in the context of the MARIE project. MARIE's objective was to improve regional public policy by supporting the dissemination and uptake of RRI among enterprises. Using interregional activities, communication and stakeholder engagement, the partners worked to develop new action plans and strategies for the implementation of RRI. The rationale behind the project was a common experience of challenges related to RRI including, for example, the complexity of the concept and lack of appropriate approaches to implement it in public policies. One of the pilots of the MARIE project was the development of open innovation platforms in one of the 6Aika cities in Finland (Tampere). The MARIE workshop thus created an opportunity for feedback on the RRI evaluation approach and supported the overall goals of MARIE in the implementation of RRI.

The participants of the workshop were unanimous that an evaluation framework for the innovation platforms and for supporting the implementation of RRI is needed, even though this is challenging due to problems related to the attribution of factors producing impacts and a lack of appropriate indicators for measuring impacts. While the developers identified the challenges of a systemic operational environment and the problems it may cause for anticipating and creating impacts, they also hoped that it would be possible to conduct long-lasting system-level evaluation of impacts and related factors. The group was in general agreement that an evaluation framework and standardized criteria for evaluation are needed, as the use of evaluation supports the implementation of RRI. Mentioned positive functions of RRI evaluation included, for instance, "to benchmark activities against other open innovation platforms" and "to identify weak points and areas". In addition, participants thought that "management without measurement is dubious" and that evaluation is needed "to improve alignment with societal needs and needs and interests of platform members".

In terms of the suggested evaluation approach, the participants thought that it would make operations and action patterns more visible and under conscious deliberation, which, in turn, would make it possible to learn from the action and develop it further. At the same time, they suggested that, in order to strengthen the approach, the criteria describing responsibility should be integrated in the general follow-up criteria, such as key performance indicators, to create incentives for "doing RRI". The more developmentally oriented evaluation would be conducted as an independent process and the two processes of setting incentives and supporting qualitative development would be separated from each other. This would create more trust towards developmental evaluation, being a major prerequisite for it.

In addition, collaboration and commitment to the shared goals necessitate trust, which can be developed only in a long-running process of working together and in honest, reciprocal dialogue. According to the participants, this makes it possible to develop a shared value base, targets and inclusive evaluation procedures. Thus, while the acknowledgement of different values was considered extremely important in co-creation processes, it was stated that, when the process is not based on trust, it might shrink to mere opinion gathering, without genuine openness, learning and critical reflexivity. Inclusion of various values in the framework of trust would also support a more holistic view of responsibility and a deeper understanding of the co-creation of activities.

Participants also stressed that societal values that facilitate innovation and democratic discussion, such as openness and transparency, capacity building, social diversity and ethics, should be emphasized more in the evaluation. According to the participants, the wider inclusion of values would, in turn, "improve quality of life and help to face a 'black swan', when it happens[11]" or "to assure that the platform is close to society". For instance, openness and transparency are strongly linked to the operation of platforms – for effective self-organization of actors and coordination of actions it is essential that all actors are visible and identifiable in terms of their capacities, opportunities and roles in the platform. Transparency and clarity of communication convey the values and interpretations of the platform and its functioning.

According to the participants, trust, inclusion and openness support the commitment of actors and the meeting of ideas, which are prerequisites for the development of sustainable and acceptable innovations in a platform. This concerns the whole innovation process, from ideation to marketing and distribution. However, the inclusion of different ideas and values may create sustainable and effective innovations only if these values and ideas for future actions, products or services are taken into account in the design and development processes.

The participants considered ensuring the effective and "fair" participation of all relevant stakeholders and actors to be a major challenge in this process. The challenge stretches from the identification of stakeholders and innovation actors to the inclusion of their values and ideas in the design or development process. Who are the stakeholders, and who should be included or left out? If all potential actors are included, is the process any more realistically "doable" and sense making to the actors? Are various perspectives and values sufficiently taken into account? And how should compromises between different perspectives and values be made? These critical questions can only be answered case by case, as the implementation context and developed technologies and services vary, along with the relevant stakeholders and the ways to engage them.

Thus, there are critical questions that need to be kept in mind and solved when the model is applied. Firstly, for motivational and trust-related reasons it might be useful to keep quantitative measurement and incentives separate from the qualitative and developmental evaluation approach. Secondly, in order to be useful the approach should be able to create genuine trust among participants and stakeholders. Otherwise, it would not optimally support the development of positive results. Thirdly, the evaluation process should be as open as possible and include widely different values. A critical issue here is that these different viewpoints must be genuinely taken into account in the process, not only as a legitimizing element to increase the social acceptability of previously made "top-down" decisions. Fourthly, the evaluation should pay careful attention to the identification and engagement of stakeholders to ensure, not only fair and wide, but also effective participation. And finally, the far from simple challenge of functional indicators and their measurement needs to be addressed to provide appropriate follow-up indicators and incentives for RRI.[12]

10.4 Conclusions

In this chapter, our premise has been that the development of responsibility in research and innovation depends on complex interactions of various actors, organizational dimensions and their dynamic co-development with the organization's external environment. This is reflected also in systems thinking in evaluation. As our societies and organizations have become increasingly complex and rapidly changing, we should not over-simplify the complexity of impact paths or the heterogeneity of factors and actors contributing to societal impacts. Heterogeneity also emphasizes a diversity of values and interests, which affect the desirability and acceptability of innovations. In addition, as innovation is by definition something that takes place in the future, we need anticipatory approaches to cast light on its social and environmental impacts.

Following from this, we have suggested an evaluation model that incorporates future orientation, a systemic view and multi-actor approach. The model emphasizes continuous development instead of outcomes. Ideally, the evaluation should be aimed at developing operations and processes from the very beginning, not afterwards. In other words, assessing the acceptability and desirability of an innovation or development takes place in a dialogic process of anticipating the impacts and outcomes of the innovation from the very beginning of the process. In the model, RRI intertwines in a stepwise process with successive innovation and organizational actions. A focal character of the model is co-creation, which helps to define values and operational goals for the innovation activity and supports the assessment of the success of operations. At its best, the process is a loop of continuous learning in which participating organizations and individuals learn from each other during the evaluation process and reflexively develop their activity. The desirability and acceptability of the innovation are supported by creating shared visions for the innovation activity.

However, implementing the model is not without its challenges. For instance, in a complex and systemic environment, the line between an organization and its environment may become blurred and the concept of stakeholder may turn out to be more or less obsolete in describing symbiotic relationships in innovation ecosystems or in global interaction. Neither is balancing between different values and avoidance of value relativism or dominance easy. In addition, not all actors may be motivated to participate in RRI activities. Motivation to participate can, however, be increased by ensuring trust among the participating actors.

As stakeholder and/or citizen participation is the "cornerstone" of our evaluation approach – and of any RRI approach – a specific critical factor for further development of the model is theoretical and practical enquiry into the stakeholder concept. Further studies are needed to clarify its theoretical relevance as well as practical usefulness in implementation situations. Correspondingly, the model as a whole needs to be tested in diverse social contexts and organizations in order to secure sufficient empirical feedback on its elements.

Notes

1 www.rri-tools.eu/
2 https://cordis.europa.eu/project/rcn/91156/reporting/en
3 https://res-agora.eu/news/
4 www.technopolis-group.com/morri/
5 http://satoriproject.eu/
6 https://6aika.fi/in-english/
7 The companies were asked to describe their current understanding of the responsibility concept in relation to their business. The company personnel were interviewed and the state of the company RRI was defined based on

these interviews. In a follow-up workshop, the results were discussed and good practices as well as potential issues for improvement were recognized. In addition, intervention activities were proposed in order to increase the companies' RRI maturity.

8 https://6aika.fi/in-english/

9 Source: What is a Stakeholder? Definitions of a Stakeholder: www.stakeholdermap.com/what-is-a-stakeholder.html#freeman

10 See: www.interregeurope.eu/marie/

11 The speaker is apparently referring to Taleb's (2010) term "black swan" symbolizing an unanticipated or unthinkable occurrence, as well as societal resilience and cohesion during these unexpected developments.

12 For an attempt to develop such indicators for industry, see: Yaghmaei (2018).

References

Adner, R. (2012) *The Wide Lens: A New Strategy for Innovation*. London: Portfolio/Penguin.

Ainasoja, M., Kaasinen, E., Vulli, E., Kulju, M., Reunanen, E., Hautala, R., & Rytövuori, S. (2012). User involvement in service innovations: Four case studies. In Chesbrough, H. & Piller, F. (eds.) *Bridging Mass Customization & Open Innovation: Proceedings of the MCPC 2011 Conference*. Raleigh, NC: Lulu, p. 40.

Alam, I. (2006). Removing the fuzziness from the fuzzy front-end of service innovations through customer interactions. *Industrial Marketing Management*, 35, 468–480.

Autio, E. & Thomas, L. D.W. (2013). Innovation ecosystems. In Dodgson, M., Gann, D.M., & Phillips, N. (eds.), *Oxford Handbook of Innovation Management*, Oxford: Oxford University Press. pp. 204–288.

Auvinen, H., Ruutu, S., Tuominen, A., Ahlqvist, T., & Oksanen, J. (2014). Process supporting strategic decision-making in systemic transitions: Towards emission-free transport in cities by 2050. *Technological Forecasting and Social Change*, 94, 97–114.

Borras, S. & Edler, J. (eds.) (2014). *The Governance of Socio-Technical Systems. Explaining Change*. Cheltenham, UK: Edward Elgar.

Cabrera, D., Colosi, L., & Lobdell, C. (2008). Systems thinking. *Evaluation and Program Planning*, 31, 299–310.

Cao, G., Clarke, S., & Lehaney, B. (2003). Diversity management in organizational change: Towards a systemic framework. *Systems Research*, 20, 231–242. doi:10.1002/sres.530

Castells, M. (1996). *The Rise of the Network Society. The Information Age: Economy, Society and Culture*, Vol. I. Oxford: Blackwell.

Chesbrough, H., Vanhaverbeke, W., & West, J. (eds.) (2006). *Open Innovation: Researching a New Paradigm*. Oxford: Oxford University Press.

Damanpour, F. (1996). Organizational complexity and innovation: Developing and testing multiple contingency models. *Management Science*, 42(5), 693–716. Retrieved from www.jstor.org/stable/2634460

Dyehouse, M., Bennett, D., Harbor, J., Childress, A., & Dark, M. (2009). A comparison of linear and systems thinking approaches for program evaluation illustrated using Indiana Interdisciplinary GK-12. *Evaluation and Program Planning*, 32, 187–196.

Eden, C. & Ackermann, F. (1998). *Making Strategy: The Journey of Strategic Management*. London: Sage.

Edquist, C. (2005). Reflections on the systems of innovation approach. *Science and Public Policy*, 31(6), 485–489.

Eoyang, G. & Holladay, R. (2013). *Adaptive Action, Leveraging Uncertainty in Your Organization*. Stanford, CA: Stanford University Press.

Forss, K., Marra, M., & Schwartz, R. (2011). *Evaluating the Complex: Attribution, Contribution, and Beyond*. London: Transaction Publishers.

Freeman, R. E. (1984). *Strategic Management: A Stakeholder Approach*. Boston, MA: Pitman.

Freeman, C. & Louca, F. (2002). *As Time Goes By: From the Industrial Revolutions to the Information Revolution*. Oxford: Oxford University Press.

Friedman, M. (2005). *The World is Flat*. Farrar: Straus and Giroux.

Friedman, B., Kahn, P. H., & Borning, A. (2006). Value sensitive design and information systems. In Zhang, P. & Galletta, D. (eds.) *Human–Computer Interaction in Management Information Systems: Foundations*. New York: M.E. Sharpe, pp. 348–371.

Geels, F. W. & Schot, J. (2007). Typology of sociotechnical transition pathways. *Research Policy*, 36, 399–417.

Georghiou, L., Cassingena Harper, J., Keenan, M., Miles, I., & Popper, R. (eds.) (2008). *The Handbook of Technology Foresight*. Cheltenham, UK: Edward Elgar.

Gertler, P. J., Martinez, S., Premand, P., Rawlings, L. B., & Vermeersch, C. M. J. (2011). *Impact Evaluation in Practice*. Washington, DC: The World Bank.

Gianni, R. (2016). *Responsibility and Freedom: The Ethical Realm of RRI*. London: John Wiley.

Gurzawska, A., Mäkinen, M., & Brey, P. (2017). Implementation of Responsible Research and Innovation (RRI) practices in industry: Providing the right incentives. *Sustainability*, 9 (6) (1036). https://doi.org/10.3390/su9061036

Helbing, D. (2013). Globally networked risks and how to respond. *Nature*, 497, 51–59.

Holland, J. H. (1995). *Hidden Order: How Adaptation Builds Complexity*. Reading, MA: Addison-Wesley.

Hughes, T. P. (1983). *Networks of Power: Electrification in Western Society, 1880–1930*. Baltimore, MD: Johns Hopkins University Press.

Hyytinen, K. (2017). *Supporting Service Innovation via Evaluation: A Future Oriented, Systemic and Multi-actor Approach*. Aalto University Publication Series Doctoral Dissertations 14/2017 & VTT Science 146. Retrieved from: www.vtt.fi/inf/pdf/science/2017/S146.pdf

João, E., Vanclay, F., & den Broeder, L. (2011). Emphasising enhancement in all forms of impact assessment: Introduction to a special issue. *Impact Assessment and Project Appraisal*, 29(3), 170–180. DOI: 10.3152/146155111X12959673796326

Kaasinen, E., Ainasoja, M., Vulli, E., Paavola, H., Hautala, R., Lehtonen, P., & Reunanen, E. (2010). *User Involvement in Service Innovations*. VTT Tiedotteita – Valtion Teknillinen Tutkimuskeskus (2552). Espoo, VTT, pp. 1–69.

Kellogg Foundation (2004). *Using Logic Models to Bring Together Planning, Evaluation and Action Logic Model Development Guide*. Battle Creek, MI: Kellogg Foundation. Retrieved from: www.wkkf.org/Pubs/Tools/Evaluation/Pub3669.pdf

Kivisaari, S., Väyrynen, E., & Saranummi, N. (2004). *Knowledge-intensive Service Activities in Health Care Innovation. Case Pirkanmaa.* VTT Research Notes 2267. Retrieved from: www.vtt.fi/inf/pdf/tiedotteet/2004/T2267.pdf

Kline, S. J. & Rosenberg, N. (1986). An overview of innovation. In Landau, R. & Rosenberg, N. (eds.) *The Positive Sum Strategy: Harnessing Technology for Economic Growth.* Washington, DC: National Academy Press, pp. 275–305.

Kurzweil, R. (1999). *The Age of Spiritual Machines: When Computers Exceed Human Intelligence.* New York: Penguin Books.

Lawrence, K. (2013). *Developing Leaders in VUCA Environment.* Retrieved from: www.growbold.com/2013/developing-leaders-in-a-vuca-environment_UNC.2013.pdf

Leikas, J. (2009). *Life-Based Design – A Holistic Approach to Designing Human–Technology Interaction.* VTT Publications 726. Espoo: VTT.

Luppicini, R. (2010). *Technoethics and the Evolving Knowledge Society: Ethical Issues in Technological Design, Research, Development, and Innovation.* Hershey: IGI Global.

Mayers, J. & Vermeulen, S. (2005). *Stakeholder Influence Mapping. Power Tools.* London: International Institute for Environment and Development (IIED).

Mayne, J. (2012). Contribution analysis: Coming of age? *Evaluation*, 18, 270–280.

Mitchell, R. K., Agle, B. R., & Wood, D. J. (1997). Toward a theory of stakeholder identification and salience: Defining the principle of who and what really counts. *The Academy of Management Review*, 22 (4), 853–886.

Moore, J. F. (1996). *The Death of Competition: Leadership and Strategy in the Age of Business Ecosystems.* New York: HarperBusiness.

MoRRI (2018). *Monitoring the Evolution and Benefits of Responsible Research and Innovation. Short Draft Final Report on Insights from Monitoring the Evolution and Benefits of RRI in Europe.* European Commission. Retrieved from: www.technopolisgroup.com/wpcontent/uploads/2018/03/D11_MoRRI_Short_Draft_Final_Report.pdf

Mowles, C. (2014). Complex, but not quite complex enough: The turn to the complexity sciences in evaluation scholarship. *Evaluation*, 20 (2), 160–175.

Nieminen, M. & Hyytinen, K. (2015). Future-oriented impact assessment: Supporting strategic decision-making in complex socio-technical environment. *Evaluation*, 21(4), 448–461.

Ortt, J. R. & van der Duin, P. A. (2008). The evolution of innovation management towards contextual innovation. *European Journal of Innovation Management*, 11(4), 522–538, https://doi.org/10.1108/14601060810911147

Owen, R., Bessant, J., & Heintz, M. (eds.) (2013). *Responsible Innovation.* Oxford: Wiley.

Patton, M. Q. (2011). *Developmental Evaluation: Applying Complexity Concepts to Enhance Innovation and Use.* New York, NY: Guilford.

Porter, M. E. & Kramer, M. R. (January–February 2011). Creating shared value. How to reinvent capitalism and unleash a wave of innovation and growth. *Harvard Business Review.* Retrieved from: https://hbr.org/2011/01/the-big-idea-creating-shared-value

Prahalad, C. K. & Ramaswamy, V. (January–February 2000). Co-opting customer competence. *Harvard Business Review.* Retrieved from: https://hbr.org/2000/01/co-opting-customer-competence

Reed, M. S., Graves, A., Dandy, N., Posthumus, H., Hubacek, K., Morris, J., Prell, C., Quinn, C. H., & Stringer, L. C. (2009). Who's in and why? A typology of stakeholder analysis methods for natural resource management. *Journal of Environmental Management*, 90(5), 1933–1949.

Reijers, W., Brey, P., Jansen, P., Rodrigues, R., Koivisto, R., & Tuominen, A. (2018). *A Common Framework for Ethical Impact Assessment. Stakeholders Acting Together on the Ethical Impact Assessment of Research and Innovation.* SATORI. D4.1. Retrieved from: http://satoriproject.eu/media/D4.1_Annex_1_EIA_Proposal.pdf

Responsible-Industry Project Consortium (2017). *Guide for the Implementation of Responsible Research and Innovation (RRI) in the Industrial Context.* Responsible Industry. Retrieved from: www.responsible-industry.eu/

Rittel, H. & Webber, M. (1973). Dilemmas in a general theory of planning. *Policy Sciences*, 4, 155–169.

Rotmans, J. & Loorbach, D. (2009). Complexity and transition management. *Journal of Industrial Ecology*, 13(2), 184–196.

Saariluoma, P., Cañas, J. J., & Leikas, J. (2016). *Designing for Life – A Human Perspective on Technology Development.* London: Palgrave MacMillan.

Sanders, E. B. & Stappers, P. (2008). Co-creation and the new landscapes of design. *CoDesign*, 4(1), 5–18.

Schot, J. & Rip, A. (1997). The past and future of constructive technology assessment. *Technological Forecasting and Social Change*, 54, 251–268.

Schot, W. J. & Steinmueller, E. (2018). Three frames for innovation policy: R&D, systems of innovation and transformative change. *Research Policy*, 47(9), 1554–1567.

Simon, J. (2016). Value-sensitive design and responsible research and innovation. In Hansson, S. O. (ed.) *The Ethics of Technology – Methods and Approaches.* London: Rowman & Littlefield, pp. 219–236.

Smallman, M. (2018). Citizen science and Responsible Research and Innovation. In Bonn, A., Haklay, M., Hecker, S., Bowser, A., Makuch, Z. & Vogel, J. (eds.) *Citizen Science – Innovation in Open Science, Society and Policy.* London: UCL Press, pp. 241–253.

Smith, K. R. (2000). Innovation as a systemic phenomenon: Rethinking the role of policy. *Enterprise and Innovation Management Studies*, 1(1), 73–102.

Stahl, B. C., Obach, M., Yaghmaei, E., Ikonen, V., Chatfield, K., & Brem, A. (2017). The Responsible Research and Innovation (RRI) maturity model: Linking theory and practice. *Sustainability*, 9,1036.

Stilgoe, J., Owen, R., & Macnaghten, P. (2013). Developing a framework for responsible innovation. *Research Policy*, 42(9), 1568–1580.

Sutcliffe, H. (2011). *A Report on Responsible Research and Innovation.* Retrieved from: https://ec.europa.eu/research/science-society/document_library/pdf_06/rri-report-hilary-sutcliffe_en.pdf

Taleb, N. N. (2010) *The Black Swan. The Impact of the Highly Improbable.* Revised edition. London: Penguin.

Teece, D. (1988). Technological change and the nature of the firm. In Dosi, G., Freeman, C., Nelson, R., Silverberg, G. & Soete, L. (eds.) *Technical Change and Economic Theory.* London: Pinter Publishers, pp. 256–281.

Tsoukas, H. & Chia, R. (2002). On organizational becoming: Rethinking organizational change. *Organization Science*, 13(5), 567–582.

Urry, J. (2005). The complexity turn. *Theory, Culture and Society*, 22(5), 1–14.

van de Poel, I., Asveld, L., Flipse, S., Klaassen, P., Scholten, V., & Yaghmaei, E. (2017). Company strategies for Responsible Research and Innovation (RRI): A conceptual model. *Sustainability*, 9(11), 2045.

van Wezel, A. P., van Lente, H., van de Sandt, J. J., Bouwmeester, H., Vandeberg, R. L., & Sips, A. J. (2018). Risk analysis and technology assessment in support of technology development: Putting responsible innovation in practice in a case study for nanotechnology. *Integrated Environmental Assessment and Management*, 14,9–16. doi:10.1002/ieam.1989

von Hippel, E. (1988). *The Sources of Innovation*. New York: Oxford University Press.

von Schomberg, R. (2011). *Towards Responsible Research and Innovation in the Information and Communication Technologies and Security Technologies Fields*. A Report from the European Commission Services. Belgium: European Commission.

Wholey, J. S., Hatry, H. P., & Newcomer, K. E. (eds.) (2010). *Handbook of Practical Program Evaluation*. Third Edition. San Francisco, CA: Jossey-Bass.

Williams, B. & Hummelbrunner, R. (2011). *Systems Concepts in Action. A Practitioner's Toolkit*. Stanford, CA: Stanford University Press.

Williams, B. & Imam, I. (eds.) (2006). *Systems Concepts in Evaluation: An Expert Anthology*. Point Reyes, CA: EdgePress of Inverness.

Yaghmaei, E. (2018). Responsible research and innovation key performance indicators in industry: A case study in the ICT domain. *Journal of Information, Communication and Ethics in Society*, 16(2), 214–234. https://doi.org/10.1108/JICES-11-2017-0066

Best practice IV PRISMA KPI analysis tool

Steven Flipse

Description of good practice example

Within the European Commission-funded Science with and for Society project PRISMA about the instalment of Responsible Research and Innovation (RRI) in innovation practice in industrial companies, we studied how an RRI approach may improve innovation processes in companies, and their outcomes. We developed a quantitative method to score performance on both RRI and general innovation management-related key performance indicators (KPIs). The method builds upon an earlier tool, the 'success factor based live innovation project scoring and evaluation' tool (Flipse *et al.* 2014), which was expanded with specific RRI elements, and for tracking the development of criteria scores over time.

Aims and goals

This tool is meant to create a basis of discussion about RRI in corporate innovation contexts, and to explore the links between RRI and innovation management-related KPIs. It is recommended to also collect data in a qualitative manner, in conjunction with this approach, for projects at the

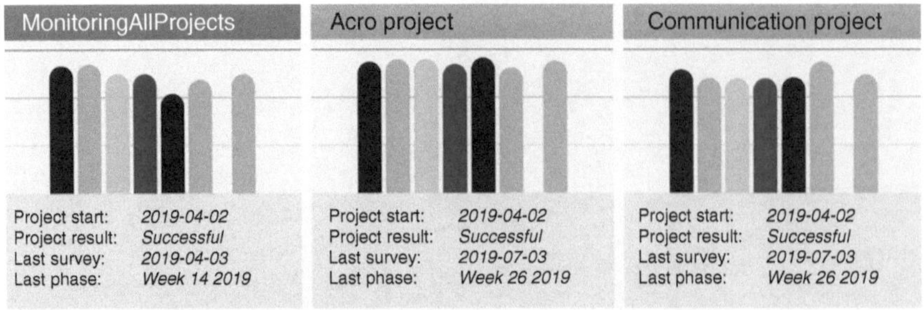

Figure IV.I Key performance indicator (KPI) performance of three projects. The six KPIs are shown on the left side of the graphs and the overall score on the right.

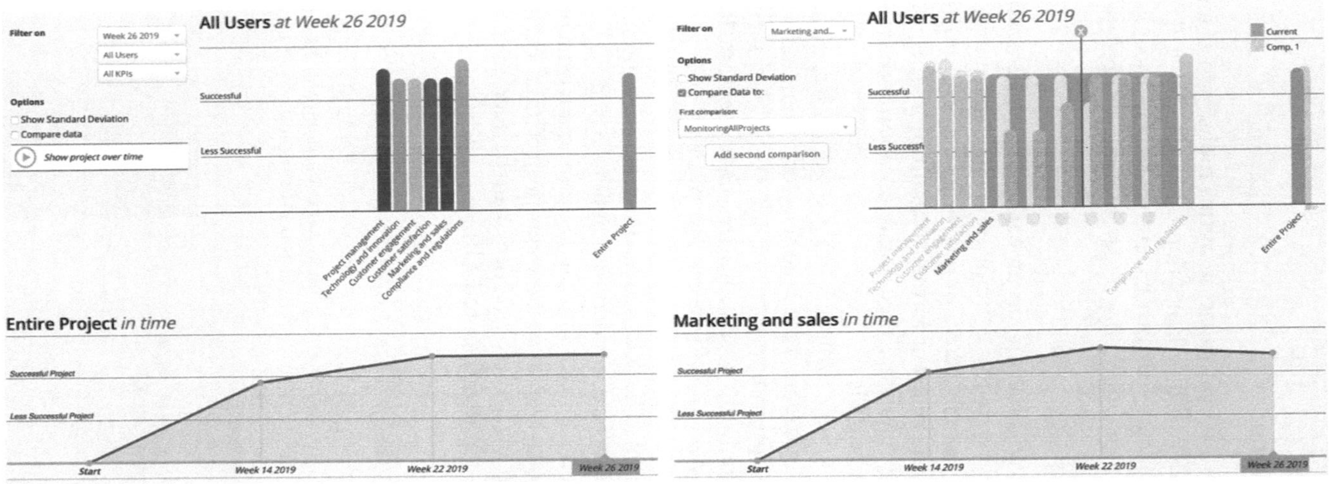

Figure IV.II (Left) Key performance indicator (KPI) scores for a project, at a given moment in time, as assessed by all users within that project team. (Right) Comparison of a project with 'MonitoringAllProjects', with one specific KPI selected (marketing and sales), showing different scores for various criteria within that one KPI.

beginning, middle and end of their runtime, in order to make sense of the scoring patterns.

Benefit for organizations

An open discussion with the information in this tool supports innovators in explicitly discussing RRI within their organization. Linking RRI to innovation management elements helps to ensure that such discussions are directly relevant to innovators' on-going research and development work. The tool works in three steps:

1. *Identification of relevant KPIs.* In an earlier study, 92 possible success criteria (both RRI-related and innovation management-related) were identified. Companies are asked, in the form of a serious card game, to select (30–50) criteria relevant to their organization, sort these into (5–8) relevant KPIs and indicate their relative importance by distributing 100 points over these KPIs.
2. *Assessment of company projects.* Innovation project team members (and possibly also outsiders) score the selected criteria on a 1–7 Likert-type scale, to assess project performance at a certain moment, using the protected online tool, for which all users have login credentials. This results in an overview of KPI scores for projects at a given moment (Figure IV.I).
3. *Discussion of results.* The KPI scores for the innovation projects are analysed, and allow for guided discussion based on (well/poorly performing) aspects of RRI and innovation management. The tool allows for comparative analysis between projects and between people within the same project: best practices can be identified and mutual learning may be facilitated (Figure IV.II).

Reference

Flipse, S.M., Van der Sanden, M.C.A., & Osseweijer, P. (2014) Improving industrial R&D practices with social and ethical aspects: Aligning key performance indicators with social and ethical aspects in food technology R&D. *Technological Forecasting & Social Change* 85: 185–197.

11 Assessing Responsible Research and Innovation (RRI) systems in the digital age

Claudia Werker

11.1 Introduction

In the past few decades the processes as well as the assessment of research and innovation have become increasingly complicated. There have been three major factors driving this development. First, despite innovation and technology sometimes displaying a dark side (Gebauer, Füller, & Pezzei, 2013), they have become core to solving commercial, economic and societal problems in a responsible way (Stilgoe, Owen, & Macnaghten, 2013). Accordingly, policy has been more and more concentrating on *Responsible Research and Innovation (RRI)*, i.e. on marrying socio-economic change with ethical concerns (European Commission, 2013). Second, the development and deployment of all-encompassing fields, such as biotechnology, information and communication technologies as well as nanotechnology, have required a deeper integration of industry, academia, civil society and government (Etzkowitz & Viale, 2010; European_Commission, 2012). Third, *digitization*, i.e. the transformation of established patterns of socio-economic life driven by processes of converting information into digital format and analysing them at large scale, has been infiltrating all walks of life (Loebbecke & Picot, 2015): it changes the way we connect to others, exchange information and knowledge, take decisions, work and shop. It also fundamentally alters research and innovation.

In this chapter, I investigate the opportunities and challenges of assessing RRI systems in the digital age. I start with showing why RRI needs a system approach and introduce the concept of *RRI systems* (Section 11.2). Then, I turn to how innovation systems have been assessed so far and summarize the key steps in a scheme (Section 11.3.1) that serves as a starting point for developing a new scheme adding responsibility to the innovation system and its assessment (Section 11.3.2). After that, I show that in the digital age big data alters the assessment of RRI systems severely (Section 11.4.1) and discuss the arduous challenges and promising solutions resulting from that (Section 11.4.2). Finally, I briefly summarize the results of my analysis, elaborate on their implications for the various stakeholders and provide a fundamental open research question emerging from them (Section 11.5).

11.2 RRI requires a system approach

RRI is at the heart of better employing and exploiting the potential of science, technology and innovation by considering and incorporating values, i.e. "things worth striving for" (Taebi, Correljé, Cuppen, Dignum, & Pesch, 2014, p. 119), societal needs as well as economic opportunities (European Commission, 2013; Taebi et al., 2014). "[I]n an ideal situation, responsible innovation can best be conceptualized as an endorsement of the relevant values during the innovation process" (Taebi et al., 2014, p. 118).

In order to realize RRI all stakeholders need to be part of the process (Taebi et al., 2014). In particular, not only innovative agents, i.e. those *actively* influencing research and innovation, are relevant for the RRI process but also all other stakeholders which might be affected by either the process or the outcome. In order to be able to understand the implications of the research and innovation processes, these stakeholders not only have to identify the purposes and trajectories of research and innovation but also their effects and alternatives (Forsberg, Shelley-Egan, Ladikas, & Owen, 2018). Based on these insights stakeholders are able to assess research and innovation in the light of their values. In fact, value implications caused by new technologies are core to processes leading to responsible innovation as they can either stimulate or undermine values (van de Poel, 2009).

RRI requires a system approach for two major reasons. The first reason points to the necessity to analyse research and innovation from a system perspective in general terms. In recent decades all-encompassing fields, such as biotechnology, information and communication technologies as well as nanotechnology, have emerged and have required a deeper integration of industry, academia, the civil society and government (Etzkowitz & Viale, 2010; European Commission, 2012). A typical issue addressed by such a deeper integration is a missing fit between academic capabilities and industrial needs. Deeper integration means that innovative agents engage in reciprocal relationships by considering the point of view of the other agents in order to enhance their performance.

The second reason for employing a system perspective emerges from the nature of RRI. Developing shared values about the process and outcomes of research and innovation requires integration of the values of all relevant stakeholders (European Commission, 2013; Taebi et al., 2014). While the values of stakeholders can substantially differ, shared values can only emerge if stakeholders eventually agree on them. Developing shared values is non-trivial, because stakeholders differ regarding both their power and their legitimacy (Mitchell, Agle, & Wood, 1997). In the context of RRI processes, the question about the power positions clarifies who depends on whom and who can influence or even pressure others to act against their own values. The question about legitimacy touches the very heart of shared values, because it clarifies what socially accepted and expected behaviour of the various stakeholders is. Therefore, the institutions guiding stakeholders' behaviour and the relationships between them are core to RRI processes. Furthermore, these processes change in time, so that the dynamics of relationships and of institutions supporting RRI are centre-stage (Forsberg et al., 2018). To find RRI, i.e. to base jointly acceptable solutions in research and innovation on shared values, entails that all stakeholders communicate and possibly collaborate. Analysing these processes means investigating all stakeholders as well as the dynamics of their relationships and the supporting institutions, i.e. using a system approach.

As shown above, analysing and realizing RRI require a system approach. While the term responsible innovation system has been used in passing to call for "a broader imaginary of a responsible innovation system involving reconfiguration of knowledge flows and institutions" (Forsberg et al., 2018, p. 10), to my knowledge, it has not been comprehensively defined or used for analysis so far. In the following, in order to capture *all relevant stakeholders of RRI and the way their values affect their activities, relationships and supporting institutions* I will use the term *RRI system.*

11.3 Assessment of RRI systems

First, I show how innovation systems have been assessed so far and provide a scheme of how to assess an innovation system (Section 11.3.1). Following that, I augment this scheme with responsibility. I do so by pointing at those elements in the assessment of RRI systems that are required in addition to the ones necessary for innovation systems alone (Section 11.3.2).

11.3.1 Assessing innovation systems: reviewing the state of the art

Innovation systems display four major features which are crucial for assessing them. *Innovative agents*, i.e. those influencing research and innovation processes, include companies, universities, private and public research organizations, civic communities as well as governmental bodies, such as ministries or city councils (Bergek, Jacobsson, Carlsson, Lindmark, & Rickne, 2008; Edquist, 2011; Klein Woolthuis, Lankhuizen, & Gilsing, 2005). Innovative agents' communication and collaboration with partners form their *relationships*. These relationships depend on how close (potential) collaboration partners are to each other in cognitive, organizational, personal, social, institutional and geographical terms (Boschma, 2005; Caniëls, Kronenberg, & Werker, 2014; Werker, Ooms, & Caniëls, 2016). Innovative agents can support and use formal (e.g. laws and regulations) and informal *institutions* (e.g. codes of conducts) which can either enable or hinder innovation processes (Bergek et al., 2008; Edquist, 2011; Werker et al., 2016). Innovative agents' *activities*, i.e. how they organize research and innovation, are crucial to understand the drivers of change in innovation systems (Bergek et al., 2008; Werker, 2001). While innovation systems display structural regularities such as management or policy processes, innovation processes themselves are inherently uncertain (Schwerin & Werker, 2003). All four major features mentioned above may change in time, thereby influencing each other's evolution.

Assessing an innovation system means identifying opportunities and problems regarding the above-mentioned four features. To come to terms with this in a practical manner, a number of schemes have been provided (Bergek et al., 2008; Edquist, 2011; Klein Woolthuis et al., 2005). All of these schemes include the structural components of innovation systems, i.e. innovative agents, their relationships as well as formal and informal institutions. Some emphasize activities (Edquist, 2011), called functions by Bergek et al. (2008), thereby also focusing on the dynamics of innovation systems, i.e. co-evolution between innovative agents, their relationships and the supporting institutions. In Figure 11.1, I provide a scheme capturing the structural elements and the processes of innovation systems comprising the following five steps:

I. identifying the structural components, i.e. innovative agents, relationships and institutions (Klein Woolthuis et al., 2005)
II. finding crucial activities, such as knowledge advance by and diffusion amongst stakeholders, entrepreneurial experimentation, legitimation, market formation, development of institutions and influence on the direction of search by different selection mechanisms, such as business models, technology development, market and institutional forces (Bergek et al., 2008; Edquist, 2011)

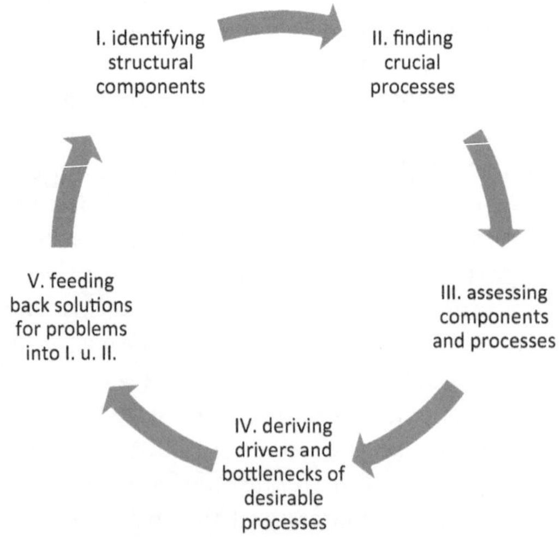

Figure 11.1 A scheme for assessing innovation systems.

III. assessing components and processes by uncovering desirable ones (Bergek et al., 2008; Edquist, 2011; Klein Woolthuis et al., 2005)
IV. deriving drivers and bottlenecks of desirable components and processes (Bergek et al., 2008; Edquist, 2011; Klein Woolthuis et al., 2005)
V. feeding back solutions for problems into the structural components (I) and processes (II) including their functioning and co-evolution.

In the list below, I summarize the potential problems for which solutions have to be fed back into the next cycle of the innovation system. They can be related to either structural components (Klein Woolthuis et al., 2005), i.e. 1–4 in the list below, or activities (Bergek et al., 2008; Edquist, 2011), i.e. 5–10 in the list below:

1. innovative agents facing lack of capability so that they are not able to obtain, create or use knowledge and innovation sufficiently
2. missing innovative agents that would be important for the functioning of the system, e.g. no business angels providing finance and advice for start-ups
3. problems in collaborating, i.e. lack of necessary relationships or too strong relationships leading to lock-ins
4. problems because of wrong or missing formal or informal institutions, e.g. missing standards
5. lack of knowledge development
6. lack of knowledge diffusion amongst stakeholders
7. missing entrepreneurial experimentation
8. no or insufficient legitimation
9. lacking market formation
10. insufficient processes to search for a sufficient variety of business models and technology solutions.

The ten problems summarized above lead agents and stakeholders of innovation systems to challenges they need to address in the future. To do so effectively and efficiently they need a qualification or quantification of the problems in the form of indicators. These indicators help them to assess the problem, develop solutions and measures as well as control whether the measures they take have effect.

For many of the potential problems, there exist rather straightforward quantitative indicators: data on patents, patent citations, licences, publications or publication citations (Nelson, 2009) can measure the knowledge capability of innovative agents (1). Moreover, when using them to compare different innovation systems they might point at a lack of knowledge development (5) or wanting diffusion amongst innovative agents (6). Problems of missing innovative agents (2) might be detected by indicators such as venture capital investment (e.g. EU Commission, 2018a). Problems of collaborating (3) might be captured by using the indicator of innovative small and medium-sized enterprises (SMEs) collaborating with others (e.g. EU Commission, 2018a). Problems of missing entrepreneurial experimentation (7) might show when looking at the indicators SMEs product/process innovations, SMEs marketing/organizational innovations and SMEs innovating in-house (e.g. EU Commission, 2018a). Insufficient processes to search for a sufficient variety of business models and technology solutions (10) could be measured by the newly defined indicators product innovation, i.e. "a new or improved good or service that differs significantly from the firm's previous goods or services and that has been introduced on the market" (OECD/Eurostat, 2018, p. 34), and business process innovation, i.e. "a new or improved business process for one or more business functions that differs significantly from the firm's previous business processes and that has been brought into use by the firm" (OECD/Eurostat, 2018, p. 34).

For other problems mentioned above the combined use of quantitative and qualitative indicators is necessary: Insufficient legitimation, i.e. not fully abiding by the innovative agents' norms and values (8), can be overcome by conscious processes of changing institutions or finding solutions within existing ones (4) (Bergek et al., 2008). Lacking market formation (9) requires the analysis of the different market phases, of the users, their preferences and their purchasing processes as well as of the institutions, particularly whether they stimulate market formation or if they need to change to do so (Bergek et al., 2008). The respective indicators are partly readily available quantitative facts on market size and on customer groups and partly qualitative data, e.g. stakeholders' strategies (Bergek et al., 2008).

Often indicators are combined and compared in order to identify whether or not individual innovative agents or whole systems, e.g. national innovation systems, perform well or not (e.g. EU Commission, 2018a). The combined use of indicators requires the weighing of the underlying indicators and might lead to tuned and unreliable rankings (Grupp & Schubert, 2010). Recently, the well-established European Innovation Scoreboard (EU Commission, 2018a) was severely criticized for not carefully reflecting on the use of input and output indicators in their analysis, because it mixes input and output innovation indicators and uses their average without defining and theoretically underpinning the meaning of innovation performance. According to Edquist, Zabala-Iturriagagoitia, Barbero, and Zofío (2018) this makes their interpretations and the ranking of European Union (EU) innovation systems misleading and worthless. So, while this combined use of indicators to rank innovation systems is well established, it has to be done with care. Otherwise it might be misleading.

11.3.2 Assessment of RRI systems: adding responsibility to the mix

To develop an RRI system approach we need to extend the innovation system approach with responsibility. In order to do so we have to include the following aspects to our analysis. First of all, we have to identify those stakeholders who are affected by research and innovation and do not have any influence on innovation processes and design. In contrast to innovative agents, those stakeholders are usually left out of the analysis of innovation systems. On the rather rare occasion those stakeholders are included in the analysis they are just described as collaterals whose values are not integrated in the process and design. In order to make innovation responsible *all stakeholders* affected by research and innovation processes *including their values* have to be included in the analysis (see also Section 11.2). So, the first step in the assessment scheme of RRI systems has to change accordingly (Figure 11.2). Second, responsible innovations are only possible when research and innovation processes lead to designs appropriately incorporating shared values (Taebi et al., 2014; van de Poel, 2009). So, developing these *shared values* has to be part of the second step of the assessment scheme. Moreover, in the third step the processes and outcomes of the RRI system would be *assessed on the basis of these shared values.* In the fourth step when the drivers and bottlenecks of desirable processes are identified *additional value-related drivers and bottlenecks* might occur. This might lead to additional opportunities as value-related drivers might lead to much more acceptance of the RRI amongst all stakeholders. Yet failure to develop shared values might lead to exactly the opposite. In the fifth step when solutions to the identified problems are chosen *shared values might help to choose between various opportunities.* In some cases it is possible to come up with a solution that is an improvement for everyone or an improvement for some with the situation being the same for the rest of the

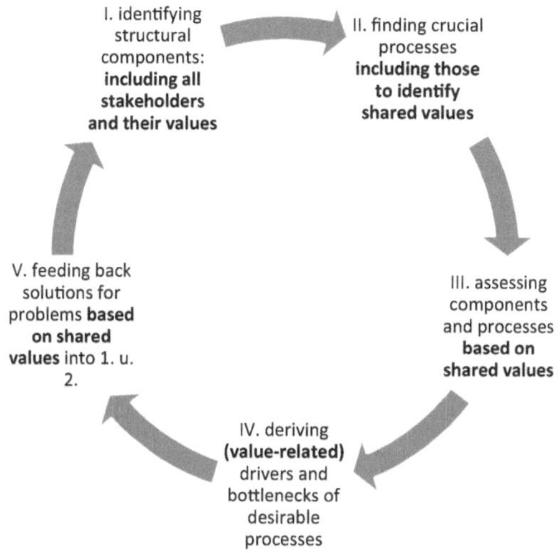

Figure 11.2 A scheme for assessing Responsible Research and Innovation (RRI) systems.

stakeholders. In other cases stakeholders need to find ways to compensate those worse off because of the innovation process (Taebi et al., 2014). In doing so the relationships between stakeholders as well as institutions might change, thereby also substantially changing the assessment of innovative processes and design.

To assess RRI systems, we have to add four more potential problems to the already existing list regarding innovation systems (Section 11.3.1). Yet these problems do also exist when not taking the RRI perspective. They are simply not acknowledged and might lead to severe acceptance problems, such as with the Dutch electronic patients' card which eventually fell through parliament because of privacy concerns that had not been sufficiently taken into consideration early on in developing these cards (Von Schomberg, 2013). The potential additional problems are:

11. problems in including all stakeholders (a) and all their values (b), e.g. accounting for environmental damage that will affect future generations
12. lack of adequate processes to communicate and collaborate in order to develop shared values
13 failure to compensate those stakeholders worse off because of an RRI
14. inability to base decisions on how to choose solutions for problems identified on shared values.

In order to detect the four additional problems summarized above as well as to develop solutions and measures that can be assessed we need additional indicators. While the information on all relevant stakeholders (11a) might be readily available – at least for experts in the field of the respective RRI – getting a clear picture on all their values (11b) might turn out to be difficult. In academic settings mid-stream modulation leads to a situation where individual researchers are more aware and reflect more on the social and ethical aspects of their work (Flipse, Sanden, & Osseweijer, 2013). This higher reflective awareness might be used to collect information on values in other settings as well. Another way to identify values was suggested by van de Poel et al. (2017). When companies translate their RRI strategy into activities resulting in RRI outcomes they could also develop company-specific RRI key performance indicators to assess their performance. The lack of adequate processes to communicate and collaborate in order to develop shared values (12) might be detected by an "embedded humanist" carrying out a mid-stream modulation (Flipse et al., 2013) but would require constant supervision by a third party. The same holds for the failure to compensate those stakeholders worse off because of an RRI (13) as well as the inability to base decisions on how to choose solutions for problems identified on shared values (14). A way out might be to choose indicators on a much more aggregated level, e.g. by uncovering whether research-performing organizations and research-funding organizations have and apply ethical guidelines (Mejlgaard, Bloch, & Madsen, 2018). Yet this does not indicate whether values are appropriately included, developed into shared ones and applied to the decision process when solving problems in the case of specific RRIs.

11.4 Assessing RRI systems in the digital age

The digital age comes with various opportunities and challenges, two of which are crucial for the assessment for RRI systems, i.e. big data and internet of things (IoT) platforms (Section 11.4.1). In Section 11.4.2 I will enhance the scheme of assessing RRI systems (see Section 11.3.2) by including opportunities and challenges of the digital age, as identified in Section 11.4.1.

11.4.1 The digital age: big data analytics and IoT platforms

There are two core elements of the digital age that have the potential to influence RRI systems substantially. The first element is big data, which changes the playing field of assessing RRI systems because of its very nature. The second is the IoT platforms which offer opportunities of connecting RRI systems.

11.4.1.1 Big data analytics

The digital age is driven by so-called big data and most particularly its analysis. Often large amounts of digitized information are defined as big data. Data, i.e. Latin for "things given" to us, stem from various sources, e.g. nature, careful observation or people (Jacobs, 2009). While chatting, shopping and connecting with platforms, we produce data (IDC, 2016; Loebbecke & Picot, 2015). Those of us wearing sensors, e.g. collecting heart rate measurement and number of steps walked, or blogging, add to this enormous amount of data even more (Porter & Heppelmann, 2014).

> What makes most big data big is repeated observations over time and/or space. The Web log records millions of visits a day to a handful of pages; the cellphone database stores time and location every 15 seconds for each of a few million phones; the retailer ... logs billions and billions of individual transactions in a year. Scientific measurements are often made at a high time resolution (thousands of samples a second in neurophysiology, far more in particle physics) .. [They] really start to get huge when they involve two or three dimensions of space as well; [e.g. in] neuroimaging.
>
> (Jacobs, 2009, p. 5)

While the sheer amount of data, i.e. its *volume*, is often used to define big data, there is much more to it. First of all, velocity and variety characterize big data as well (Erevelles, Fukawa, & Swayne, 2016; Galbraith, 2014).

Velocity points at the quick availability sometimes connected with big data. *Variety* captures the fact that big data usually stems from numerous sources, because it often combines both structured transactional data, e.g. from files or sensors, with unstructured behavioural data, e.g. from blogs or text messages. Second, to use big data stakeholders need big data analytics, i.e. "analytical techniques in applications that are so large (from terabytes to exabytes) and complex (from sensor to social media data) that they require advanced and unique data storage, management, analysis, and visualization technologies" (Chen, Chiang, & Storey, 2012, p. 1166).

Using big data analytics comes with numerous challenges both in practical terms and in terms of values. In practical terms the challenges emerge from the necessity to *digitize*, *align*, *analyse and store big data* (Galbraith, 2014; Kshetri, 2014). Stakeholders in RRI systems have to acknowledge and overcome these practical challenges to understand big data and the consequences of its use so that they can possibly use it for their own purposes. In terms of integrating all stakeholders of RRI systems into the processes of research and innovation this might become an additional challenge, as some of the stakeholders might be much better equipped with the capacities and financial means to engage in big data analytics than others, giving them an information advantage in the RRI processes.

In terms of values the characteristics of big data and the use of big data analytics lead to challenges with regard to privacy and security issues as well as concerns regarding welfare, discrimination and strategic behaviour. Analysing big data means that huge amounts of data from different sources, including unstructured data, e.g. from blogs, often collected and processed in real time are used to act and react (cf. this and the following Kshetri, 2014; Sogeti, 2013). Most organizations lack mechanisms of using and protecting unstructured data sufficiently, thereby creating *privacy and security issues*. Outsourcing data to cloud service suppliers comes with even more privacy and security issues, because it makes the process of data collection, storage and use even more complex. In addition, the process of how big data is collected, analysed and used is often not sufficiently transparent. So, the owners of the data, often private entities such as consumers or patients, do not get the chance to give informed consent to participate. Frequently they are even not aware that the combination of various kinds of sources makes it more likely that personal identifiable information about them is revealed. This can lead to *welfare losses*. While nowadays consumers have the chance to hide how much they are willing to pay for a specific product and are usually asked the same price as everyone else buying the product, in the digital world they might undergo price discrimination, because the seller of the product collected, combined and used a lot of data to find out how much they are willing to pay.

The consequences of personal identifiable information falling into the wrong hands can be even more severe. In cases where location-based real-time personal information is used this might lead to stalking in real time.

Moreover, while data-driven technologies have often been viewed as positive, they might enhance *discriminatory biases*, e.g. by embedding algorithms in data-driven technologies used for screening potential incoming students, discriminating against women or people with non-European names (Niklas & Pena Gangadharan, 2018). In addition, the nature of big data might inherently alter the exchange between stakeholders. It is known that *stakeholders might react strategically to assessment* (Freeman & Soete, 2009; van de Poel et al., 2017). As big data is often available in real time (Galbraith, 2014), this problem becomes much bigger, because we might even lose the information of how non-strategic behaviour would have looked like. It will become a challenge to determine shared values under such circumstances.

While big data analytics has been subject to severe concerns, at the same time, it might offer new ways of assessing RRI systems. Nowadays, *big data analytics often drives new solutions in research and innovation*. In these cases it also provides additional and appropriate ways in creating *new indicators to identify opportunities and challenges in the respective RRI systems*. Take the health sector where the well-established gender bias in treatment of women still has to be overcome (Hamberg, 2008). While there have been numerous studies on gender bias regarding cardiovascular disease, follow-up protocols about medical treatment and outcome by gender still need to be evaluated more carefully. By using big data analytics physicians and pharmacists can assess huge volumes of information on who is getting and using which drugs under what circumstances (healthworkscollective.com, 2014). Based on this they can design and implement optimal drug therapies – also for women. So, big data, in this example, provides the stakeholders in the RRI system with a solution for a problem identified as the shared value of gender equity in health care (step V along the lines of the scheme depicted in Figure 11.3).

Another example is autonomous driving. While huge amounts of data from different sources will help to make traffic much safer (PWC & Flanders Make, 2017), it does not make it 100% safe. So, in principle the well-established trolley problem of whom to hurt or kill – when you do not have the option of hitting no one, remains (Nyholm & Smids, 2016). Yet big data does not only help advancing safety in traffic it also provides those involved in the respective RRI system with much better and more detailed information to make the value choices involved. This example shows that big data might help to identify shared values in step II and use them along the lines of the scheme depicted in Figure 11.3.

While big data might help to identify shared values and deal with them in the RRI process, there need to be both appropriate data available and institutions in place to guard their use in the RRI process. Data about technologies that are almost readily available will usually suffice to support RRI processes. Yet for technologies in the beginning of their development this might not be the case. Moreover, the values of minorities, of relatively powerless groups of stakeholders or of future generations might be underrepresented. It is a typical task of governmental agencies to either

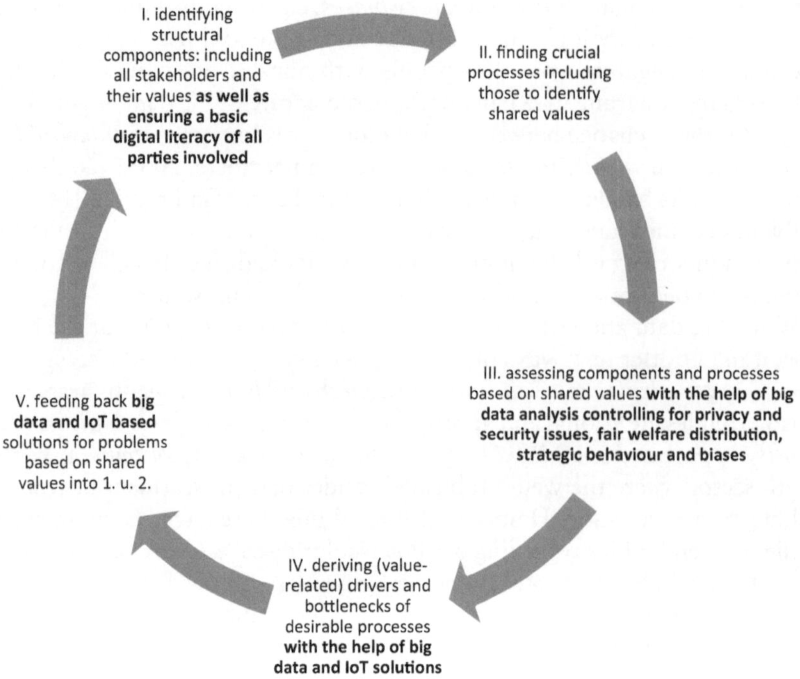

Figure 11.3 A scheme for assessing Responsible Research and Innovation (RRI) systems in the digital age.
Note: IoT, internet of things.

represent them or to make sure that the institutions suffice to have them represented appropriately.

11.4.1.2 IoT platforms

Important elements of the digital age are the *IoT platforms* which allow "communication between humans as well as machines in Cyber-Physical-Systems (CPS) throughout large networks" (Brettel, Friederichsen, Keller, & Rosenberg, 2014, p. 37). This does not only change industrial production processes, often subsumed under the title industry 4.0 (Brettel et al., 2014), but has the potential to change the situation for all stakeholders in RRI systems. Already modern communication had changed the way collaboration partners organize and maintain ties (Genoni, Merrick, & Willson, 2005; Haythornthewait, 2002; Rychen & Zimmermann, 2008). Putting IoT platforms to use to organize RRI processes, particularly taking on board all stakeholders of the RRI system, would give ample opportunities to change communication patterns. To give an example, it would be much more difficult to justify why stakeholders other than innovative agents would not be included in the process as IoT platforms would make it much easier. The EU Commission and other governmental bodies are already making use of this possibility by providing webpages where interested parties can give their input for new initiatives (EU Commission, 2018b). In addition, IoT platforms could be used to supervise RRI processes by following the communication partners on these platforms.

While using communication data might lead to strategic behaviour and might come with privacy or security concerns (see discussion above), it also means that the person or organization mediating the RRI process about how and when stakeholders contact each other and what kind of information they exchange has much more information. This might give them much more detailed insights into the technological, innovative and ethical aspects of the exchanges, thereby giving them the opportunity to influence the communication within RRI systems to overcome a lack of adequate processes to communicate and collaborate in order to develop shared values (problem 12 as identified in Section 11.3.2). This would mean that *the values of all stakeholders* would be *better included* and better aligned to technological opportunities emerging.

11.4.2 Assessing RRI systems in the digital age: a scheme

Assessing RRI systems in the digital age requires to consider its challenges and to use its opportunities. In Figure 11.3, I add the major opportunities and challenges of the digital age to the scheme provided for the assessment of RRI systems in Figure 11.2.

In the course of the digital age all stakeholders have to get some understanding of how the digitization, alignment, analysis and storage of big data take place and of how IoT platforms work (see Section 11.4.1). Stakeholders less digitally educated, less wealthy and less powerful than other stakeholders in an RRI system will have a hard time dealing with the opportunities and challenges of the digital age and will most likely fall behind (Bloem et al., 2013). This opens ample opportunity for governmental, academic and civic agents to step up by educating and involving these disadvantaged stakeholders (see step I in Figure 11.3). As big data has been driving big science, i.e. data-driven solutions in research, e.g. at CERN (Bloem et al., 2013), we might expect that the values emerging from, in this case, the academic sector, might already include goals of inclusiveness and enabling people by educating them. As long as an RRI system is not dominated by profit-oriented organizations only there is a good chance that the RRI process will lead to shared values, providing a level playing field in the RRI system.

When using big data analytics in RRI systems stakeholders have to be aware of privacy and security issues as well as concerns regarding welfare, discrimination and strategic behaviour (see Section 11.4.1). After having determined the structural components and the processes of the RRI system at hand it is therefore crucial to address these potential issue in the assessment of them in step III of Figure 11.3.

Currently, big data and its analytical tools are often the driver of new solutions in RRI systems. They come with new indicators to identify the opportunities and challenges in the RRI system they drive (see Section 11.4.1). Moreover, IoT platforms could be used to better include all stakeholders of RRI systems into the RRI process as well as better mediating this process (see Section 11.4.1), e.g. identifying capability problems of innovative agents (problem 1) or missing entrepreneurial experimentation (problem 7). As many of the problems mentioned in Section 11.3 could be identified and potentially solved by big data analytics and IoT solutions, they can be important elements in deriving (value-related) drivers and bottlenecks of desirable process in step IV of Figure 11.3. The same holds for feeding back possible solutions into the next round of the scheme in step V of Figure 11.3, which might also benefit from big data and IoT solutions. While big data and IoT solutions come with numerous potential concerns, including them in the RRI process should help to overcome them. For example, companies have the opportunity to offer big data-driven solutions that protect personally identifiable information by privacy by design, i.e. being transparent about what they do, complying with the wishes of the customers regarding privacy and explaining as much as possible about what they do with their data (Bloem et al., 2013).

11.5 Conclusions

As the digital transformation of our societies including RRI systems will take off even more in coming years, more consequences might emerge in the future. In order to assess RRI in the digital age we need a system approach, because RRI emerges from the co-evolution of stakeholders, their relationships and activities as well as the supporting institutions (see Sections 11.2 and 11.3). Moreover, we have to acknowledge and use the opportunities and challenges emerging from the digital transformation which changes the playing field of RRI systems substantially (see Section 11.4).

So far, we can derive three major lessons for the assessment of RRI systems in the digital age. First of all, in order to create a level playing field in the RRI process it is important that all stakeholders have at least some understanding of how the digitization, alignment, analysis and storage of big data take place and of how IoT platforms work. This will help them to assess their situation and act accordingly. Second, all stakeholders have to understand that ignoring or suppressing the values of other stakeholders involved will lead to complications along the process. So, it is important to be aware of and to acknowledge the privacy and security issues as well as concerns regarding welfare, discrimination and strategic behaviour of all stakeholders involved. Third, big data analytics and IoT platforms are more often than not giving the tools for assessing and maintaining RRI systems.

Big data analytics often drive new solutions in RRI systems and provide indicators to assess the related opportunities and challenges. Moreover, IoT platforms could be used to better include all stakeholders of RRI systems into the RRI process as well as better mediating this. Therefore, big data analytics and IoT solutions can be important in governing RRI processes.

Values are core to the evolution and assessment of RRI systems. Therefore, it is crucial that stakeholders from all sectors, i.e. government, academia, industry and civic society, have a clear picture of their own values, how they change in time, what the values of other stakeholders in their RRI system look like and what opportunities and ways there are to develop shared ones. In the digital age research and innovation offer an abundance of opportunities to solve commercial, economic and societal problems responsibly. How stakeholders of RRI systems assess them and how they deal with the challenges will to a large extent determine how our society will look in the future.

By using shared values as the cornerstone to assess the components and activities of RRI systems the question of which activities are welcome or not in a society can be answered on the micro-level. The use of shared values requires that all stakeholders eventually agree on them. Yet it opens up the more fundamental question of how the direction and intensity of innovative activities (Cantner & Vannuccini, 2018) are related to the development of shared values. As long as we have a clear view on the various options RRI opens up in the future – which is a non-trivial problem as such – shared values can guide us in finding jointly acceptable directions of innovative activities. Yet shared values do not provide any objective criteria of the "right" intensity of innovative activities – a problem possibly only solvable by experimentation. Clarifying the relationship between shared values on the one hand and the direction and intensity of innovation activities on the other hand might open up a perspective on innovation policy that goes beyond market or system failure approaches. This is subject to further research.

References

Bergek, A., Jacobsson, S., Carlsson, B., Lindmark, S., & Rickne, A. (2008). Analyzing the functional dynamics of technological innovation systems: A scheme of analysis. *Research Policy*, *37*(3), 407–429. doi:10.1016/j.respol.2007.12.003

Bloem, J., van Doorn, M., Duivestein, S., van Manen. T., van Ommeren, E., & Sachdeva, S. (2013). *No More Secrets with Big Data Analytics* (pp. 206). Groningen: Sogeti. ISBN: 9789075414738.

Boschma, R. (2005). Proximity and innovation: A critical assessment. *Regional Studies*, *39*(1), 61–74. doi:10.1080/0034340052000320887

Brettel, M., Friederichsen, N., Keller, M., & Rosenberg, M. (2014). How Virtualization, Decentralization and Network Building Change the Manufacturing Landscape: An Industry 4.0 Perspective. Retrieved from: http://waset.org/publications/9997144

Caniëls, M. C. J., Kronenberg, K., & Werker, C. (2014). Conceptualizing proximity in research collaborations. In R. Rutten, P. Benneworth, D. Irawati, & F. Boekma

(Eds.), *The Social Dynamics of Innovation Networks* (pp. 221–238). London: Routledge.

Cantner, U., & Vannuccini, S. (2018). Elements of a Schumpeterian catalytic research and innovation policy. *Industrial and Corporate Change*, 27(5), 833–850. doi:10.1093/icc/dty028

Chen, H., Chiang, R. H. L., & Storey, V. C. (2012). Business intelligence and analytics: From big data to big impact. *MIS Quarterly*, *36*(4), 1165–1188.

Edquist, C. (2011). Design of innovation policy through diagnostic analysis: Identification of systemic problems (or failures). *Industrial and Corporate Change*, *20*(6), 1725–1753. doi:10.1093/icc/dtr060

Edquist, C., Zabala-Iturriagagoitia, J. M., Barbero, J., & Zofío, J. L. (2018). On the meaning of innovation performance: Is the synthetic indicator of the Innovation Union Scoreboard flawed? *Research Evaluation*, 27(3), 196–211. doi:10.1093/reseval/rvy011

Erevelles, S., Fukawa, N., & Swayne, L. (2016). Big data consumer analytics and the transformation of marketing. *Journal of Business Research*, *69*(2), 897–904. doi:10.1016/j.jbusres.2015.07.001

Etzkowitz, H., & Viale, R. (2010). Polyvalent knowledge and the entrepreneurial university: A third academic revolution? *Critical Sociology*, *36*(4), 595–609. doi:10.1177/0896920510365921

EU Commission. (2018a). *European Innovation Scoreboard*. Retrieved from: https://op.europa.eu/en/publication-detail/-/publication/8e458033-74fc-11e8-9483-01aa75ed71a1/language-en/format-PDF/source-99539237

EU Commission. (2018b). Have your Say. Retrieved from: https://ec.europa.eu/info/law/better-regulation/have-your-say_en

European Commission. (2012). A European Strategy for Key Enabling Technologies – A Bridge to Growth and Jobs. Retrieved from: http://eur-lex.europa.eu/LexUriServ/LexUriServ.do?uri=COM:2012:0341:FIN:EN:PDF

European Commission. (2013). *Options for Strengthening Responsible Research and Innovation*. Luxembourg: European Commission.

Flipse, S., Sanden, M. A., & Osseweijer, P. (2013). Midstream modulation in biotechnology industry: Redefining what is 'part of the job' of researchers in industry. *Science and Engineering Ethics*, *19*(3), 1141–1164. doi:10.1007/s11948-012-9411-6

Forsberg, E.-M., Shelley-Egan, C., Ladikas, M., & Owen, R. (2018). Implementing Responsible Research and Innovation in research funding and research conducting organisations – what have we learned so far? In F. Ferri, N. Dwyer, S. Raicevich, P. Grifoni, H. Altiok, H. T. Andersen, Y. Laouris, & C. Silvestri (Eds.), *Governance and Sustainability of Responsible Research and Innovation Processes* (pp. 1–11). Springers Brief in Research and Innovation Governanc, Springer, Cham (Switzerland). https://doi.org/10.1007/978-3-319-73105-6_1

Freeman, C., & Soete, L. (2009). Developing science, technology and innovation indicators: What we can learn from the past. *Research Policy*, *38*, 583–589. doi:10.1016/j.respol.2009.01.018

Galbraith, J. R. (2014). Organizational design challenges resulting from big data. *Journal of Organization Design*, *3*(1), 2. doi:10.7146/jod.8856

Gebauer, J., Füller, J., & Pezzei, R. (2013). The dark and the bright side of co-creation: Triggers of member behavior in online innovation communities.

Journal of Business Research, *66*(9), 1516–1527. doi: https://doi.org/10.1016/j.jbusres.2012.09.013

Genoni, P., Merrick, H., & Willson, M. (2005). The use of the internet to activate latent ties in scholarly communities. *First Monday*, *10*(12). doi: http://dx.doi.org/10.5210/fm.v10i12.1301

Grupp, H., & Schubert, T. (2010). Review and new evidence on composite innovation indicators for evaluating national performance. *Research Policy*, *39*(1), 67–78. doi:10.1016/j.respol.2009.10.002

Hamberg, K. (2008). Gender bias in medicine. *Women's Health*, *4*(3), 237–243. Retrieved from: http://journals.sagepub.com/doi/pdf/10.2217/17455057.4.3.237

Haythornthewait, C. (2002). Strong, weak, and latent ties and the impact of new media. *The Information Society*, *18*, 385–401. doi:10.1080/01972240290108l9

healthworkscollective.com. (2014). Five Ways Big Data is Improving Patient Outcomes. Retrieved from: www.healthworkscollective.com/ways-big-data-healthcare-improving-patient-outcomes/

IDC. (2016). *Industrial Data Platforms – Key Enablers of Industry Digitization*. Retrieved from: www.key4biz.it/wp-content/uploads/2018/04/SMART20130063_Final-Report_030417_2.pdf

Jacobs, A. (2009). The pathologies of big data. *ACM Queue*, *2014*, 1–12. Retrieved from: http://queue.acm.org/detail.cfm?id=1563874

Klein Woolthuis, R., Lankhuizen, M., & Gilsing, V. (2005). A system failure framework for innovation policy design. *Technovation*, *25*(6), 609–619. doi:10.1016/j.technovation.2003.11.002

Kshetri, N. (2014). Big data's impact on privacy, security and consumer welfare. *Telecommunications Policy*, *38*(11), 1134–1145. doi:10.1016/j.telpol.2014.10.002

Loebbecke, C., & Picot, A. (2015). Reflections on societal and business model transformation arising from digitization and big data analytics: A research agenda. *Journal of Strategic Information Systems*, *24*, 149–157. doi:10.1016/j.jsis.2015.08.002

Mejlgaard, N., Bloch, C., & Madsen, E. B. (2018). Responsible research and innovation in Europe: A cross-country comparative analysis. *Science and Public Policy*. doi:10.1093/scipol/scy048

Mitchell, R. K., Agle, B. R., & Wood, D. J. (1997). Toward a theory of stakeholder identification and salience: Defining the principle of who and what really counts. *Academy of Management Review*, *22*(4), 853–886. doi:10.5465/amr.1997.9711022105

Nelson, A. J. (2009). Measuring knowledge spillovers: What patents, licenses and publications reveal about innovation diffusion. *Research Policy*, *38*(6), 994–1005. doi:10.1016/j.respol.2009.01.023

Niklas, J., & Pena Gangadharan, S. (2018). *Data-driven Discrimination: A New Challenge for Civil Society*. London: London School of Economics Impact Blog.

Nyholm, S., & Smids, J. (2016). The ethics of accident-algorithms for self-driving cars: An applied trolley problem? *Ethical Theory and Moral Practice*, *19*(5), 1275–1289. doi:10.1007/s10677-016-9745-2

OECD/Eurostat. (2018). *Oslo Manual 2018*. Retrieved from www.oecd-ilibrary.org/docserver/9789264304604-en.pdf?expires=1599730730&id=id&accname=guest&checksum=3CEF6403AC7338FF492B28799B5117BF

Porter, M. E., & Heppelmann, J. E. (2014). How smart, connected products are transforming competition. *Harvard Business Review*(November), 1–23.

PWC, & Flanders Make. (2017). *Industry 4.0: Hype or Reality?* Retrieved from: www.pwc.be/en/news-publications/publications/2017/industry-hype-or-reality.html

Rychen, F., & Zimmermann, J.-B. (2008). Clusters in the global knowledge-based economy: Knowledge gatekeepers and temporary proximity. *Regional Studies*, *42*(6), 767–776. doi:10.1080/00343400802088300

Schwerin, J., & Werker, C. (2003). Learning innovation policy based on historical experience. *Structural Change and Economic Dynamics*, *14*(4), 385–404. doi:10.1016/s0954-349x(03)00031-6

Stilgoe, J., Owen, R., & Macnaghten, P. (2013). Developing a framework for responsible innovation. *Research Policy*. doi:10.1016/j.respol.2013.05.008

Taebi, B., Correljé, A., Cuppen, E., Dignum, M., & Pesch, U. (2014). Responsible innovation as an endorsement of public values: The need for interdisciplinary research. *Journal of Responsible Innovation*, *1*(1), 118–124. doi:10.1080/23299460.2014.882072

van de Poel, I. R. (2009). Values in engineering design. In A. Meijers (Ed.), *Philosophy of Technology and Engineering Sciences* (pp. 973–1006). Amsterdam: Elsevier. www.sciencedirect.com/science/article/pii/B9780444516671500409

van de Poel, I., Asveld, L., Flipse, S., Klaassen, P., Scholten, V., & Yaghmaei, E. (2017). Company strategies for Responsible Research and Innovation (RRI): A conceptual model. *Sustainability*, *9*(11), 2045. doi:10.3390/su9112045

Von Schomberg, R. (2013). A vision of Responsible Research and Innovation. In R. Owen, M. Heintz, & J. Bessant (Eds.), *Responsible Innovation* (51–74). London: John Wiley.

Werker, C. (2001). Knowledge and organisation strategies in innovation systems. *International Journal of Innovation Management*, *5*(1), 105–127. https://doi.org/10.1142/S1363919601000300

Werker, C., Ooms, W., & Caniëls, M. C. J. (2016). Personal and related kinds of proximity driving collaborations: a multi-case study of Dutch nanotechnology researchers. *SpringerPlus*, *5*(1). doi:10.1186/s40064-016-3445-1

Best practice V Managing social impacts and ethical issues of research and innovation: the CEN/WS 105 guidelines to innovate responsibly

Andrea Porcari and Elena Mocchio

Aims and goals

The standard under development (formally named as workshop CEN/WS[1] 105) aims to provide practical guidelines for organisations to approach and evaluate the most significant ethical, legal and social impacts (ELSI) expected by the development of their innovative products, and to identify a Responsible Research and Innovation (RRI) vision and operative RRI actions (roadmap) to address them.

Description of good practice example

Recent experiences suggest that RRI implementation can support anticipation of impacts of technological innovation and prevent externalities from the early stages of product or service development. For this reason, a roadmap for RRI implementation in the industrial context is under development at the European Committee for Standardisation (CEN),[2] starting from the experience of the PRISMA project[3] regarding RRI uptake in companies dealing with transformative technologies, such as nanomaterials, synthetic biotechnologies, Internet of Things and autonomous vehicles.

The roadmap is meant as a support tool for strategy setting, integrating technological and responsibility aspects. It is addressed to organisations involved in planning and performing technological development, with a focus on enabling technologies. It defines a set of acknowledged principles based on three key actions:

1. Integrate the analysis of ELSI from the early stages of product/service development (reflection and anticipation)

Figure V.I European Committee for Standardisation (CEN) Workshop (CEN/WS) 105: methodological steps for the design of the Responsible Research and Innovation (RRI) roadmap.

2. Perform stakeholder engagement (inclusiveness)
3. Integrate monitoring, learning and adaptive mechanisms to address public and social values and normative principles (responsiveness).

These actions are considered as a minimum requirement for RRI implementation. The methodology follows a circular process, based on six steps (Figure V.I and Table V.I). The final step is the design of an RRI roadmap, using the architecture of Innovation Policy Road-mapping Methodologies (Ahlqvist et al., 2012).

The guidelines take into account terminology and approaches of ISO and CEN standards on management systems in the areas of social responsibility, sustainability, innovation, quality and risk assessment (e.g. ISO 26000, ISO 31000 , EN ISO 9001 and series ISO 56000).

PRISMA experience shows that RRI implementation is context-sensitive: what exactly it entails and what it can deliver depend on the sector, technology, type of company/organisation and business. RRI actions involve different functions and could have both tangible and intangible,

Table V.I CEN/WS 105: methodological steps for the design of the Responsible Research and Innovation (RRI) roadmap

Top management committment and leadership	Ensure endorsement by the organisation of RRI values and approach
Context analysis	Analyse the organisation, the R&I product(s) and technologies on which to focus; identify ethical, social and legal impacts of the product and stakeholders of the product innovation eco-system
Materiality	Identify and prioritise: drivers and challenges for RRI; risks and barriers to overcome; stakeholders to work with; significant RRI actions to pursue
Experiment and engage	Perform exploratory/pilot RRI actions, engaging with stakeholders to inform the RRI roadmap
Validate	Evaluate impact of the roadmap on both the product development and the organisation (key performance indicators)
Roadmap design	Consolidate and visualise the long-term RRI strategy, covering all the R&I value chain (time to market) and product life-cycle

Note: R&I, research and innovation.

short and long-term impacts at different levels (product/service, company, stakeholders, society).

The responsible innovation guidelines provide a flexible and modular approach, that could be adapted to any size and type of organisation and implemented both through informal and formal procedures, to guide organisations to integrate responsibility aspects all along the research and innovation stages and product development.

Benefits for the organisation

These guidelines could help organisations to introduce structural changes in their usual business practices, toward more anticipatory, inclusive, responsive research and innovation practices, anticipating social or market trends or requirements, technological scenarios, possible regulatory changes and building trust and legitimacy.

Notes

1 A CEN Workshop (CEN/WS) is a flexible working platform open to the participation of any company, organization or interested expert for rapid elaboration of CEN Workshop Agreements (CWA). A CWA is a technical document agreed by the participants of a CEN/WS, which can be the basis for future standardization works.

2 www.cen.eu.

3 PRISMA project, Horizon 2020 research and innovation programme, grant agreement no. 710059 – www.rri-prisma.eu

References

Ahlqvist, T., Valovirta, V., & Loikkanen, T. (2012). Innovation policy roadmapping as a systemic instrument for forward-looking policy design. Science and Public Policy, 39(2), 178–190.

CEN (2018). *Innovation Management*. Series CEN/TS 16555. Retrieved from https://standards.cen.eu/dyn/www/f?p=204:110:0::::FSP_PROJECT,FSP_ORG_ID:35932,671850&cs=13A816A57184977C465944D2F2E2C5645

ISO (2010). *Guidance on Social Responsibility.* ISO 26000: 2010. Retrieved from www.iso.org/iso-26000-social-responsibility.html

ISO (2010). *Using ISO 26000 in Management Systems*. IWA 26. Retrieved from www.iso.org/standard/72669.html

ISO (2015). *Quality management systems – Requirements*. EN ISO 9001. Retrieved from www.iso.org/standard/62085.html

ISO (2018). *Risk Management – Guidelines*. ISO 31000: 2018. Retrieved from www.iso.org/standard/65694.html

ISO (2019). *Innovation Management -- Innovation management system – Guidance*. ISO 56002: 2019. Retrieved from www.iso.org/Standard/68221.html

UNI (2016). *Social Responsibility in Organizations – Guidance to the Application of UNI ISO 26000*. UNI/PdR 18. Retrieved from http://store.uni.com/catalogo/index.php/uni-pdr-18–2016.html

UNI (2017). *Guidelines for Management and Processes Development for Responsible Innovation*. UNI/PdR 27. Retrieved from http://store.uni.com/catalogo/index.php/uni-pdr-27–2017.html

12 RRI intensity

A proposed method of assessing the requirement for responsible innovation in ICT projects

Martin De Heaver, Marina Jirotka, Margherita Nulli, Bernd Carsten Stahl, and Carolyn Ten Holter

12.1 Introduction

The rapid pace of technological progress in recent decades has not, it is now widely acknowledged, been matched by a similar increase in society's ability to manage the level and variety of change that this progress has brought (Wadhwa, 2014). Although ICT technologies such as machine learning and robotics have vast potential, alongside nano- and bio-technologies, they are deeply pervasive and their effects – both positive and negative – are only just beginning to be understood.

12.1.1 Societal impact

The acceleration of technology-enabled transformation proliferating through society creates an ever-decreasing window of opportunity for society to adapt. Technologies evolve at a much greater pace than societies can, and so change has time to spread before its true systematic impact can be recognized, quantified and assimilated. In addition, ICT is now deeply embedded within society and its omnipresence – as well as the way in which it has supplanted older forms of technology (for example, navigation apps have transformed the way we consume geospatial data) – means that the issues it brings with it are also ubiquitous.

12.1.2 Possible responses: before and after impact occurs

An *ex post* response – reacting to impact that may already have occurred – is only one way of dealing with the broader implications of technology-enabled transformation. It is the main method that is open to governments, regulatory bodies and other agencies that consider the effects of technology on the societies of which they are a part. However, regulatory frameworks, laws and the passage of cases through the courts all take time. These mechanisms are often slow, unwieldy and may only be able to react a long time after impacts have already occurred. Even institutional bodies set up to examine particular issues, such as the Ada Lovelace Institute[1] and the Centre for Data Ethics and Innovation,[2] are responding only slowly to events. In the case of the Cambridge Analytica use of Facebook's data, the implications were only beginning to become clear over two years after the event, and as of December 2018, no prosecutions have taken place (though Facebook was fined £500,000 by the UK's Information Commissioner in October 2018 – the maximum possible[3]).

An alternative to mechanisms that are predicated on actual impact are mechanisms that act *ex ante*. This approach aims to act to anticipate impact. There are many ways of potentially doing this, but many focus on the actions of researchers, developers and other innovators at the design and development stages of new technologies. These approaches are often drawn together collectively under the banner of responsible innovation (RI) or Responsible Research and Innovation (RRI). However, it is crucial to consider that RRI is not just about maximising benefits and managing risks, but about integrating societal preferences and actors into research and innovation (R&I).

The UK's Research Councils, which fund much of the UK's academic research, have adopted the AREA (anticipate, reflect, engage, act) framework approach to RRI created by Stilgoe, Owen, and Macnaghten in 2013. RRI is sometimes regarded as a type of risk management, but its fundamental aim is to bring science and society closer together and to foster mutual learning by involving a range of stakeholders in the R&I process.

In this paper we go beyond the *'why'* of RRI to consider the qualitative features and quantitative parameters which could help to guide researchers in making a case for *how* and *how much* RRI should be undertaken to create the best outcome for society. We introduce the concept of RRI intensity level (RIL) as a method of evaluating what type of and how much RRI activity is meaningful, useful, sensible and effective (MUSE). We calculate the RIL based upon assessments of two parameters – the anticipated technology readiness level (TRL) and the novel concept of innovation potential. We define innovation potential here as an indicator of the potential breadth and depth of societal impact of a technology, and suggest that this is a crucial parameter to consider when determining the choice and scale of the RRI activities to be undertaken.

In calculating this anticipated impact, it is necessary to use some agreed goal against which to measure the impact, and we utilise the United Nations (UN) sustainable development goals (SDGs), as these represent the best current approximation for what the representatives of a significant proportion of humanity regard as important.

12.1.3 Research gap

The academic discourse around RRI is now well established, with numerous academic papers (e.g. Jirotka, Grimpe, Stahl, Eden & Hartswood, 2017;

Martinuzzi, Blok, Brem, Stahl, & Schönherr, 2018; Stahl, Obach, Yaghmaei, Ikonen, Chatfield & Brem, 2017; von Schomberg, 2013) and even a dedicated academic journal (the *Journal of Responsible Innovation*). There are also many dedicated projects funded by the European Commission (EC) and others looking at RRI in different disciplines and fields of industry. These include: education (Ark of Enquiry, HEIRRI, IRRESISTABLE); industry (Responsible-Industry, COMPASS); research-funding organisations (RRI-Practice) or the application of RRI in specific areas of R&I, such as synthetic biology (Zhang, Marris, & Rose, 2011) or ICT (Jirotka, Grimpe, Stahl, Eden, & Hartswood, 2017). This volume of research work demonstrates that RRI has taken its place as an established and reputable means of maximising the benefits of technology transformation whilst minimising the potentially harmful associated outcomes.

What is less clear, however, is the degree to which it is necessary to realise and implement RRI to optimise its beneficial effects. We therefore use this chapter to discuss the following research question:

- How can the level of RRI activity be optimised to suit the characteristics of an individual project?

Given the variety of projects where RRI is mandated by funding bodies, the range of technology stages that may be covered and the relative experience of the project investigators in implementing RRI, this is a critical question that needs to be answered appropriately in order for practical RRI activities to be developed and implemented in a manner that suits the project variables.

In order to answer this research question, we first review the definitions and current discourse on RRI. This is followed by the introduction of the concept of RIL. We show how this concept can be extended from the literature and which factors would be important to consider. We then develop this into a formal structure of RIL. The chapter closes by suggesting further research that would help to validate this idea.

12.2 Responsible Research and Innovation (RRI)

12.2.1 Definitions and foundation

RRI has been described as a "way of thinking more systematically about the public benefits of science and technology-based research" (Nuffield Council on Bioethics, 2013), though in the same document the council points out the number of other definitions and factors to be considered. Since then, the RRI community has largely coalesced around this broad view, with a lengthier definition, from von Schomberg, now being widely cited. This sees RRI as a:

> transparent, interactive process by which societal actors and innovators become mutually responsive to each other with a view to the (ethical)

> acceptability, sustainability and societal desirability of the innovation process and its marketable products (in order to allow a proper embedding of scientific and technological advances in our society).
>
> (von Schomberg, 2013, p. 63)

This definition is in accordance with the aforementioned broad description, but it is crucial to consider that RRI is not just about managing risks and maximising benefits, but about integrating societal preferences and actors into the R&I process.

In the same vein, Stilgoe, Owen, and Macnaghten (2013) proposed a framework for RRI that focuses on four integrated dimensions to be adopted when carrying out responsible research activities. These are anticipation, reflexivity, inclusion and responsiveness. This framework was adapted and adopted by the UK Engineering and Physical Sciences Research Council (EPSRC) to form the AREA framework (Owen, 2014).

Another highly influential contribution to the RRI discourse comes from the EC that, as the largest funder of research in Europe, has a significant influence on research policy. The EC has adopted RRI as a cross-cutting activity that informs all aspects of its Horizon 2020 research framework programme in the years 2014 to 2020. In the EC's view, RRI consists of a number of pillars that correspond to established research policy areas. These are civil society engagement, gender equality, science education, open access, ethics compliance and governance.

12.2.2 Development and implementation

The overall aim for RRI is to promote the consideration of the effects of R&I on the environment and society before, during and after a particular project. While several studies have focused on RRI in the broad theoretical sense, few have detailed the specifics of the ways in which those RRI concepts can be applied to actual projects. The practical reflections in this chapter will enrich the further development of the RRI framework, thereby contributing to its wider use.

RRI principles allude to Latour's observation that "there [is] no direct connection between scientific results and the larger context of society, which could do no more than slow down and speed up the advancement of an autonomous science" (Latour, 1998, p. 208). The discourse around RRI has now reached a significant level of maturity. Following initial attempts to define the term and promote it as a relevant aspect of research governance, there is now growing agreement on key aspects of RRI and possible means of implementation. This does not mean, however, that RRI is unproblematic. There are deep-seated contradictions and problems that can limit the potential success of RRI. One example is the desire of RRI to broaden the conversation about R&I to include wider stakeholder groups; this sits very uneasily with the traditionally tightly focused definition of excellence in research. Another challenge for RRI is that much of the academic discourse centres around the carrying out of publicly funded research. However, a significant amount of research, and a large proportion of innovation activities, are conducted by privately funded organisations. It is not immediately obvious whether the arguments in favour of RRI are relevant to companies, or whether the incentive structures that are being put in place resonate with companies' R&I processes. There is, therefore, a significant need for investigation of the relevance and potential role for RRI in industry.

Using the core concept of responsibility in RRI as a point of departure, Stahl (2013, p. 3) suggests that RRI could be further refined "as a higher-level responsibility or meta-responsibility that aims to shape, maintain, develop, coordinate and align existing and novel research and innovation-related processes, actors and responsibilities with a view to ensuring desirable and acceptable research outcomes". Stahl also viewed RRI as a space defined by activities, actors and norms, thus emphasising the need for a systemic approach to determine the acceptability and desirability of R&I.

12.2.3 RRI as an extension of prior methods

For innovations in ICT, the introduction of RRI may be viewed as a reinforcement and extension of existing models and conceptual frameworks, such as user acceptance models (Davis, 1989) and participatory action research approaches (Baskerville, 1999). While these models emphasise user involvement for increasing the acceptance of technology products and services, the Scandinavian research on information system development stresses workplace democracy as a driver for involving users and giving them power over design decisions (Bjerknes and Bratteteig, 1995). However, the discussions around user involvement and participation have been limited to the organisational level. Participatory approaches have also been used mainly to promote the traditional view of using participation for "right impact" and risk mitigation (Owen, Macnaghten, & Stilgoe, 2012). The concept of engagement in RRI expands the discourse of user engagement to the societal level for critically investigating the social, economic and ethical dilemmas and opportunities surrounding the decisions, actions and outcomes related to R&I. RRI, however, is more than a theoretical debate. It has become an overarching practical R&I governance approach for ensuring the desirability, acceptability and sustainability of ICT-based innovations (Jirotka et al., 2017).

12.2.4 Balancing creativity and compliance

The discourse on the nature, justification and outcomes which might be expected of RRI has stimulated many research projects. However, when it comes to the practicalities of implementing RRI as part of an R&I activity the discourse has bifurcated into two distinct strands: one which favours detailed methodologies akin to those developed in projects such as COMPASS and RRI Tools, and the other which suggests that researchers looking to implement RRI in their projects draw upon their innate creativity in defining how RRI should be implemented for a particular project in a particular field.

We recognise that the discourse on RRI has, with the best of intentions, focused on avoiding being seen as a constraint on innovation and curiosity, but this has arguably come at the cost of making the scoping, design, management, costing and justification of RRI activities in research projects difficult for researchers not experienced in RRI. Thus, the appeal to creativity rather than compliance has stimulated innovation, but could be seen as a contributing factor to the inconsistent approach in the adoption of RRI in ICT projects in particular.

We recognise that an overly prescriptive and formulaic approach to how RRI is implemented could be seen as confining the freedom of the ICT research community; however in other fields where structured processes have been implemented for simple elements of highly complex tasks, these simple procedures have been found to yield significant benefits. A good example of this is the use of checklists in surgical procedures. Such checklists were trialled in surgical procedures and found to significantly reduce clinical error, and improve patient outcomes, despite the initial resistance and scepticism of the surgical community (Gawande, 2010). Checklists have also been used in the aviation sector for many years, for example in pre-flight checks, as a way of avoiding simple mistakes within the context of a complex set of tasks. These findings demonstrate that highly experienced professionals can avoid simple but important errors by the use of prescriptive procedures which, rather than constraining their creativity, liberate them from the risks associated with systematic error, and allow them to focus on areas where creativity rather than compliance will yield the greatest outcome.

We are not suggesting that RRI needs to be implemented in the form of checklists. The example of such checklists is used to show the importance of taking seriously the question of implementation of generally accepted, socially beneficial processes, and how a relatively structured approach can be valuable. While there has been much discussion of ways of "doing" RRI in the literature and while there are numerous case studies, there is at present no clear view on which RRI-related processes or activities would be required or appropriate in a particular situation. This is a clear gap in the literature, as it is intuitively plausible that different types, levels and sizes of projects will require different responses. So far, however, there has been little discussion of how a decision on the appropriate implementation of RRI in a particular case should be made. In order to address this gap, we propose the idea of RIL.

12.3 Responsible Research and Innovation intensity level

This methodology is intended to be applied during a project definition process, most likely as part of an application for funding. The RIL process is an

RRI activity in its own right in as much as it helps to define the intensity of RRI which should take place in the future, should the project be carried out.

Any assessment of RI requirements must gauge not only what RRI must be undertaken (what is *necessary*) but also how much is enough (*sufficient*). These two questions must both be answered correctly in order to ensure that the right balance is found. An appropriate pathway must be followed between a zealous approach to RRI that could be seen to be a constraint on curiosity as well as potentially a misuse of public research funding, or an overly *laissez-faire* approach that could put society at risk from innovation that has not been carefully considered.

The questions of *whether* and *why* RRI is carried out must therefore be extended to include the concepts of when, *how* and *how much* RRI should be undertaken. Utilising and integrating the concept of the RRI maturity model proposed by Stahl et al. (2017), the authors are therefore proposing this new concept aimed at the assessment of this measurement. This method seeks to gauge how much RRI is both *necessary* and *sufficient*. This assessment is a crucial precursor to the actual carrying out of RRI activities, which cannot be undertaken without reference to the context and development of a particular project.

This novel method is being termed RIL. It is a method of evaluating a research project (with a particular focus on ICT projects) and then selecting how much and what type of RRI activity is feasible, appropriate and effective.

We propose an approach to establishing the RIL based upon a matrix that takes into account its position along two axes: the TRL, and the innovation potential level. Innovation potential is a hybrid scale intended to indicate in what manner and to what degree a technology might affect wider humanity. The result of a given project's position along these two axes results in an RIL score. The score provides a measurement of when, how and how much RRI should be undertaken before and during the project. For example, a high RIL score may indicate very significant stakeholder involvement, perhaps even at policy-making level, as well as other governance measures such as a series of agreed "stage-gates" (cf. Stilgoe et al., 2013). Conversely, a low RIL may only require involvement of the researcher's broader team or department and sufficient departmental or supervisory oversight.

12.3.1 Technology readiness and innovation potential

The selection of these two dimensions is based on the recognition that RRI practice is related to at least two aspects of a research or innovation project.

The first of these is the project's *readiness*. This links to Collingridge's famous dilemma of control (Collingridge, 1981). As Collingridge showed, at an early stage of development it is relatively easy to change the trajectory and thus the outcomes of a particular technology. However, at such an early stage it is difficult to predict what exactly the outcomes will be. During the development process, as a piece of innovation or technology matures, its consequences become clearer but the trajectory is more difficult to change. Another analogy would be that of stem cells, that in their initial state may develop into many different forms, but once that development matures, the earlier options become closed off. This shows the need for RRI along the entire life cycle of an innovation but also demonstrates that more mature technologies need to be covered in more depth, as their consequences can be understood better, not only by their developers but by other stakeholders in society. Therefore, in order to measure this developmental stage of a project, we propose using a well-established indicator (Olechowski et al., 2015) of maturity of a technology, the TRL. TRLs were originally developed by NASA[4] as a method of measuring the maturity of space exploration technology. A variety of different industries now use the same approach and utilising this measure provides a cross-platform, consensus-based measure that can be understood across both academia and industry. It did not appear to be valuable to create a new measure of assessing a project's developmental phase.

The second aspect that affects RRI decisions is the *impact* and *reach* of a project. Technology readiness is important, but it is easy to see that on its own it is not sufficient to provide accurate information about the amount of effort that needs to be invested into RRI in a particular instance. A technology may have a high TRL and be close to market, but could have very little societal impact, thus the TRL measure alone is insufficient to determine the level of RRI activities which are appropriate. We therefore need an indication of the *type* (impact) and *level* (reach) of social change that the technology may create. There are various indicators for the level of disruption that can result from the deployment of a technology, such as those from KPMG,[5] Gartner,[6] Cisco[7] and other professional service organisations, but these do not concern themselves with social disruption. Their focus is on disruptive impact on business and therefore seek to find ways to manage risk. These may, therefore, be sufficient for the risk management part of RRI.

However, it is crucial to consider that RRI is not just about managing risks and maximising benefits, but about integrating societal preferences and actors into R&I. If RRI is to incorporate and take into account societal values, then it is vital to understand the level to which these are likely to be affected by an innovation, in order to determine the scope and scale of RRI-related efforts. There is much debate within the RRI community about the types and instantiations of these values that R&I are meant to promote. Suggestions include philosophical ethics, more applied approaches such as research ethics or corporate social responsibility, or more politically oriented candidates such as human rights (Oftedal, 2014). We believe that all of these are valuable, but lack the specificity required to guide practical action. We therefore propose to rely on a set of normative aims that are now well established, build on and incorporate many of the earlier more abstract values, and that are internationally discussed and agreed. These are the UN's SDGs (United Nations, 2015), as we will show in more detail below. Relevance of a piece of research, innovation or technology to the

SDGs is therefore the second measure that we consider when determining the level of RRI that would be *necessary* and *sufficient* to ensure that societal needs and preferences are expressed.

12.3.2 Step 1: Axis 1 – technology readiness level

To be effective, RRI needs to be embedded in the research process, but we posit, at an appropriate level of intensity, proportional to the stage that the research is attempting to reach. For example, undertaking public engagement activities for basic research would pose significant challenges in defining any realistic scenarios for potential societal impacts, comprehensively identifying stakeholders or creating actionable insights. Broad public engagement work would, however, be entirely appropriate for work that anticipated a substantial impact and an imminently marketable result.

TRL is one development pathway for new technologies. These give a common reference point when considering the level of development a given technology has achieved. The journey of any given technology through the technology development pathway is rarely a simple linear progression; there are likely to be frequent forks in the path, dead ends or feedback loops where activities generate new avenues of research, invention and iteration, or where components of systems at higher readiness levels in their own right are considered to be at a lower level when integrated into a new system.

The TRL scale has its own challenges when it comes to using it as a predictor of how technologies may interact with marketplaces. The science 'push' model, which is predicated upon the suggestion that science leads market adoption, has limited predictive validity and ignores the vast range of activity that will never reach a marketplace. Conversely the technology 'pull' model – that the market pulls desired new technologies out of the laboratory – is also too simplistic. Neither model takes into account the range of complex ways in which new technologies may reach fruition. In addition, this has a consequentialist approach to harms which conflicts with the virtue-ethics approach that suggests responsible methods should be undertaken even if the development may never progress to a final, public, form. Pathways to widespread adoption may also vary by industry sector and technology domain.

Table 12.1 shows the increments of the TRL score. The TRL is the score that is aimed at for the *end* of the project. As noted above, the technology may develop through loops, or may not reach its targeted readiness level, so the score will undoubtedly be less than 100% accurate, but it is suggested that this is sufficient for our purposes.

12.3.3 Step 2: Axis 2 – innovation potential

As discussed above in considering the fundamental principles of RRI, one of the critical success factors for an RI approach is the consideration of societal

Table 12.1 Technology readiness levels (TRLs)

TRL 1	Basic principles observed
TRL 2	Technology concept formulated
TRL 3	Experimental proof of concept
TRL 4	Technology validated in lab
TRL 5	Technology validated in relevant environment (industrially relevant environment in the case of key enabling technologies)
TRL 6	Technology demonstrated in relevant environment (industrially relevant environment in the case of key enabling technologies)
TRL 7	System prototype demonstration in operational environment
TRL 8	System complete and qualified
TRL 9	Actual system proven in operational environment

Source: NASA (2012).

impact – the manner and the level of impact are key pieces of evidence. The inclusion of civil society and a wide range of stakeholders is an integral part of the RRI process. It is, however, clearly neither necessary nor possible to include public engagement in every project. In addition to the practical issues, different societies will assign different levels of importance to societal goals. There is therefore a need, when proposing a new method for assessing RI, for a single, widely agreed metric against which to measure the interests of society in the proposed research, innovation or development.

The metric we propose for this axis of the RIL matrix is therefore achieved by creating a score on how and in what way the technology may affect society. Given the difficulty of formulating widely agreed, generally applicable societal measures, and subject to the reservations expressed previously, we are here proposing the use of the UN SDGs, and the governance schemes which surround them (Voegtlin & Scherer, 2017).

12.3.3.1 *Sustainable development goals*

The SDGs were created by the UN to be universal goals, targets and indicators that UN member states are expected to use to frame their agendas and political policies over the next 12 years. The SDGs follow and expand on the millennium development goals (MDGs), which were agreed by governments in 2001. Every country is expected to work towards achieving the SDGs. The goals are:

1. End poverty in all its forms everywhere.
2. End hunger, achieve food security and improved nutrition and promote sustainable agriculture.
3. Ensure healthy lives and promote wellbeing for all at all ages.
4. Ensure inclusive and equitable quality education and promote lifelong learning opportunities for all.

5. Achieve gender equality and empower all women and girls.
6. Ensure availability and sustainable management of water and sanitation for all.
7. Ensure access to affordable, reliable, sustainable and modern energy for all.
8. Promote sustained, inclusive and sustainable economic growth, full and productive employment and decent work for all.
9. Build resilient infrastructure, promote inclusive and sustainable industrialisation and foster innovation.
10. Reduce inequality within and among countries.
11. Make cities and human settlements inclusive, safe, resilient and sustainable.
12. Ensure sustainable consumption and production patterns.
13. Take urgent action to combat climate change and its impacts (taking note of agreements made by the UN Framework Convention on Climate Change forum).
14. Conserve and sustainably use the oceans, seas and marine resources for sustainable development.
15. Protect, restore and promote sustainable use of terrestrial ecosystems, sustainably manage forests, combat desertification and halt and reverse land degradation and halt biodiversity loss.
16. Promote peaceful and inclusive societies for sustainable development, provide access to justice for all and build effective, accountable and inclusive institutions at all levels.
17. Strengthen the means of implementation and revitalise the global partnership for sustainable development.

Within the goals are 169 targets that expand on the main overall goal to provide more specific and measurable targets for countries to aim at. Targets under goal 1, for example, include reducing by at least half the number of people living in poverty by 2030, and eradicating extreme poverty (people living on less than $1.25 a day). The targets thus lend themselves to being tailored to the specific sets of circumstances prevailing in individual countries.

This proposal does not seek to argue that the SDGs are a perfect set of societal goals. The RIL method uses them as a widely adopted measure of social "goods" and therefore a useful metric against which to measure the potential impact of a technology and assess the level of required RI work. The "impact" encompasses both positive and negative impact; for example, genetically modified crops that are beneficial for increasing food supply but may have a negative impact on the environment.

12.3.3.2 Establishing the innovation potential score

Recognizing that there is potential for different ICT technologies to impact either none or many SDG indicators we needed to introduce a measure of the consolidated impact, to permit comparison, and ultimately to gauge granularity in scaling of appropriate RRI activities. This was given the title innovation potential. The innovation potential score is formulated based on the impact on the SDGs. It is necessary for the researcher to assess the size of the potential impact of a technology on one or more of the SDGs. This can range from *no clear impact* to formally *quantifiable impact*.

12.3.3.2.1 EXAMPLE

The technology for the purposes of this example is a tool to allow parents, teachers and care-givers to find out how many data trackers or harvesters are contained within a tablet game aimed at children. This could have an impact against SDG 3 – promoting wellbeing for all at all ages. This type of tool will clearly have some impact, given the problematic nature of data-harvesting, storage and re-use and its effects on people's lives, but it could be difficult to quantify without other research. We therefore use Table 12.2 to gather the innovation potential score. One SDG is affected, and there is an unquantifiable impact, giving an innovation potential score of 2.

12.3.4 Step 3: assessing RIL

Assessing a project's RIL is the final stage in the assessment process. Once the TRL and the innovation potential are established, the project's RIL can be found using the matrix in Table 12.3. The higher the innovation potential, and the higher the TRL, the higher the RIL. Using our example of the parental tool above, which would need very little additional development to reach the marketplace (and therefore is being given a TRL of 8), we can see, using the matrix, that this gives an overall RIL of 4.

RIL is founded on the principle that, in order to be effective, RRI needs to be appropriate to the stage that the research has reached. For example, undertaking public engagement activities for basic research would pose significant challenges in terms of defining any potential societal impacts, comprehensively identifying stakeholders or creating actionable insights. This would be especially true when research is being undertaken for the sake of pure knowledge. This method is predicated upon assumptions such as

Table 12.2 Innovation potential (IP)

Number of sustainable development goals affected	*Impact of the technology*		
	No clear impact	*Unquantified impact*	*Quantified impact*
1	IP1	IP2	IP4
2 or more	IP1	IP3	IP5

Table 12.3 Finding the Responsible Research and Innovation intensity level (RIL)

Technology readiness level	9	RIL3	RIL4	RIL5	RIL5	RIL5
	8	RIL2	RIL4	RIL4	RIL4	RIL5
	7	RIL2	RIL3	RIL4	RIL4	RIL5
	6	RIL2	RIL3	RIL4	RIL4	RIL5
	5	RIL2	RIL3	RIL4	RIL4	RIL5
	4	RIL2	RIL2	RIL3	RIL3	RIL5
	3	RIL2	RIL2	RIL3	RIL3	RIL5
	2	RIL1	RIL1	RIL2	RIL2	RIL5
	1	RIL1	RIL1	RIL2	RIL2	RIL5
		1	2	3	4	5
	Innovation potential					

the concept that any kind of transformative technology or research should *always* have a high RIL. By contrast, very basic or conceptual research that will have little direct impact on society should still have some RI work, but this can be relatively low key.

Following this through with our worked example, its RIL score of 4 might indicate RRI activities such as parents' and children's focus groups or juries during the development process, as well as possible co-ordination with games manufacturers. The inclusion of an appropriate group of stakeholders is a key part of all RRI activity.

12.3.5 Operationalising RRI intensity level

We have proposed RIL as a scale of the intensity of RRI activities for a particular ICT research project, but we must then consider the characteristics of the organisation proposing to undertake such a project, and the experience and capabilities necessary in the design and delivery of RRI in an ICT research context.

When considering suitable candidates, the work done by Stahl et al. on RRI maturity modelling (Stahl et al., 2017) presents a body of evidence of the observed RRI capabilities of organisations delivering ICT R&I projects. This work proposes a scale of organisational maturity in the undertaking of RRI activities based on a substantial body of evidence, which can be applied as a structured approach to evaluating an organisation's alignment to RRI practices, and their suitability to deliver different levels of RRI intensity.

Maturity modelling has its origins in software engineering, but has evolved into other applications such as project management, with a notable example being P3M3. The Portfolio, Programme and Project Management Maturity Model (Young & Conboy, 2013) was developed by the UK's Office of Government Commerce. The purpose of P3M3 is to provide a frame of reference that can be used to baseline an organisation's capabilities in project, programme and portfolio management.

With P3M3 the intention is to use past performance in project delivery as a predictor of the future outcome of projects, and to then use this measure as an important factor in evaluating a particular organisation's suitability to deliver projects of a particular type or level of importance.

Once the RIL required for a particular project has been established by a research actor, be they prospective researcher or funding body, the question then comes as to what type of research project organisation has the right characteristics to deliver such an initiative. Clearly this evaluation covers a very wide range of parameters associated with research excellence, and so forth, but we propose that these parameters be extended to include due consideration of the RRI maturity level of the organisation.

The objectives of RRI are to ensure the acceptability, desirability and sustainability of research outcomes, and reaffirming the assertion (Stahl et al., 2017) that RRI should be considered a meta-responsibility, we propose the use of RRI maturity level as a justifiable proxy for RIL.

This approach could go some way to answering the question: "From an RRI perspective, what type of research organisation is best suited to undertake research of different RILs?" Table 12.4 proposes a conversion from RRI maturity to RILs.

12.3.6 *Other factors*

12.3.6.1 *Research portfolios*

For projects with a low RIL but which form part of a portfolio of research (such as quantum computing), the researcher must ensure that at the portfolio

Table 12.4 Conversion table for Responsible Research and Innovation intensity level (RIL) to Responsible Research and Innovation (RRI) maturity

RRI maturity level	*Maturity definition*	*RRI intensity level*
1	Organisation is unaware of RRI or its components and does not incorporate it in its processes	RIL 1
2	Organisation reacts to external pressures concerning aspects of RRI and experiments concerning appropriate processes	RIL2
3	Organisation has a definition of RRI and has integrated these into its processes	RIL3
4	Organisation realises the benefits of RRI and seeks to proactively and increasingly integrate these into its processes	RIL4
5	Organisation has adopted RRI as a component of its strategic framework and aims to ensure all R&D activities cover all RRI components	RIL5

Note: R&D, research and development.

level an RRI exercise has been undertaken, and that the implications of their work on that portfolio have been properly considered. To do this the researcher should produce a written assessment of the potential implications of their work on the portfolio as a portfolio impact report for the portfolio manager, and get positive assurance that the implications of this have been considered in the portfolio-level RRI work.

If the portfolio-level RRI activity has not been considered then the RIL for the assessed research should be increased to take account of this lack in the overall portfolio, and provision needs to be made for the appropriate cost and time implications.

12.3.6.2 Significant risks

There are some technologies and developments that – regardless of how many SDGs they affect and whether that impact is quantifiable – must be subject to the maximum level of RRI and therefore incur an RIL 5 score. This includes technologies that are only in relatively early stages such as quantum computing. It is anticipated that relatively few technologies will fall into this "significant risk" category.

12.3.6.3 Alternative applications

Where an alternative application for the research with a higher innovation potential has been identified, but this alternative application is outside the expertise of the principal investigator, then external advice should be sought from an RRI adviser.

12.4 Conclusions

It will be noted that there is no RIL score of zero. This is because it is the position of these authors that *every* project requires RRI consideration. There is no stage of research or type of development that requires nil reflection, consideration of stakeholders or assessment of societal impact. An RRI mindset has already partially permeated the strata of research practice bedrock in many parts of the world – an ethical thread that runs throughout the research at every stage.

Once the appropriate RIL is established for a project, the next stage envisaged by the method proposed here is to use the EPSRCAREA framework[8] to formulate a range of activity types suitable to the individual requirements of the project. Formulating this range of activities is beyond the scope of this chapter, but the intention is to allow researchers to choose or create the most suitable RRI activities for their project, establish the type of team members to undertake those activities and the types of tools and resources which will be required to undertake them effectively. The aim for these activities is to optimise the value of RRI for that project, but equally to identify where significant uncertainties should and do exist which will require further analysis. It is envisaged that this self-critical and reflective process can only improve the project's work and its outcomes.

We have here outlined a method that aims to address the question identified as a lacuna within the current research. It will be recalled that our research question at the outset was "*How can the level of RRI activity be aligned with the requirements of the project?*" and we have attempted to demonstrate a method that we believe allows researchers to shape an appropriate level and type of RRI activity to the needs of their project. They can then move on to design RRI activities that incorporate the need for reflection and stakeholder involvement.

It is our belief that this method makes a much-needed contribution to the discourse on RRI and supplies a critical step in the process of applied (rather than theoretical) RRI.

12.4.1 *Limitations of the method*

The number of RILs is currently set at five; however it may prove in practice that this does not offer sufficiently nuanced grading and that there need to be more levels to respond not just to projects at different stages of the technology-readiness metric but also that affect more of the SDGs in possibly variable ways.

Secondly, funding of early-stage research may have little realistic prospect of correctly anticipating outcomes; however work that is highly conceptual in nature should not be regarded as being exempt from RI work. This is recognised currently in the EPSRC's People at the Heart of ICT[9] requirement which applies to all projects, even the strictly theoretical.

The limitations of the SDGs as one of the metrics for innovation potential are acknowledged, particularly at the granular level obtaining below the top-line measures.

The use of RRI maturity as a proxy for RIL shows some promise, but does have the limitation of looking at past RRI activities to determine maturity. Considering that RRI emerged as a term comparatively recently (2010), many research organisations will have limited track records.

12.5 Further research

The approach proposed here is as yet untested in real-world scenarios, although the authors are undertaking research to evaluate the method. It is also clear that, as well as evaluation of the method, real-world case studies would provide a valuable addition to the theory in this field.

We also propose developing the MUSE test for project RRI activities along the lines of the following definitions:

- Meaningful: Can a sufficiently broad range of stakeholders sufficiently appreciate what the potential impact of a novel technology is, based on its stage of development?
- Useful: Are we stifling creativity by demanding a statement of instrumental value too early?
- Sensible: What level of resources is it sensible to expend on RRI (based on stage, impact, importance, incremental/transformational)?
- Effective: Can we create sufficient evidence of potential impact to convince or effect changes in society or stimulate anticipatory regulation?

Further evaluation of the alignment between RRI maturity and RIL is required, as well as addressing the identified limitations of this approach.

Notes

1 www.adalovelaceinstitute.org/
2 www.gov.uk/government/consultations/consultation-on-the-centre-for-data-ethics-and-innovation/centre-for-data-ethics-and-innovation-consultation
3 https://ico.org.uk/about-the-ico/news-and-events/news-and-blogs/2018/10/facebook-issued-with-maximum-500-000-fine/
4 www.nasa.gov/directorates/heo/scan/engineering/technology/txt_accordion1.html
5 https://assets.kpmg.com/content/dam/kpmg/xx/pdf/2016/11/disruptive-technologies-barometer-tech-report.pdf
6 www.gartner.com/imagesrv/books/digital-disruption/pdf/digital_disruption_ebook.pdf
7 www.cisco.com/c/dam/en/us/solutions/collateral/industry-solutions/digital-vortex-report.pdf
8 https://epsrc.ukri.org/research/framework/area/
9 https://epsrc.ukri.org/research/ourportfolio/themes/ict/introduction/crossictpriorities/peopleattheheart/

References

Baskerville, R. L. (1999). Investigating information systems with action research. Communications of the Association for Information Systems, Vol. 2, Article 19. https://doi.org/10.17705/1CAIS.00219

Bjerknes, G., & Bratteteig, T. (1995) User participation and democracy: A discussion of Scandinavian research on system development. *Scandinavian Journal of Information Systems*, 7(1), 73–98, Article 1.

Collingridge, D. (1981). *The Social Control of Technology*. Palgrave Macmillan.

Davis, F. D. (1989). Perceived usefulness, perceived ease of use, and user acceptance of information technology. *MIS Quarterly*, *13*(3), 319. https://doi.org/10.2307/249008

Gawande, A. (2010). *The Checklist Manifesto: How to Get Things Right*. New York, NY: Metropolitan Books.

Jirotka, M., Grimpe, B., Stahl, B., Eden, G., & Hartswood, M. (2017). Responsible research and innovation in the digital age. *Communications of the ACM, 60*(5), 62–68. https://doi.org/10.1145/3064940

Latour, B. (1998). From the world of science to the world of research? *Science, 280*, 208–209. https://doi.org/10.1126/science.280.5361.208

Martinuzzi, A., Blok, V., Brem, A., Stahl, B., & Schönherr, N. (2018). Responsible Research and Innovation in industry – challenges, insights and perspectives. *Sustainability*, 10(3), 702. https://doi.org/10.3390/su10030702

NASA. (2012). *Technology Readiness Level.* Retrieved from https://www.nasa.gov/directorates/heo/scan/engineering/technology/txt_accordion1.html

Nuffield Council on Bioethics. (2013). *Novel Neurotechnologies : Intervening in the Brain. Intervening in the Brain : a Guide to the Report.* Retrieved from http://books.google.de/books?id=mihzngEACAAJ&dq=intitle:Novel+neurotechnologies&hl=&cd=1&source=gbs_api%0Apapers3://publication/uuid/9969CF40-6E91-4BDD-9DD4-509F8B30EEB8

Oftedal, G. (2014). The role of philosophy of science in Responsible Research and Innovation (RRI): The case of nanomedicine. *Life Sciences, Society and Policy*, 10(5). doi:10.1186/s40504-014-0005-8

Olechowski, A., Eppinger, S. D., & Joglekar, N. (2015). Technology readiness levels at 40: A study of state-of-the-art use, challenges, and opportunities. 2015 Portland International Conference on Management of Engineering and Technology (PICMET), August 2–6 2015, Institute of Electrical and Electronics Engineers (IEEE), September 2015.

Owen, R. (2014). The UK Engineering and Physical Sciences Research Council's commitment to a framework for responsible innovation. *Journal of Responsible Innovation*, *1*(1), 113–117. https://doi.org/10.1080/23299460.2014.882065

Owen, R., Macnaghten, P., & Stilgoe, J. (2012). Responsible Research and Innovation: From Science in Society to Science for Society, with Society. Retrieved from www.wur.nl/upload_mm/1/e/a/0542fde4-acaa-455e-899a-e9680820ccd8_Responsible Research and Innovation- from Science in Society to Science for Society%2C with Society_Owen et al 2012_pre-pub copy.pdf

Stahl, B.C. (2013). Responsible research and innovation: The role of privacy in an emerging framework. *Science and Public Policy, 40*, 708–716. https://doi.org/10.1093/scipol/sct067

Stahl, B., Obach, M., Yaghmaei, E., Ikonen, V., Chatfield, K., & Brem, A. (2017). The Responsible Research and Innovation (RRI) maturity model: Linking theory and practice. *Sustainability*, *9*(6), 1036. https://doi.org/10.3390/su9061036

Stilgoe, J., Owen, R., & Macnaghten, P. (2013). Developing a framework for responsible innovation. *Research Policy*, *42*(9), 1568–1580. https://doi.org/http://dx.doi.org/10.1016/j.respol.2013.05.008

United Nations. (2015). Sustainable Development Goals – United Nations, United Nations Sustainable Development §. https://sdgs.un.org/goals

Voegtlin, C., & Scherer, A. G. (2017). Responsible innovation and the innovation of responsibility: Governing sustainable development in a globalized world. *Journal of Business Ethics*, *143*(2), 227–243. https://doi.org/10.1007/s10551-015-2769-z

von Schomberg, R. (2013). A Vision of Responsible Research and Innovation, 51–74. Retrieved from http://onlinelibrary.wiley.com/doi/10.1002/9781118551424.ch3/summary

Wadhwa, V. (2014) Law and ethics can't keep pace with technology. MIT Technology Review. Retrieved from www.technologyreview.com/s/526401/laws-and-ethics-cant-keep-pace-with-technology/

Young, M., & Conboy, K. (2013). Contemporary project portfolio management: Reflections on the development of an Australian Competency Standard for Project Portfolio Management. *International Journal of Project Management*. http://dx.doi.org/10.1016/j.ijproman.2013.03.005

Zhang, J. Y., Marris, C., & Rose, N. (2011). *The Transnational Governance of Synthetic Biology – Scientific Uncertainty, Cross-borderness and the 'Art' of Governance*. BIOS working paper no. 4. London: London School of Economics and Political Science.

Best practice VI Benchmarking for a better world: assessing corporate performance on the SDGs

Lisanne Urlings

Description of the good practice example

The global challenges we face, from climate disaster to gender inequality, unsustainable food systems to digital exclusion, are interconnected and affect us all. There is a perfect storm brewing with growing pressure to deliver on the United Nations' Sustainable Development Goals (SDGs) and the clock is ticking. The private sector has enormous potential to drive change. But there is currently no global accountability mechanism to understand where companies are today and where they need to get to. The World Benchmarking Alliance (WBA) has set out to develop a range of free and publicly available benchmarks to assess the progress of 2,000 keystone companies against seven systems transformations needed to achieve the SDGs (Figure VI.I).

'Business as usual' or incremental progress will not deliver the transformations needed to achieve the SDGs. It will not encourage companies to develop the innovations needed to address the challenges in the SDGs. We need business to move beyond pure financial performance and an 'avoid doing harm' mentality, to real measurable positive impact. Together with allies from government, business platforms, financial institutions, academia and civil society, WBA is developing transformative methodologies and benchmarks to measure and rank the world's most influential companies on the SDGs. WBA benchmarks produce insights, data and public rankings of these keystone companies so that investors, governments, civil society, individuals and the companies themselves better understand corporate progress and can more effectively hold companies accountable for their role in advancing the SDGs.

Companies often deal with a wide range of stakeholders with diverging expectations and priorities. Through an extensive multi-stakeholder process, benchmarks identify common ground among stakeholders and work to build consensus around these expectations. Benchmark methodologies translate these expectations into clear metrics, providing companies with

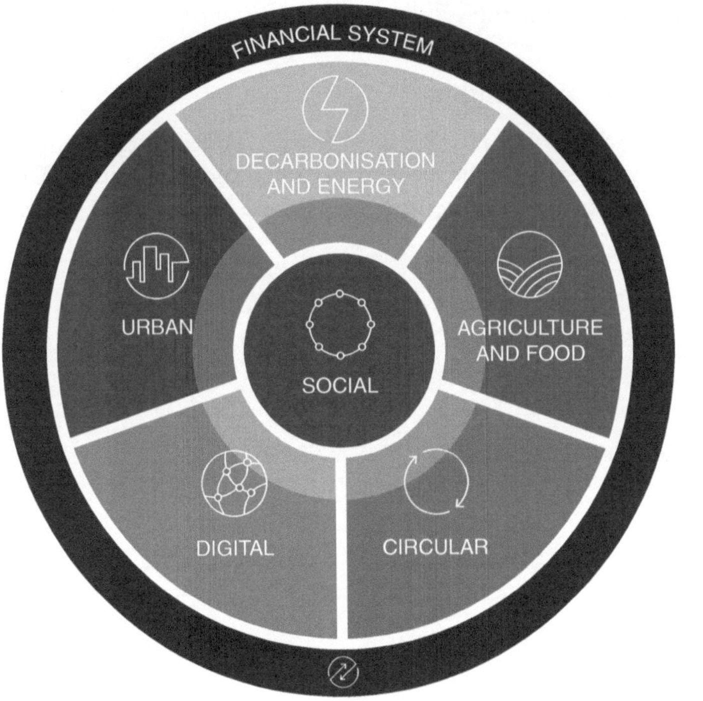

Figure VI.1 World Benchmarking Alliance (WBA) seven systems transformations.

a path forward. In addition, methodologies are informed by best available science and build on existing norms and standards, frameworks and initiatives.

WBA benchmarks reveal to both companies and stakeholders where each company stands compared to its peers, where it can improve and where urgent action is needed to deliver on the SDGs in its business strategies, operations, supply chains and product and service portfolios. These benchmarks are free for everyone to use and continually improved through an open, multi-stakeholder dialogue. By virtue of being public and due to the way in which the data are presented, benchmarks empower all stakeholders, from consumers and investors to employees and business leaders, with key data and insights to encourage sustainable business practices across all sectors.

Aims and goals

WBA aims to drive the private sector's engagement in the SDGs through its benchmarks, envisioning a future where companies, investors, governments, civil society and individuals can quickly and easily compare businesses and motivate a 'race to the top'.

Benefit for the organisation

The benchmark results show how companies perform, both positively and negatively, in regard to the transformations and associated SDGs, show what best practice looks like, pinpoint where more action is needed and inspire innovation.

13 The responsible side of innovation

Towards the measurement of a new construct

Robert Verburg, Laurens Rook, and Udo Pesch

13.1 Introduction

Innovation relates to developing new products and services, production methods and procedures, production technologies, and to administrative changes. Not only does innovation allow for the general improvement of the quality of life, it is also the motor of economic progress, which is motivated by the desire of companies to gain competitive advantage. As such, innovation is essential for many companies (Amabile, Conti, Coon, Lazenby, & Herron, 1996). In a business context, innovation relates to the intentional adoption of new ideas, processes, products or procedures within organizational units in order to improve the current organization or wider society (West & Farr, 1990). As Damanpour (1991) notes, the field of innovation is broad, and distinctions can be made between the diffusion and adoption of innovations in the market versus organizational innovating and innovativeness. Here, we regard innovation as 'a process of turning opportunity into new ideas and of putting these into widely used practice' (Tidd, Bessant, & Pavitt, 2005, p. 66).

For the last two decades, the ethical dimension of innovation has attracted the attention of various scholars; it features as an important topic of discussion under the heading of Responsible Research and Innovation (RRI, e.g. Guston, 2014; Guston et al., 2014; Hellström, 2003; Lubberink et al., 2017; Van den Hoven, 2007; Van de Poel, 2014). In spite of its growing importance and relevance for the practice of innovation, however, RRI seems almost invisible in business contexts (Blok & Lemmens, 2015; Lubberink et al., 2017). The current literature on RRI is dominated by conceptual pieces on the nature of RRI (e.g. Stilgoe et al., 2013) and writings on its societal importance (e.g. Van den Hoven, Lokhorst, & Van de Poel, 2012), while less work seems to be done on applying RRI as business practices in organizations. In this chapter, we aim to address this gap by providing a measurement scale for assessing RRI on the level of employees in business contexts, particularly in business contexts where innovations are most likely to occur (Baregheh, Rowley, & Sambrook, 2009; Foley et al., 2016).

In innovation-driven firms, employees are instrumental in order to attain organizational success. They do so through their ability to generate ideas, and use their creative self-efficacy to develop new and better products, services and work processes. In other words, it is through their innovative work behaviour that employees attain innovation outcomes (e.g. De Jong & Den Hartog, 2010; Richter, Hirst, Van Knippenberg, & Baer, 2012). Janssen (2000, p. 228) defines innovative work behaviour 'as the intentiona creation, introduction and application of new ideas within a work role, group or organization, in order to benefit role performance, the group, or the organization'. Ideas are generally the building blocks of innovation. Therefore, innovative work behaviour includes behaviours related to the generation of ideas, idea promotion and the realization or implementation of ideas.

In addition, and related to innovative work behaviour, the more general proactivity of employees affects organizational success (e.g. Botero & Van Dyne. 2009; Parker, Williams & Turner, 2006). Especially in innovative organizations, in which novel and unpredictable situations are likely to occur, employees need to proactively solve problems and speak up when needed. Proactivity in a corporate setting is defined as constructive and anticipatory work behaviour which is intended to be functional (Crant, 2000). Proactive employees show goal-oriented behaviour at work, explore opportunities and speak up with concerns and suggestions for improvement (Crant, 2000; Grant, 2013; Grant & Ashford, 2008). While innovative work behaviour is specifically focused on generating, championing and implementing new ideas, we view proactivity as taking charge, which is defined as the voluntary and constructive efforts of employees to effect organizationally functional change in how work is executed within the context of their jobs, work units or organizations (Morrison & Phelps, 1999).

Since employees differ in their ability to explore novel and useful ideas, a number of individual differences measures to assess innovation performance on the individual level were developed. Much of this work highlights creative achievements (e.g. Carson, Peterson, & Higgins, 2005; Runco, Plucker, & Lim, 2001) or focuses on innovative behaviour (e.g. De Jong & Den Hartog, 2010; Jansen, 2000), without considering the ethical dimensions as stressed by the RRI literature. Here, we aim to provide such an elaboration of RRI while still focusing on behavioural aspects or the attributes of individual employees. The chapter is structured as follows. First, we will explore RRI in relation to business activities on the basis of a more elaborate literature review. Next, we will translate the findings of this literature study into relevant items for assessing RRI. Then, we will compose a measurement scale and refine this measure through an empirical study among a sample of $n = 244$ students. We will conclude the chapter with the implications of the RRI measures for both theory and practice.

13.1.1 Creativity and responsibility

Creativity is an important and established research domain in contemporary behavioural research, as expressed in handbooks (cf. Kaufman &

Sternberg, 2006, 2010; Paulus & Nijstad, 2003; Smith, Ward, & Finke, 1995; Sternberg, 1999; Zhou & Shalley, 2008) and even an encyclopaedia (Runco & Pritzker, 2011). Most researchers in the field follow the standard definition of creativity, which considers creativity as consisting of novelty (or originality) and usefulness (fit or appropriateness). This standard definition was first coined by Stein in 1953 (see Runco & Jaeger, 2012), but has primarily become popular through the seminal work of Harvard professor Teresa Amabile (1983, 1996).

Over the past 25 years, scholars have increasingly sought to gain insights on creativity in the moral domain. A special issue in the *Creativity Research Journal* in 1993 inspired a research agenda aimed at increasing the understanding of moral creativity as a responsible creativity aimed at responding to societal, scientific and technological dilemmas with moral integrity (Gruber, 1993; Runco & Nemiro, 2003). In this special issue, several authors pointed to the existence of a darker side of creativity, characterized by a focus on societally undesirable novel outcomes (cf. McLaren, 1993). These contributions set the stage for exploration of negative – or even malevolent – creativity as the intentional production of novel solutions that deliberately aim at causing harm or immoral results (Clark & James, 1999; Cropley, Kaufman, & Cropley, 2008; Harris, Reiter-Palmon, & Kaufman, 2013; James, Clark, & Cropanzano, 1999).

The 1993 special issue also inspired a research agenda aimed at increasing the understanding of moral creativity as a *responsible creativity* focused on responding to societal, scientific and technological dilemmas with moral integrity. McLaren (1993) described it as: '[creativity] cannot be fully understood, any more than ... any other human endeavor, apart from its sociomoral context and intent' (p. 142). Scholars on moral creativity, indeed, all point to the need to account for goal-directedness (Clark & James, 1999; James, Clark, & Cropanzano, 1999) or intentionality (Runco & Nemiro, 2003) during the creative act. Haste (1993) argued that a moral creative act is about 'an unusual vision of moral seriousness of the situation [that fuels] the individual's own sense of efficacy and personal responsibility' (p. 3). Thus, it is a person's intention to make a difference in a responsible way, together with that person's *efficacy* – i.e., the self-perceived belief of being capable of setting in motion some intended course of action (Bandura, 1997) – that constitute the behavioural underpinnings of moral creativity. In addition, James and colleagues (1999) suggest that the type of (positive or negative) creativity a person will be inclined toward probably depends on individual differences in a person's personality (see Harris et al., 2013; Lee & Dow, 2011, for a similar point). Taken together, these considerations suggest that a scale for measuring the behavioural aspects of RRI practices can be established by reworking existing scales for creativity – especially those containing an element of self-efficacy together with creativity – in such a way that they also explicitly take into account ethical concerns at work. However, insights from the current RRI literature could be relevant to our aim to design a measurement scale to assess RRI. Therefore, we will start by extracting the business implications in relation to the dimensions of RRI in order to explore its relevance for the design of our measure.

13.2 The relevance of the dimensions of Responsible Research and Innovation for moral creativity

The current literature on RRI focuses strongly on the creation of new knowledge rather than on the overall innovation process in business contexts where new products, services or business models are developed by people in organizations. Here, we will extract the business implications in relation to the dimensions of RRI – i.e., anticipation, reflexivity, inclusion, and responsiveness as put forward by Owen et al. (2013). The rationale for this approach is that these four dimensions comprise the most commonly used framework in the field of RRI.

13.2.1 Anticipation

'Anticipation' refers to the ability to foresee possible consequences of innovation (Asante, Owen, & Williamson, 2014; Owen et al., 2013; Stilgoe, Owen, & Macnaghten, 2013) and implies reflection about both the desired and undesired (technical, political and environmental) impacts that may result from innovations (cf. Guston, 2014; Pesch, 2015). Anticipation does not only articulate promising narratives of expectation, but also prompts scientists and innovators to think of 'what if ...' and 'what else might it do?' questions. Therefore, anticipation is needed to 'articulate and assess expectations in science and technology' and see 'the difference between looking into the future and looking at the future' (Borup, Brown, Konrad, & Van Lente, 2006, p. 296).

In a business context, anticipation implies that a company needs to determine the impact of innovation activities. According to Lubberink et al. (2017), this requires a company to engage in activities that will improve knowledge about the innovation context by monitoring the innovation environment (Chadha, 2011; Gaziulusoy, Boyle, & McDowall, 2013). Next, a company needs to determine the impact of innovation, including its unanticipated consequences. Again, the company can reduce the uncertainties of innovation by monitoring the innovation environment (Berker, 2010; Biondi, Iraldo, & Meredith, 2002; Chadha, 2011; Rohrbeck, Konnertz, & Knab, 2013). This enables a company to assess the risks, uncertainties and impact of innovations as perceived by the stakeholders, such as the regulatory risks (e.g. corruptive culture of bureaucratic procedures and regulation instability), political risks (e.g. low political instability and lack of support from local government) and even *force majeure* (e.g. natural disaster and terrorism). Such risks and uncertainties are usually assessed by elaborate decision-making models (Wang, Chan, Yee, & Diaz-Rainey, 2012).

This short overview of business aspects related to the dimension of anticipation shows a strong emphasis on the (undesirable) impacts of innovation processes. Subsequently, the suggested measures involve the way in which both foreseeable and unforeseeable impacts may be effectively coped with by installing appropriate organizational structures and activities. These measures involve tools and models, but do not include any behavioural characteristics of the people who engage in a responsible way.

13.2.2 Reflexivity

The reflexivity dimension is aimed at making values and beliefs in innovation processes more explicit in order to enable adaptation when necessary. Reflexivity usually requires collaboration approaches, such as the involvement of the public (Asante et al., 2014; Flipse, Van Dam, Stragier, Oude Vrielink, & Van der Sanden, 2015; Macnaghten & Chilvers, 2014). In order to evaluate current and previous actions with regard to achievement of the desired innovation outcomes formal evaluations, third-party critical appraisal or an informal (self-) assessment should be ensured. Also, employee empowerment is needed in order to create innovations that are more in line with environmental needs by having regular staff meetings, managing internal selection and enabling new knowledge creation (Muduli, Govindan, Barve, Kannan, & Geng, 2013).

When a company is involved with innovation, conflicting values may arise (Lubberink et al., 2017). It is essential to prioritize values and motivations before thinking about the impact of these on innovation. For example, if a company values sustainability, different innovation governances, such as a democratic organizational structure, innovation workshops or an innovation task force, may be needed in order to gain the desired sustainability outcomes (Ayuso, Ángel Rodríguez, & Enric Ricart, 2006). A third aspect of reflexivity concerns the way in which a company may critically assess the way its knowledge could impact innovation processes (Lubberink et al., 2017). Companies should carefully analyse the presence, absence and subjectivity of information. In other words, a company ought to reflect and reframe perceived realities. This also relates to the perceived realities of different actors, which calls for reconciling the conflicting information resulting from these different realities.

By involving stakeholders with different perspectives, a company can expand its reflexive capacities and also further improve capabilities to sustain growth (Vickers & Lyon, 2014). Also for the dimension of reflexivity, a re-articulation into elements relevant for a business context shows a strong emphasis on process characteristics. These characteristics are found beneficial for creating responsible innovation outcomes. In its core, collaborative structures aimed at facilitating interaction among relevant stakeholders within a corporate setting are suggested as the designated route towards reflexivity.

13.2.3 Inclusion

Often, the notion of inclusion is used interchangeably with that of deliberation. In the context of businesses, these two notions are sometimes separated (Pellé & Reber, 2015), as stakeholder inclusion can have conflicting goals with stakeholder deliberation (Papadopoulos & Warin, 2007). The twin dimensions of inclusiveness and deliberation pertain to the interrelations with other agents. The dimensions respectively serve to identify the values which with an innovation might become more societally acceptable or responsive and to the entertainment of stakeholder dialogues. As such, a company's 'licence to operate' may be supported.

In this, 'inclusion' is aimed at engaging different stakeholders in the early stage of innovation processes, such as in the involvement of public values (Bozeman, Rimes, & Youtie, 2015; Taebi et al., 2014). Public values would encompass the outcomes that are important to the public at large. In a business context, inclusion implies the involvement of various stakeholders, such as the wider public. Also, a company may be encouraged to attract and collaborate with experts (cf. Baba & Walsh, 2010), but it could also team up with governmental agencies (Carrillo-Hermosilla, del Río, & Könnölä, 2010).

An important aspect of inclusion is to raise stakeholders' commitment and contribution (Le Ber & Branzei, 2010). Transparency is central here and requires a clear communication about the costs and benefits of the different stakeholders (De Chiara & Russo Spena, 2011). Clear and open communication implies several costs, such as guarding business secrets, reducing community entry barriers, giving up control and organizational inertia (Stuermer, Spaeth, & Von Krogh, 2009). Therefore, many companies selectively apply openness as they are aware that such openness would not always improve innovation performance (Balka, Raasch, & Herstatt, 2014).

'Deliberation' is defined by Van de Kerkhof (2006, p. 282) as 'a process of argumentation and communication in which the participants engage into an open process in which they exchange opinions and viewpoints, weigh and balance arguments, and offer reflections and associations'. This implies that stakeholder inclusion emphasizes questions about which actors should be involved, when to involve them and whether the actors represent the relevant stakeholders. Deliberation does not take much attention of the challenges of inclusivity and stakeholders' representativeness. Being deliberative refers to a sustained interaction between stakeholders in the whole innovation process (Jasanoff, 2003; Parkhill, Pidgeon, Corner, & Vaughan, 2013).

In order to implement deliberative governance, companies could engage in various ways of dialogue, such as crowdsourcing (Franke, Keinz, & Klausberger, 2013), focus group discussions (Dossa & Kaeufer, 2014) or community visits (Chadha, 2011). Companies might develop so-called dialogical relations by formalizing the process of deliberation governance (Asif,

Searcy, Santos, & Kensah, 2013; Harrisson, Chaari, & Comeau-Vallée, 2012). It is important to note that the dialogue should not be seen as an attempt to influence or coerce one another, but it emphasizes that one party should listen to other parties with empathy and should focus on the common goals with the intention of coming up with solutions (Ayuso et al., 2006).

13.2.4 Responsiveness

The 'responsiveness' dimension concerns the response to new knowledge, different emerging perspectives, views and norms (Stilgoe et al., 2013). Responsiveness is needed because the efforts to involve the public in innovation processes are limited. Therefore, the risk that the innovation could lead to undesired impacts is growing. Von Schomberg (2013) argues that the challenge of responsible innovation is to improve responsiveness to societal challenges. Also, Maynard (2015, p. 199) explains that innovators are likely to be blinkered by their optimism – especially regarding the 'seemingly speculative future health, environmental and other societal impacts [that] look like an ill-affordable luxury'.

While innovating, a company may obtain new information from the external environment that requires adjustments to the innovation process. This could imply that the company will try to avoid unintended consequences through incremental improvement (Berker, 2010). In this case, the company applies careful adaptation to achieve the best possible results. The development of a platform strategy is also increasingly proposed as a way to create product architecture that allows for many (customizable) design variants (Evans, Partidário, & Lambert, 2007). By implementing such a customization strategy, the company can maintain the production costs, the quality and the delivery of the products based on consumers' needs (Trentin, Perin, & Forza, 2012). Like the other dimensions of RRI, the recommendations distilled here suggest a focus on how to structure organizational processes in such a way that they contribute to an innovation outcome that is desirable.

13.3 Findings

Our review of RRI literature highlights the need to account for novelty and ethical aspects in innovation (Barron & Harrington, 1981; King, Walker, & Broyles, 1996) and strengthens our goal to develop a suitable measure for people involved with the creation of new products, services or business models in organizations. Although our findings show some clear business implications in relation to the dimensions of RRI, our review of the RRI literature indicates that these suggested links are primarily focused on *process* characteristics. Research in this emerging paradigm has primarily addressed the question about how to make innovations responsible (Koops, 2015) – i.e., the main focus in the community is on the steps through which an innovative end product eventually becomes 'what it ought to be'. RRI can therefore be seen as a cognitive precursor to the ethically driven novelty-producing behaviours of different actors in the innovation process.

The strong focus does seem to neglect a number of other relevant concerns (see Cuppen et al., 2019 for an overview). For business contexts the most notable neglect may be that employees are important drivers of innovation, but the role of behavioural factors of such employees (attitudes and competencies) is almost completely neglected by the work on the RRI dimensions. Therefore, we are motivated to derive the relevant behavioural characteristics that can be connected to responsible innovation. In the next section, we will present the construct of 'responsible creativity and innovation' and develop a measure to highlight the responsible side of innovation in organizations.

13.4 Development of the responsible creativity and innovation scale

In order to develop a measurement instrument for assessing responsible creativity and innovation on the level of employees in business contexts, we started from the definition of responsible innovation, and supplemented this with a literature on moral creativity at work (as outlined in the previous section). Below, we will first describe how these two literatures informed the development of a preliminary pool of items. Next, we will discuss how we arrived at a validated final measurement instrument.

As discussed above, creativity theoretically refers to the originality and usefulness, fit or appropriateness of a product or idea (Amabile, 1983, 1996; Runco & Jaeger, 2012). In line with this standard definition of creativity, researchers have developed several individual difference measures to explore to what extent a person performs more creatively in creative task settings (cf. the Creative Achievement Questionnaire and the Ideational Behavior Scale; Carson, Peterson, & Higgins, 2005; Runco, Plucker, & Lim, 2001). Most measures focus on the first stage of the creative process – idea generation – thus overlooking the implementation stage, where creative ideas are transformed into innovation (West, 2002), such as novel organizational arrangements, products, services or business models. This is also true for the three-item creative self-efficacy (CSE) construct, which was explicitly developed to capture an employee's self-reported motivation and ability to perform creatively at work (Tierney & Farmer, 2002). CSE is a predictor of individual creativity at work (cf. Tierney & Farmer, 2004, 2011), but has also been shown to predict individual creativity within the broader corporate setting of larger research and development teams (Richter et al., 2012).

Janssen (2000) takes a broader perspective and defines innovative work behaviour as related to idea generation, idea promotion and eventually the overall implementation of such ideas. First evidence exists that creativity and innovation are related to ethical business practices of individuals and groups (Valentine, Godkin, Fleischman, & Kidwell, 2010). Inspired by the economical nature of the CSE, the added focus on goal-directedness / intentionality in moral creativity (see above), as well as the work on innovative work behaviour, we generated items that explicitly captured the ethical appropriateness of novel ideas and their applications. Next, we formulated items to assess a person's moral considerations during the implementation of creative ideas in organizations. This led to a preliminary pool of 12 items intended to capture an employee's self-efficacy for responsible creativity and innovation at work. Each item was measured on a seven-point scale, ranging from 1 (very strongly disagree) to 7 (very strongly agree).

The preliminary pool of 12 items was tested among a sample of bachelor students, who were enrolled in an applied statistics course at a business school in The Netherlands. Participating students did not receive any reward (no class credit, no financial benefits) for participating in this online survey. Participation was voluntary (they could withdraw from the study at any point in time) and their confidentiality was assured. A total of 244 students (90 male; 154 female; age = 21.09 years; SD = 1.38) participated. Following the recommendations by Reise, Waller and Comrey (2000), we specifically targeted bachelor students for our pilot test, since the standard pool of undergraduates tends to be characterized by heterogeneity in terms of traits and background.

The following steps were taken in order to refine our measure. First, we explored the inter-correlations between the 12 preliminary items. Table 13.1 offers the descriptive statistics and correlations for each item, and indicates that each variable had a large number of significant inter-correlations (with the exception of item 11). This tentatively hinted at the possibility that most of the items belong to an overarching construct (Hair et al., 2014). In order to ensure that the items are part of an overarching construct, we applied Bartlett's test of sphericity. This test allows for checking the assumption that 'the correlation matrix has significant correlations among at least some of the variables' (Hair et al., 2014, p. 102). The significant result in our case, χ^2 (66) = 1036.48, $p < 0.0001$, rendered it appropriate for factor analysis in the remaining steps.

Next, we applied a random split half method on the dataset in order to determine a stable factor structure and to enable item reduction for the measured construct. The random split half method assesses the internal consistency of a construct (Hair et al., 2014). We ran a series of 50 random half-sample splits on the full dataset to determine the measure of sampling adequacy. The mean Kaiser–Meyer–Olkin (KMO) value after 50 splits was 0.82, SD = 0.03. This value was well within the range that is considered meritorious for further exploratory analysis (KMO ≥ 0.80; Kaiser, 1974). Following the procedure by Costello and Osborne (2005), we calculated coefficient comparability in a stepwise series of 50 random half-sample splits. This reduced the items from 12 to 8.

In order to refine the factor structure, we conducted an exploratory factor analysis (EFA) on a random half split dataset to derive the using principal

Table 13.1 Descriptive statistics and correlations of 12 preliminary items

	M	*SD*	*1.*	*2.*	*3.*	*4.*	*5.*	*6.*	*7.*	*8.*	*9.*	*10.*	*11.*	*12.*
1. Item 1	4.71	0.97	–											
2. Item 2	4.55	1.04	0.62**	–										
3. Item 3	3.95	1.41	0.21**	0.34**	–									
4. Item 4	4.20	1.16	0.26**	0.25**	–0.01	–								
5. Item 5	4.25	1.15	0.41**	0.27**	–0.14*	0.48**	–							
6. Item 6	4.43	1.12	0.43**	0.43**	–0.16*	0.47**	0.67**	–						
7. Item 7	4.36	1.13	0.36**	0.31**	–0.04	0.37**	0.38**	0.49**	–					
8. Item 8	4.54	1.24	0.24**	0.20**	0.20**	0.54**	0.36**	0.44**	0.40**	–				
9. Item 9	4.58	1.21	0.34**	0.27**	–0.11	0.29**	0.34**	0.44**	0.23**	0.36**	–			
10. Item 10	4.62	1.08	0.43**	0.36**	–0.12	0.36**	0.47**	0.57**	0.42**	0.38**	0.54**	–		
11. Item 11	4.58	1.40	–0.03	–0.03	0.41**	0.10	–0.07	0.00	–0.09	0.25**	0.09	0.02	–	
12. Item 12	4.95	1.09	0.33**	0.34**	–0.05	0.31**	0.30**	0.37**	0.24**	0.39**	0.46**	0.36**	0.20**	–

Note: $n = 244$. $^{*}p < 0.05$. $^{**}p < 0.01$.

component analysis with an oblique (direct Oblimin) rotation. This approach is recommended when dealing with behavioural factors, which are typically assumed to correlate (Clark & Watson, 1995). The EFA produced a two-factor solution with Eigenvalues > 1. The Eigenvalue of the first factor was 3.97 (49.68% of variance). All items loaded on the primary factor, but a single item also loaded strongly on the secondary factor (factor loading > 0.54). We removed this item, which resulted in a stable, single-factor, measurement instrument with a total of seven items. This scale has good overall reliability (Cronbach $\alpha = 0.83$). Table 13.2 offers the final items and descriptive statistics for the responsible creativity and innovation scale.

13.5 Discussion and conclusion

In spite of its growing importance for the practice of innovation, RRI seems almost invisible in business contexts. In order to create awareness among managers and innovators in industry of their responsibility towards society, a clear understanding of the RRI concept is necessary. To provide such an elaboration of RRI, we focused on the attributes of individual employees by paying attention to innovative and ethical behaviour. The aim of this chapter was to show the development of a measure that enables the assessment of responsible innovation on the level of employees who are engaged in work leading to innovation outcomes. Although there are some good measures to evaluate the innovative behaviour of employees, such measures do not include the responsibility dimension, but focus on a person's ability to generate ideas and to develop new and better products, services or business models. One could argue that it would be easy to just add a responsibility measure on top of one of the existing innovative behaviour measures rather than to develop a new scale to capture the RRI of employees. However, the reason why we decided to develop such a new scale is rooted in the fact that we wish to capture the ethical appropriateness of novel ideas and their applications rather than to segregate innovations from responsibility. As such, responsibility should not be entered after the intentional creation, introduction and application of new ideas, but should be embedded into these processes right from the very start. For example, employees involved with innovation should not just be good at generating ideas, but should also be able to capture the responsible side of innovation at the same time.

With the RRI measure, more research can be done in bridging the gap between intended business strategies involving RRI and what employees actually do when creating novel and useful products, services or business models. Such multilevel approaches to study RRI are now possible by applying the RRI measure. It could be very relevant to study RRI of employees in international contexts and to compare employees from different levels and backgrounds. The RRI measure could further feature in studies on the relationship between management practices or leadership styles and innovation outcomes. As such, the RRI measure could also be applied in practice

Table 13.2 Descriptive statistics and correlations of the final measurement instrument

	M	*SD*	1.	2.	3.	4.	5.	6.	7.
1. I am good at coming up with ideas that are novel, but also right	4.67	0.96	–						
2. I find it easy to generate original solutions that reflect how the problem at hand ought to be solved	4.59	1.06	0.58**	–					
3. Producing large numbers of solutions for a better future comes naturally to me	4.33	1.08	0.46**	0.28**	–				
4. I am good at generating many ideas that capture the responsible side of innovation	4.56	1.07	0.54**	0.43**	0.65**	–			
5. It is important to me to explore the various ethical aspects of my ideas	4.69	1.16	0.29**	0.25**	0.24**	0.41**	–		
6. Transforming new ideas into useful applications (such as new products, services or business models that make the world a better place) means a lot to me	4.61	1.16	0.47**	0.34**	0.35*	0.45**	0.32**	–	
7. I am good at introducing responsible solutions to a meaningful problem into new organizational arrangements	4.67	1.03	0.56**	0.42**	0.40**	0.57**	0.35**	0.56**	–

Note: $n = 121$. $^{*}p < 0.05$. $^{**}p < 0.01$.

as this will enable companies to compare different departments on their level of RRI by using aggregated data. The measure could also feature as an outcome for training and development efforts in the area of RRI or be used for the assessment of employees or managers.

Although the properties of the measure are promising, some important next steps are necessary before the RRI measure can be applied in both research and practice. First of all, the measure needs to be further refined. A large-scale follow-up study would be useful to establish the validity and reliability of the measure. Second, although we deliberately used bachelor students in order to secure sufficient heterogeneity of our sample, there could still be the possibility of response bias. Therefore, the final measure needs to be tested among employees who are actually involved with innovation. Third, further analysis is needed in order to establish whether the overall construct of 'responsible creativity and innovation' should be treated as a formative rather than as a reflective (second-order) construct.

In conclusion, our results highlight the importance of an RRI scale for both future research and practice. In spite of its growing importance for the practice of innovation, RRI seems almost invisible in business contexts. By presenting a measure for assessing RRI of employees in business contexts, we open the road for future studies that may further our understanding of the responsible innovation process in practice.

Acknowledgement

The authors would like to express their thanks to Mr Hakim Agung Ramadhan, who first explored our idea to develop a responsible innovation measure in his master thesis in the area of Management of Technology at Delft University of Technology. Through exploring our dataset, Hakim has helped us in refining our thinking and enabled us to take the next steps in the development of the responsible innovation measurement scale.

References

Amabile, T. M. (1983). *The Social Psychology of Creativity*. New York (NY): Springer.

Amabile, T. M. (1996). *Creativity in Context*. Boulder (CO): Westview.

Amabile, T. M., Conti, R., Coon, H., Lazenby, J., & Herron, M. (1996). Assessing the work environment for creativity. *Academy of Management Journal*, 39(5), 1154–1184.

Asante, K., Owen, R., & Williamson, G. (2014). Governance of new product development and perceptions of responsible innovation in the financial sector: Insights from an ethnographic case study. *Journal of Responsible Innovation*, 1(1), 9–30.

Asif, M., Searcy, C., Santos, P. d., & Kensah, D. (2013). A review of Dutch corporate sustainable development reports. *Corporate Social Responsibility and Environmental Management*, *20*(6), 321–339.

Ayuso, S., Ángel Rodríguez, M., & Enric Ricart, J. (2006). Using stakeholder dialogue as a source for new ideas: A dynamic capability underlying sustainable innovation. *Corporate Governance: The International Journal of Business in Society*, 6(4), 475–490.

Baba, Y., & Walsh, J. P. (2010). Embeddedness, social epistemology and breakthrough innovation: The case of the development of statins. *Research Policy*, 39(4), 511–522.

Balka, K., Raasch, C., & Herstatt, C. (2014). The effect of selective openness on value creation in user innovation communities. *Journal of Product Innovation Management*, 31(2), 392–407.

Bandura, A. (1997). *Self-efficacy: The Exercise of Control*. New York: Freeman.

Baregheh, A., Rowley, J., & Sambrook, S. (2009). Towards a multidisciplinary definition of innovation. *Management Decision*, 47(8), 1323–1339.

Barron, F., & Harrington, D. M. (1981). Creativity, intelligence, and personality. *Annual Review of Psychology*, 32(1), 439–476.

Berker, T. (2010). Dealing with uncertainty in sustainable innovation: mainstreaming and substitution. *International Journal of Innovation and Sustainable Development*, 5(1), 66–79.

Biondi, V., Iraldo, F., & Meredith, S. (2002). Achieving sustainability through environmental innovation: The role of SMEs. *International Journal of Technology Management*, 24(5–6), 612–626.

Blok, V., & Lemmens, P. (2015). The emerging concept of responsible innovation. Three reasons why it is questionable and calls for a radical transformation of the concept of innovation. In B. Koops, I. Oosterlaken, H. Romijn, T. Swierstra & J. Van den Hoven (Eds.), *Responsible Innovation 2. Concepts, Approaches, Applications* (pp. 19–35). Cham: Springer.

Borup, M., Brown, N., Konrad, K., & Van Lente, H. (2006). The sociology of expectations in science and technology. *Technology Analysis & Strategic Management*, 18(3–4), 285–298.

Botero, I. C., & Van Dyne, L. (2009). Predicting voice: Interactive effects of LMX and power distance in the U.S. and Colombia. *Management Communication Quarterly*, 23 (1), 84–104.

Bozeman, B., Rimes, H., & Youtie, J. (2015). The evolving state-of-the-art in technology transfer research: Revisiting the contingent effectiveness model. *Research Policy*, 44(1), 34–49.

Carrillo-Hermosilla, J., del Río, P., & Könnölä, T. (2010). Diversity of eco-innovations: Reflections from selected case studies. *Journal of Cleaner Production*, 18(10), 1073–1083.

Carson, S. H., Peterson, J. B., & Higgins, D. M. (2005). Reliability, validity, and factor structure of the creative achievement questionnaire. *Creativity Research Journal*, 17(1), 37–50.

Chadha, A. (2011). Overcoming competence lock-in for the development of radical eco-innovations: The case of biopolymer technology. *Industry and Innovation*, 18(3), 335–350.

Clark, K., & James, K. (1999). Justice and positive and negative creativity. *Creativity Research Journal*, 12(3), 311–320.

Clark, L. A., & Watson, D. (1995). Constructing validity: Basic issues in objective scale development. *Psychological Assessment*, 7(3), 12–25.

Costello, A. B., & Osborne, J. W. (2005). Best practices in exploratory factor analysis: Four recommendations for getting the most from your analysis. *Practical Assessment, Research & Evaluation*, 10(7), 1–9.

Crant, J. M. (2000). Proactive behavior in organizations. *Journal of Management*, 26(3), 435–462.

Cropley, D. H., Kaufman, J. C., & Cropley, A. J. (2008). Malevolent creativity: A functional model of creativity in terrorism and crime. *Creativity Research Journal*, 20(2), 105–115.

Cuppen, E., Van de Grift, E., & Pesch, U. (2019). Reviewing Responsible Research and Innovation: Lessons for a sustainable innovation research agenda? In F. Boons & A. McMeekin (Eds.), *Handbook of Sustainable Innovation*. Cheltenham: Edward Elgar.

Damanpour, F. (1991). Organizational innovation: A meta-analysis of effects of determinants and moderators. *Academy of Management Journal*, 34(3), 555–590.

De Chiara, A., & Russo Spena, T. (2011). CSR strategy in multinational firms: Focus on human resources, suppliers and community. *Journal of Global Responsibility*, 2(1), 60–74.

De Jong, J. P. J. & Den Hartog, D. N. (2010). Measuring innovative work behaviour. *Creativity and Innovation Management*, 19(1), 23–36.

Dossa, Z., & Kaeufer, K. (2014). Understanding sustainability innovations through positive ethical networks. *Journal of Business Ethics*, 119(4), 543–559.

Evans, S., Partidário, P. J., & Lambert, J. (2007). Industrialization as a key element of sustainable product-service solutions. *International Journal of Production Research*, *45*(18–19), 4225–4246.

Flipse, S., Van Dam, K., Stragier, J., Oude Vrielink, T., & Van der Sanden, M. (2015). Operationalizing responsible research and innovation in industry through decision support in innovation practice. *Journal on Chain and Network Science*, 15(2), 135–146.

Foley, R. W., Bernstein, M. J., & Wiek, A. (2016). Towards an alignment of activities, aspirations and stakeholders for responsible innovation. *Journal of Responsible Innovation*, 3(3), 209–232.

Franke, N., Keinz, P., & Klausberger, K. (2013). "Does this sound like a fair deal?": Antecedents and consequences of fairness expectations in the individual's decision to participate in firm innovation. *Organization Science*, 24(5), 1495–1516.

Gaziulusoy, A. İ., Boyle, C., & McDowall, R. (2013). System innovation for sustainability: A systemic double-flow scenario method for companies. *Journal of Cleaner Production*, *45*, 104–116.

Grant, A. M. (2013). Rocking the boat but keeping it steady: The role of emotion regulation in employee voice. *Academy of Management Journal*, 56(6), 1703 –1723.

Grant, A. M. & Ashford, S. J. (2008). The dynamics of proactivity at work. *Research in Organizational Behavior*, 28, 3–34.

Gruber, H. E. (1993). Creativity in the moral domain: Ought implies Can implies Create. *Creativity Research Journal*, 6 (1–2), 3–16.

Guston, D. H. (2014). Understanding 'anticipatory governance'. *Social Studies of Science*, 44(2), 218–242.

Guston, D. H., Fisher, E., Grunwald, A., Owen, R., Swierstra, T., & Van der Burg, S. (2014). Responsible innovation: Motivations for a new journal. *Journal of Responsible Innovation* 1(1), 1–8.

Hair, J. F., Black, W. C., Babin, B. J., Anderson, R. E., & Tatham, R. L. (2014). *Multivariate Data Analysis*. Harlow: Pearson.

Harris, D. J., Reiter-Palmon, R., & Kaufman, J. C. (2013). The effect of emotional intelligence and task type on malevolent creativity. *Psychology of Aesthetics, Creativity, and the Arts*, 7(3), 237–244.

Harrisson, D., Chaari, N., & Comeau-Vallée, M. (2012). Intersectoral alliance and social innovation: When corporations meet civil society. *Annals of Public and Cooperative Economics*, 83(1), 1–24.

Haste, H. (1993). Moral creativity and education for citizenship. *Creativity Research Journal*, 6(1–2), 153–164.

Hellström, T. (2003). Systemic innovation and risk: Technology assessment and the challenge of responsible innovation. *Technology in Society*, 25(3), 369–384.

James, K., Clark, K., & Cropanzano, R. (1999). Positive and negative creativity in groups, institutions, and organizations: A model and theoretical extension. *Creativity Research Journal*, 12(3), 211–226.

Jansen, O. (2000). Job demands and innovative work behavior. *Journal of Occupational and Organizational Psychology*, 73(3), 287–302.

Jasanoff, S. (2003). Technologies of humility: Citizen participation in governing science. *Minerva*, 41(3), 223–244.

Kaiser, H. F. (1974). An index of factorial simplicity. *Psychometrika*, 39(1), 31–36.

Kaufman, J. C., & Sternberg, R. J. (2006). *International Handbook of Creativity Research*. New York (NY): Cambridge University.

Kaufman, J. C., & Sternberg, R. J. (2010). *Cambridge Handbook of Creativity*. New York (NY): Cambridge University.

King, L. A., Walker, L. M., & Broyles, S. J. (1996). Creativity and the five-factor model. *Journal of Research in Personality*, 30(2), 189–203.

Koops, B.-J. (2015). The concepts, approaches, and applications of responsible innovation. In B.-J. Koops, I. Oosterlaken, H. Romijn, T. Swierstra & J. van den Hoven (Eds.), *Responsible Innovation 2: Concepts, Approaches, and Applications* (pp. 1–15). Cham: Springer.

Le Ber, M. J., & Branzei, O. (2010). Forming strategic cross-sector partnerships: Relational processes of social innovation. *Business & Society*, 49(1), 140–172.

Lee, S., & Dow, G. (2011). Malevolent creativity: Does personality influence malicious divergent thinking? *Creativity Research Journal*, 23(2), 73–82.

Lubberink, R., Blok, V., van Ophem, J., & Omta, O. (2017). Lessons for responsible innovation in the business context: A systematic literature review of responsible, social and sustainable innovation practices. *Sustainability*, 9(5), 721–752.

Macnaghten, P., & Chilvers, J. (2014). The future of science governance: Publics, policies, practices. *Environment and Planning C: Government and Policy*, 32(3), 530–548.

Maynard, A. D. (2015). The (nano) entrepreneur's dilemma. *Nature Nanotechnology*, 10(3), 199–200.

McLaren, R. (1993). The dark side of creativity. *Creativity Research Journal*, 6(1–2), 137–144.

Morrison, E. W., & Phelps, C. C. (1999). Taking charge at work: Extra-role efforts to initiate workplace change. *Academy of Management Journal*, 42(4), 403–419.

Muduli, K., Govindan, K., Barve, A., Kannan, D., & Geng, Y. (2013). Role of behavioural factors in green supply chain management implementation in Indian mining industries. *Resources, Conservation and Recycling*, 76, 50–60.

Owen, R., Stilgoe, J., Macnaghten, P., Gorman, M., Fisher, E., & Guston, D. H. (2013). A framework for responsible innovation. In R. Owen, J. Bessant & M. Heintz (Eds.), *Responsible Innovation: Managing the Responsible Emergence of Science and Innovation in Society* (pp. 27–50). Chichester: Wiley.

Papadopoulos, Y., & Warin, P. (2007). Are innovative, participatory and deliberative procedures in policy making democratic and effective? *European Journal of Political Research*, 46(4), 445–472.

Parker, S. K., Williams, H. M., & Turner, N. (2006). Modeling the antecedents of proactive behavior at work. *Journal of Applied Psychology*, 91(3), 636–652.

Parkhill, K., Pidgeon, N., Corner, A., & Vaughan, N. (2013). Deliberation and responsible innovation: A geoengineering case study. In R. Owen, J. Bessant & M. Heintz (Eds.), *Responsible Innovation: Managing the Responsible Emergence of Science and Innovation in Society* (pp. 219–240). Chichester: Wiley.

Paulus, P. B., & Nijstad, B. A. (2003). *Group Creativity: Innovation Through Collaboration*. Oxford (UK): Oxford University.

Pellé, S., & Reber, B. (2015). Responsible innovation in the light of moral responsibility. *Journal on Chain and Network Science*, 15(2), 107–117.

Pesch, U. (2015). Engineers and active responsibility. *Science and Engineering Ethics*, 21(4), 925–939.

Reise, S. P., Waller, N. G., & Comrey, A. L. (2000). Factor analysis and scale revision. *Psychological Assessment*, 12(3), 287–297.

Richter, A. W., Hirst, G., Van Knippenberg, D., & Baer, M. (2012). Creative self-efficacy and individual creativity in team contexts: Cross-level interactions with team informational resources. *Journal of Applied Psychology*, 97(6), 1282–1290.

Rohrbeck, R., Konnertz, L., & Knab, S. (2013). Collaborative business modelling for systemic and sustainability innovations. *International Journal of Technology Management*, *63*(1–2), 4–23.

Runco, M. A., & Jaeger, G. J. (2012). The standard definition of creativity. *Creativity Research Journal*, 24(1), 92–96.

Runco, M. A., & Nemiro, J. (2003). Creativity in the moral domain: Integration and implications. *Creativity Research Journal*, 15(1), 91–105.

Runco, M. A., Plucker, J. A., & Lim, W. (2001). Development and psychometric integrity of a measure of ideational behavior. *Creativity Research Journal*, 13(3–4), 393–400.

Runco, M. A., & Pritzker, S. (2011). *Encyclopedia of Creativity*. Oxford (UK): Elsevier.

Smith, S. M., Ward, T. B., & Finke, R. A. (1995). *The Creative Cognition Approach*. Cambridge (MA): MIT.

Sternberg, R. J. (1999). *Handbook of Creativity*. New York (NY): Cambridge University.

Stilgoe, J., Owen, R., & Macnaghten, P. (2013). Developing a framework for responsible innovation. *Research Policy*, 42(9), 1568–1580.

Stuermer, M., Spaeth, S., & Von Krogh, G. (2009). Extending private-collective innovation: A case study. *R&D Management*, 39(2), 170–191.

Taebi, B., Correljé, A. F., Cuppen, E., Dignum, M., & Pesch, U. (2014). Responsible innovation as an endorsement of public values: The need for interdisciplinary research. *Journal of Responsible Innovation*, 1(1), 118–124.

Tidd, J., Bessant, J., & Pavitt, K. (2005). *Managing Innovation* (Vol. 3). West Sussex (UK): Wiley.

Tierney, P., & Farmer, S. M. (2002). Creative self-efficacy: Its potential antecedents and relationship to creative performance. *Academy of Management Journal*, 45(6), 1137–1148.

Tierney, P., & Farmer, S. M. (2004). The Pygmalion process and employee creativity. *Journal of Management*, *30*(*3*), 413–432.

Tierney, P., & Farmer, S. M. (2011). Creative self-efficacy development and creative performance over time. *Journal of Applied Psychology*, *96*(2), 277–293.
Trentin, A., Perin, E., & Forza, C. (2012). Product configurator impact on product quality. *International Journal of Production Economics*, *135*(2), 850–859.
Valentine, S., Godkin, L., Fleischman, G. M., & Kidwell, R. (2010). Corporate ethical values, group creativity, job satisfaction and turnover attention: The impact of work context on work response. *Journal of Business Ethics*, 98(3), 353–372.
Van de Kerkhof, M. (2006). Making a difference: On the constraints of consensus building and the relevance of deliberation in stakeholder dialogues. *Policy Sciences*, 39(3), 279–299.
Van de Poel, I. (2014). Translating values into design requirements. In D. P. Michelfelder, N. McCarthy & D. E. Goldberg (Eds.), *Philosophy and Engineering: Reflections on Practice, Principles and Process* (pp. 253–266). Dordrecht: Springer.
Van den Hoven, J. (2007). ICT and value sensitive design. In P. Goujon, S. Lavelle, P. Duquenoy, K. Kimppa & V. Laurent (Eds.), *The Information Society: Innovation, Legitimacy, Ethics and Democracy In honor of Professor Jacques Berleur sj* (pp. 67–72). New York: Springer.
Van den Hoven, J., Lokhorst, G.-J., & Van de Poel, I. (2012). Engineering and the problem of moral overload. *Science and Engineering Ethics*, 18(1), 143–155.
Vickers, I., & Lyon, F. (2014). Beyond green niches? Growth strategies of environmentally-motivated social enterprises. *International Small Business Journal*, 32(4), 449–470.
Von Schomberg, R. (2013). A vision of responsible research and innovation. Responsible innovation. In R. Owen, J. Bessant & M. Heintz (Eds.), *Responsible Innovation: Managing the Responsible Emergence of Science and Innovation in Society* (pp. 51–74). Chichester: Wiley
Wang, X., Chan, H. K., Yee, R. W., & Diaz-Rainey, I. (2012). A two-stage fuzzy-AHP model for risk assessment of implementing green initiatives in the fashion supply chain. *International Journal of Production Economics*, 135(2), 595–606.
West, M. A. (2002). Sparkling fountains or stagnant ponds: An integrative model of creativity and innovation implementation in work groups. *Applied Psychology*, *51*(3), 355–387.
West, M. & Farr, J. (Eds.) (1990). *Innovation and Creativity at Work*. Chichester: Wiley.
Zhou, J., & Shalley, C. E. (2008). *Handbook of Organizational Creativity*. Philadelphia (PA): Taylor & Francis.

Best practice VII Enabling the private sector to manage its impact on the SDGs

Charlotte Portier

The 17 sustainable development goals (SDGs) are anticipated to generate over US$12 trillion worth of business opportunities by 2030, which makes thousands of small and large companies around the world well incentivized to work towards achieving the goals. But until now, the lack of a uniform methodology for business to measure and report on their contributions to the global goals had become a stumbling block. Developed by Global Reporting Initiative (GRI) and the United Nations Global Compact with support from Shift and PwC — and as part of the Action Platform on Reporting on the SDGs — the publication, *Integrating the SDGs into Corporate Reporting: A Practical Guide*[1] completes a toolkit that aims to boost and improve businesses' reporting on the global goals and their efforts to achieve the 2030 agenda.

This new practical guide helps companies of all sizes to prioritize SDG targets to act and report on, set related business objectives and measure and report on progress.

The newly published guide specifically addresses the current challenges in the reporting arena, namely the need for harmonization and the prioritization process. As such it does not create a new reporting framework but builds on the Ten Principles of the UN Global Compact, the UN Guiding Principles on Business and Human Rights, and the GRI Sustainability Reporting Standards. It discourages businesses from selecting SDGs and related targets based on what is easiest for them to report on. Rather, companies should focus on what accounts for the highest priorities and, therefore, is material for their business. The report calls for companies to report on both their positive and negative impacts on the SDGs.

The publication is meant to be used together with other relevant tools released by GRI, the UN Global Compact and their partners as part of a company's regular reporting cycle, such as *Business Reporting on the SDGs: An Analysis of the Goals and Targets*[2], which lists qualitative and quantitative disclosures for SDG targets that are likely to be relevant to businesses regardless of size, sector or operating location. It also lists illustrative actions that businesses can take to contribute to the achievement of the targets.

Aims and goals

GRI's aim is to enable business to disclose SDG-related information and manage their impact on sustainable development. Business can measure their contribution to the global goals to take informed decisions and maximize their positive contributions.

Benefit for organizations

By empowering business to report on their impact on sustainable development and the SDGs, GRI realizes its mission: to empower decisions that create social, environmental and economic benefits for everyone.

Notes

1 GRI and UN Global Compact (2018). Integrating the SDGs into Corporate Reporting: A Practical Guide. Retrieved from: https://www.globalreporting.org/public-policy-partnerships/sustainable-development/integrating-sdgs-into-sustainability-reporting/.

2 GRI and UN Global Compact (2017). Business Reporting on the SDGs: An Analysis of the Goals and Targets. Retrieved from: https://www.globalreporting.org/public-policy-partnerships/sustainable-development/integrating-sdgs-into-sustainability-reporting/.

14 RRI measurement and assessment

Some pitfalls and a proposed way forward

Ibo van de Poel

14.1 Introduction

In the last decade, there has been increasing attention on responsible innovation, or Responsible Research and Innovation (RRI), as a way to increase the responsible development and deployment of new technology in society. RRI refers to a more reflexive form of research and development (R&D) and innovation that is aimed at better aligning research and innovation with the "values, needs and expectations of society" (European Commission, 2014: 73).

Uptake of RRI by industry and research organizations appears to be a slow and gradual process. One underlying factor is that the vocabulary of RRI is sometimes perceived as academic or "foreign" by innovators (Dreyer et al., 2017). This does, however, not mean that the underlying rationale and motives are not recognized. Many of the actions and activities that are now promoted as RRI are already undertaken by companies and research organizations, albeit often under different heading such as corporate social responsibility (CSR), social innovation and sustainable innovation (Lubberink, Blok, van Ophem, & Omta, 2017; van de Poel et al., 2017). RRI may perhaps be best construed as an attempt to broaden and systematize such activities. What RRI seems to add is particularly a broadening of values considered (e.g. not only safety, sustainability and privacy), more attention for stakeholder involvement and public engagement and a more proactive stance where such issues are already addressed during the early phases of R&D and innovation (Stilgoe, Owen, & Macnaghten, 2013; Van den Hoven, 2013). Some have also argued that responsible innovation starts from societal challenges and needs rather than from technical opportunities, as in traditional innovation (Von Schomberg, 2019).

It has been pointed out that the uptake of RRI is constrained by the fact that current incentives for research organizations and companies point in other directions. This has led to pleas for incentivizing RRI (Gurzawska, Mäkinen, & Brey, 2017). Such incentivizing may take a multiplicity of forms, from legal requirements to financial incentives. Examples are the development of quality marks for RRI, or governments requiring RRI compliance as a condition for allowing new products on to the market.

The above considerations have led to an increasing attention for the development of methods and tools to measure, assess and monitor RRI performance. Examples can be found in the various chapters in this book (e.g. Tharani, Jarmai, Schönherr, & Urban, Chapter 8; Klaassen, Verwoerd, Kupper, & Regeer, Chapter 9; Verburg, Rook, & Pesch, Chapter 13). These tools may, for example, be used to assess where organizations stand with respect to RRI activities and awareness, even if such activities may not always be defined as RRI by the organization itself. Such assessment tools may point out possibilities for improvement, but they may also play a role in incentivizing schemes intended to promote the uptake of RRI.

If tools for RRI measurement, assessment and incentivizing are to be effective, they need not only to result in a measurement of RRI performance that is reliable and valid, but also to contribute to behavior incentives that contribute to an increased uptake of RRI and its underlying aims. However, this is by no means straightforward. Performance measurement in general is a messy and difficult process, and it may occasionally result in incentives and behavior effects that are contrary to what was intended (De Bruijn, 2007).

The aim of this chapter therefore is to highlight some of the potential pitfalls of RRI measurement, assessment and incentivizing. As we will see, avoiding these pitfalls may not be so easy (although not necessarily impossible) and there may be trade-offs between avoiding different types of pitfalls. In order to better deal with these pitfalls and dilemmas, I will propose a potential way forward.

The chapter starts with sketching the constellation of actors that may be typically involved in RRI measurement, assessment and incentivizing (Section 14.2). Next, I discuss typical motivations that may exist for doing RRI assessment (Section 14.3). These first two sections not only provide an introduction to the topic, but also provide the analytical tools that can be used to analyze in more detail concrete situations. Such analysis is a prerequisite to uncovering potential pitfalls of RRI measurement, assessment and incentivizing. Section 14.4 gives a brief overview of existing RRI assessment methods, which will also be used to illustrate some of the pitfalls in the next sections. Section 14.5 focuses on pitfalls due to measurement problems and Section 14.6 focuses on pitfalls due to behavior effects. In Section 14.7, I provide a more general discussion of the pitfalls and dilemmas in RRI assessment and I argue that the dilemmas are to a large extent due to the fact that there are different, conflicting rationales for doing RRI assessment. I suggest that getting the rationale for an RRI assessment method clear is a first step to better navigating the earlier distinguished pitfalls and dilemmas.

14.2 The constellation of actors

Before discussing the motivations behind RRI assessment and some of its potential pitfalls, it is useful to sketch an ideal-typical constellation of actors

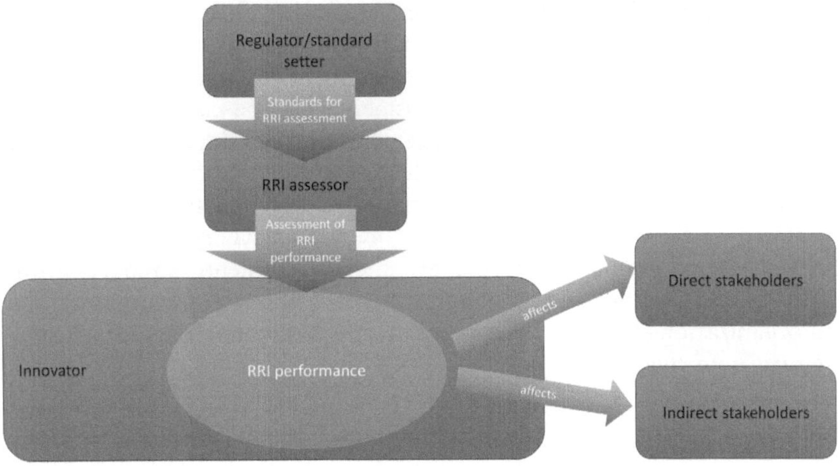

Figure 14.1 Actor roles in Responsible Research and Innovation (RRI) assessment.

that is somehow involved in, or affected by RRI assessment. Figure 14.1 lays out the main actor roles that I propose to distinguish. I discuss each of these briefly below.

14.2.1 Innovator

This is the actor that is the object of the RRI assessment. I assume here that RRI assessment is applied to a specific organization rather than to the entire knowledge or innovation system. Given the nature of RRI, I will assume that this is an organization that is developing innovative "products" (including knowledge) and that is doing R&D-like activities. I take products here broadly; basically any kind of output that can be used by another actor for another end. These products need to be somehow new and innovative, although I will refrain here from providing a precise definition of "innovation". The innovation needs to be the result of some deliberate knowledge generation activity, i.e. from R&D, although I will assume that the organization doing RRI can also focus only on research, or only on development. This first actor role of an innovator is typically played by organizations such as research laboratories, universities and companies.

14.2.2 Regulator or standard setter

This is the actor that is regulating or setting (RRI) standards for the innovator. I understand here regulation and standards very broadly. It refers to any kinds of requirements for the RRI behavior and performance of the assessed organization. Such standards can be compulsory (e.g. legal requirements by the government) or voluntary (e.g. a non-binding code of ethics), or in between (e.g. required to receive a quality mark). They can relate to the innovative products developed by the innovator, but also to organizational requirements or procedures (e.g. the requirement to have a safety officer); they can also concern how RRI assessment is to be carried out and by whom. Although I focus here on requirements and standards related to RRI, this does not always mean that they themselves have to explicitly mention RRI. Think for example of requirements for safety and sustainability that are clearly related to RRI but need not mention RRI. Typical actors that can fulfill this second actor role are the government, standard-setting organizations like ISO and its national counterparts, branch organizations and organizations formulating quality marks. Also innovators may formulate their own rules and standards.

14.2.3 RRI assessor

This is the actor doing the actual RRI assessment. Since RRI is still in its infancy, in many cases this is not yet a specialized role that is played by a separate actor. Instead, this role is often played either by the regulator or standard setter or by the innovator itself. It may currently also be played by a temporary organization, like a European RRI project that is involved in developing (tools for) RRI assessment. However, this role can in principle also be played by separate actors like, for example, a consultancy firm or an independent organization set up for such purposes.

14.2.4 Direct stakeholders

I will here understand the direct stakeholders as those actors that use the (innovative) products of the assessed organizations.[1] Presumably, the direct stakeholders have an interest in the RRI performance of the innovator. This will, for example, reassure (or even guarantee) them that the products they use from the assessed organization meet certain safety or sustainability standards or do not invade the privacy of certain groups.

14.2.5 Indirect stakeholders

Indirect stakeholders are actors that are somehow affected by the activities of the innovator or by the use of the innovative products developed by the innovator (and used by the direct stakeholders). Indirect stakeholders thus include actors further along the value chain of an innovative product (who do not directly use the product itself) or so-called bystanders. Given the understanding of RRI as developing products that meet "the needs and values of society" (European Commission, 2014: 73), the indirect stakeholders are an important category in RRI assessment, as the ultimate aim of RRI would seem to be to assure that also values and needs of those indirect stakeholders are properly addressed by the innovator in its innovation and R&D activities.

The above description of actors' roles is useful to better understand some of the motivations behind RRI assessment and some of its potential pitfalls. Figure 14.1, however, merely describes a conceptual framework with ideal-typical actor roles. It does not yet describe the situation in a specific case, with specific actors fulfilling specific actor roles. This specific constellation will be different from case to case, and describing it may occasionally also require distinguishing additional actor roles. Also, as already alluded to, in some cases actors' roles may be combined by one actor. For example, the innovator may set its own standards and assess them itself (i.e. some form of self-regulation).

Detailing Figure 14.1 for a specific situation (like a specific innovator or for a specific industry branch or technological domain) is useful because it helps to see where roles and interests of the various actors are overlapping, complementary or (potentially) conflicting. This in turn may point to certain (potential) weaknesses and pitfalls of RRI assessment in the specific situation.

14.3 Potential motivations for doing RRI assessment

There are various potential motivations for doing RRI assessment, and these may be different for different actors. I will here discuss some main motivations for doing RRI assessment without claiming to be complete. The motivations I will discuss are: (1) compliance; (2) increasing transparency, accountability and trust; (3) improving (RRI) performance; and (4) the desire to avoid (unnecessary) regulation.

14.3.1 Compliance

A first motivation for doing RRI assessment might be to check whether the innovator is compliant with existing rules and standards that somehow pertain to RRI issues. As indicated, such rules and standards need not be formulated in straightforward RRI language (as that is still rare), but could pertain to issues such as safety, sustainability, privacy, transparency, integrity, and so on. For the company, compliance may be a way to show to regulators and direct and indirect stakeholders that they meet the relevant standards. For the regulator, RRI assessment may be a means to check whether an innovator indeed is compliant, and for direct and indirect stakeholders it may be a means to ensure that they can rely on the products of the innovator.

14.3.2 Accountability, transparency and trust

Compliance requires explicit rules and standards against which compliance can be checked. In many cases, such rules and standards will be lacking for innovative products. One important reason for this is that regulation (and standardization) tend to lag behind the development of new technologies. RRI was in fact in part proposed to deal with issues that are not (yet) regulated. So while RRI assessment may be used to check compliance, in many cases RRI will be applied in situations in which (new) rules and standards have not been set yet or in which they are still being established or evolving. In such situations, RRI assessment may still be relevant to show not just that RRI procedures are in place with the innovator but also that these are functioning reasonably well. This may then make the innovation processes of the innovator more transparent, contribute to the accountability of the innovator and so create trust among other actors, like regulators and stakeholders.

14.3.3 *Learning and improving performance*

A third motivation for doing RRI assessment may to learn from it and to improve (RRI) performance. This is primarily a motive for the actor doing RRI, i.e. the innovator, but learning and improved (RRI) performance is likely also in the interest of the other actors, like regulators and stakeholders. The reasons why RRI assessment can contribute to learning and help to improve performance are quite straightforward. RRI assessment will show on which RRI parameters the innovator is performing well and on which not so well, which indicates possibilities for improving performance. Moreover, RRI assessment may help to gain insight in the (cost)effectiveness of various RRI measures and procedures and so may help to increase performance without necessarily increasing the costs of RRI.

14.3.4 *Avoiding regulation*

A fourth, more controversial, motivation for RRI assessment may be to avoid, or postpone, regulation. By doing RRI assessment innovators might want to show that they are taking their responsibility and that no government regulation is needed. One might argue that this is not a proper reason for RRI assessment but rather amounts to a pitfall. It is indeed true that innovators may misuse RRI assessment for purposes like avoiding regulation (cf. the discussion below on window dressing). However, it should be noted that regulation is not always desirable or in the interests of stakeholders (and the regulator). Government regulation comes with its own pitfalls, and if RRI can be achieved without explicit government regulation it may be more desirable than having regulation. The point, of course, is that from a societal point of view, regulation is sometimes needed.

Understanding the motivation for RRI assessment in a concrete situation is important because it will to an important extent determine whether, and which of, the pitfalls that I discuss below will actually materialize. I would therefore suggest that an analysis of concrete situations should start with both a mapping of the actors involved (Section 14.2) and their motivations behind doing RRI assessment (Section 14.3), in order to analyze which pitfalls of RRI assessment can be expected (Sections 14.5 and 14.6) and should be prevented if possible. But before we turn to the pitfalls, it is useful to briefly discuss some existing RRI assessment methods.

14.4 An overview of some existing RRI metrics and assessment methods

Table 14.1 provides an overview of some existing RRI metrics and assessment methods (see Chapters 8, 9 and 13; Flipse, Dam, Stragier, Vrielink, & Sanden, 2015; Ravn, Nielsen, & Mejlgaard, 2015; Stahl et al., 2017; Strand et al., 2015; Wickson & Carew, 2014). The dimensions have been filled out by the author on the basis of the mentioned references; in most cases this was rather straightforward; but in a number of cases it required some interpretation.[2] As the table shows, existing RRI assessment methods have a range of different aims, including monitoring, measuring RRI levels, comparison and learning. In terms of the motivations for doing RRI assessment that I discussed in Section 14.2, these aims are mainly related to the second (accountability) and third (learning and improving performance) motivation mentioned. It is not very surprising that none is related to compliance, since there are not (yet) legal requirements or quality marks for RRI, which can serve as a basis for compliance. Similarly, it is not surprising that avoiding regulation is not an (official) aim of the proposed methods. This does not rule out that use of these methods may sometimes be motivated by the desire to avoid regulation, but it is obviously not an aim of the methods.

The different methods are geared towards different objects of assessment, from employees to countries; most are however aimed at either the organizational level (company) or the project level. What is further striking is that most are based on some form of self-assessment. This is probably explained by two factors. One is that RRI is a complex notion with many dimensions; I will further reflect on this feature in the next section. The other is that RRI assessment, and in the sense the whole field of RRI, is a relatively recent endeavor and it usually takes time to operationalize complex notions like RRI into features or items that are objectively measurable; although approaches 2 and 3 in the table are clearly attempts to do so.

I will reflect on these and other features of the existing RRI assessment in the next section when I discuss typical measurement problems that the notion of RRI gives rise to.

14.5 Measurement problems

Since RRI is a complex notion, RRI (performance) cannot be directly measured. Moreover, RRI is a normative or value-laden notion, i.e. it expresses what is a (more) desirable form of innovation, which may make it

Table 14.1 Overview of some Responsible Research and Innovation (RRI) assessment methods proposed in the literature

	Reference	Unit of assessment	Assessor	Aim of assessment	Based on 6 EU keys[a]	Based on AREA framework[b]	Structure	Type of measurement	Aggregate score
1	Wickson and Carew (2014)	Project	Multiple	Multiple	No	Yes	7 criteria, rubric for each criterion	Judgement	No
2	Ravn, Nielsen, and Mejlgaard (2015)	Country	Independent assessor	Monitoring; comparison	Yes	No	6 dimensions, 36 indicators	Objective	No
3	Strand et al. (2015)	RRI initiative	Independent assessor	Monitor and assess the impacts of RRI initiatives	Yes	Yes	6 dimensions, each with performance (process and product) indicators and perception indicators[c]	Objective	No
4	Stahl et al. (2017)	Company	RRI researchers; self-assessment[d]	Assessing RRI level, monitoring	No	Yes	3 RRI categories, 14 RRI components, each scored on 5-point scale (maturity levels)	Judgement	No

5	Flipse et al. (2015)	Project (within a company)	Self-assessment	Monitoring; decision support for managers	No	Yes	8 Key performance indicators (KPIs), each consisting of several items	Judgement	No
6	Tharani et al. (Chapter 8, this book)	Company	Self-assessment	Learning	No	Yes	4 sections; 43 questions	Judgement	No
7	Klaassen et al. (Chapter 9, this book)	Project	Self-assessment	Learning	No	Yes	4 process dimensions, each with criteria, subcriteria and inviting questions	Judgement	No
8	Verburg, Rook, and Pesch (Chapter 13, this book)	Employee (in a company)	Self-assessment	Assessing RRI level	No	Yes	7 items	Judgement	no

Notes

a The six RRI keys of the EU are: engagement; gender equality; science education; ethics; open access; governance(European Commission, 2012).

b This framework identifies four procedural dimensions for RRI: anticipation, reflexivity, engagement (or inclusiveness), action (or responsiveness) (EPSRC, 2019; Owen et al., 2013).

c Based on Table 3.1 in the report with the prioritized indicators. The report suggests two more potential dimensions (and indicators).

d In the publication it is applied by the RRI researcher but self-assessment is suggested as a possible way to scale up the method.

even more difficult to measure. This means that the notion of RRI, and RRI performance, first needs to be operationalized before it can be measured.

Operationalization of RRI performance may, for example, follow a two-step procedure (cf. Keeney, 1992; Kroes & van de Poel, 2015). First, a number of dimensions (or evaluation criteria or objectives) are associated with RRI, which as such may also not be directly measurable. To make these dimensions measurable, a number of measurable items (attributes, indices or questions) may be associated with each dimension. Often, these items will not exactly measure the relevant dimensions, but only by approximation, so that they are best seen as proxies for the chosen RRI construct.

The need to operationalize RRI is clearly visible from Table 14.1. As the table shows, in six out of eight RRI assessment methods (nos. 2–7) RRI is at least a two-level construct (in one case (no. 7) it even has four levels), with typically at least ten items at the lowest level. Only two methods (nos. 1 and 8) have only one level and fewer than ten items.

If RRI performance is operationalized by defining a range of items (attributes, indices or questions), as is the case with all methods summarized in Table 14.1, a next issue is whether – and if so, how – to aggregate scores on these into an overall score. One option is to limit oneself to scores on individual items, or a number of (aggregated) indices, and to refrain from an overall score. Indeed all existing RRI assessment approaches considered here refrain from calculating an overall score. This avoids the problem of aggregation but may make it difficult to judge whether RRI performance has improved (over time), particularly if performance on some items increases while it decreases on other items (cf. Bradburn, Cartwright, & Fuller, 2017). Some of the methods propose the use of spider diagrams that can help to show which dimensions of RRI improve over time, and which dimensions decline.

Two general concerns may arise with respect to the measurement of RRI performance, namely reliability and validity (Carmines & Zeller, 1979). Reliability means that the measurement measures correctly, i.e. that it measures the "real value" of an item. This is often understood as implying that if the measurement were done again it would measure the same value for an item, or – if the measurement is done by somebody else – it will result in the same value. It also typically implies that if two items intend to measure the same phenomenon, their measured value should be the same.

There are a number of reasons why metrics for RRI may result in unreliable measurements. One is that the attributes may often not be objectively measurable, but rather are items or questions that require a judgment by the one filling out the questionnaire. Indeed, only two of the assessment methods in Table 14.1 (i.e. nos. 2 and 3) make use of objectively measurable indicators; all others require some form of judgment. The subjectivity of judgment may be further aggravated by the fact that items or questions sometimes contain vague or ambivalent terms like, for example, "sufficient", "better" and "relevant". These terms may be interpreted differently by different persons. And even one and the same person may interpret these terms differently at different points in time, which may lead to unreliable results.

Take for example a question like "Are there procedures in place to address the relevant ethical issues raised by the innovation?" A growing awareness of ethical issues may result in that the respondent answers at T_1 yes and at T_2 no, not because fewer ethical issues are addressed at T_2, but simply because the respondent has become aware of more ethical issues. So, while it may be argued that actual RRI performance has increased from T_1 to T_2 because there is more awareness of ethical issues, and because more issues are addressed at T_2 than T_1, on the basis of the answers to the indicated question the suggestion may arise that the actual RRI performance has decreased.

One way to reduce the subjectivity of judgment in RRI assessment is to provide a rubrics that gives guidance how to score questions or items, as the first method in Table 14.1 indeed does (Wickson & Carew, 2014). Another way in which subjectivity may be diminished is by involving more people in scoring items, or using questions that are a starting point for discussion, as seems to be the aim in methods 6 and 7 in Table 14.1,[3] but which may also work with some of the other methods.

A related potential source of unreliability is that RRI assessment is often a form of self-assessment, i.e. the innovator him- or herself has to fill out a questionnaire, or has to score items on a scale. This is indeed the case for six out of eight of the RRI assessment methods listed in Table 14.1. This may result in biased measurements, in particular when the innovator has a strategic interest in the outcome of the measurement. (This issue will be further discussed in the next section.)

Apart from reliability, validity may be an issue. A measurement is usually seen as valid if it measures what it intends to measure (rather than something else). As we have seen, RRI, or RRI performance, is not directly measurable. We first need to operationalize it to make it measurable. However, it is very conceivable that the chosen operationalization does not exactly or completely cover the underlying concept. The issue here is one of content or construct validity (Carmines & Zeller, 1979). By operationalizing RRI (performance), we create a construct that is measurable, but this construct may lack content or construct validity. This may be particularly so because in choosing a particular construct we may have good reasons to choose it in such a way that it is reliably measurable. However, the attributes that are most reliably measurable may not be the ones that are also most relevant for RRI, so diminishing construct validity.

Of course, the problem of content or construct validity is not unique to RRI; it applies to any complex social (or psychological) notion that we intend to measure (Carmines & Zeller, 1979). Examples are notions like intelligence or well-being. However, there seem to be a few underlying reasons why RRI may be particularly difficult to measure.

A first reason is that there is no consensus on the definition of RRI. Some definitions of RRI for example are based on the so-called AREA framework and they stress four procedural criteria for RRI: anticipation, reflection, engagement (or inclusiveness) and action (or responsiveness) (EPSRC, 2019; Owen et al., 2013; Stilgoe et al., 2013). But one may also place more emphasis on the six keys (engagement; gender equality; science education; ethics; open access; governance) for RRI defined by the European Union (EU) (European Commission, 2012). Yet other definitions of RRI stress the outcome or product dimension of RRI, for example emphasizing that innovative products should respect certain values (Van den Hoven, 2013). Or they may place emphasis on whether innovations contribute to the sustainable developments goals of the United Nations.

A look at Table 14.1 shows that all considered approaches were somehow inspired by the four procedural criteria for RRI; only two used the six EU keys for RRI. However, what is perhaps most remarkable is that the RRI constructs for the eight methods shown in Table 14.1 are very different. Methods 2 and 3 use the same six dimensions (i.e. the EU RRI keys) but somewhat different indicators. All the others RRI constructs are rather different from each other, already at the highest level, but certainly in terms of more detailed items. One might wonder how what is supposedly one concept can lead to such diverse constructs.

The underlying reason here seems to be that RRI is what Bradburn et al. (2017) call a *Ballung* concept and which they distinguish from a *pinpoint* concept. *Ballung* refers to the German word for congestion, which is used because a lot is packed into a concept. Bradburn et al. (2017: 76) say about such concepts:

> There is often no central core without which one does not merit the label, different clusterings of features among the congestion (*Ballung*) can matter for different uses, and whether a feature counts as being or outside the concept – and how far outside – is context and use dependent.

This indeed seems true of RRI. Some authors have noted that the more specific content of RRI is largely left open (Oftedal, 2014). As we have also seen, different definitions of RRI have been proposed stressing different features and some features (like science education) that are inside RRI according to some definitions (e.g. the six keys to RRI of the EU) are outside the concept in other definitions.

Ballung concepts are notoriously difficult to operationalize, also because operationalization requires first of all a clear definition. However, this does not imply that it is necessarily impossible to measure *specific* notions of RRI. Rather it becomes mandatory to make explicit what notion of RRI is operationalized and measured. Another consequence is that assessments based on different notions of RRI cannot be compared with each other. This is exactly what surfaces from Table 14.1. Different authors come with different operationalizations of RRI depending on the specific aims they have in mind (implicitly or explicitly) for RRI assessment and depending on what context and unit of assessment they assume. The consequence is also clearly that RRI levels measured with one method cannot be compared with measurements from other methods (with the possible exception of methods 2 and 3).

One may also wonder whether it is always productive to exactly pinpoint more exact notions of responsible innovation.[4] While this may make the notion easier to measure, it might also mean that it loses its function as communicative device among diverse groups. The reason is that responsible innovation often functions as what has been called a boundary object (Star & Griesemer, 1989). Boundary objects are concepts that have a common meaning among groups, or across sites, but also have some interpretative flexibility, so that different actors can adapt them to their specific local needs. Due to this combination, they can help to foster communication and cooperation between groups. As a *Ballung* concept, responsible innovation may well function as a boundary object. However, attempts to completely specify its meaning might mean that it loses some of its interpretative flexibility and that some of the involved groups may no longer subscribe to it.

A second reason why RRI may be particularly difficult to measure and assess is that RRI interventions seem often aimed at somehow improving the innovation process from a societal point of view, rather than at attaining some predefined "absolute" level of "responsibility". Of course, for an innovation process to be "responsible", some common criteria apply, like the need to include stakeholders. But what is the "right" level of stakeholder involvement may not be the same for every innovation, or for each technical domain.

Two issues are at stake here. One is that what counts as more responsible (in terms of RRI) may depend on context. In one particular RRI project, stakeholder involvement may be a main way to improve RRI performance while in other cases the emphasis is on anticipation or responsiveness. Another one is that often RRI seems to be aimed at improving performance, and hence seems to be a moving target rather than a completely predefined notion. It seems like methods 6 and 7 in particular try to address this issue. They both contain questions, and are primarily aimed at learning rather than measuring predefined levels of RRI.

A third reason why RRI may be particularly difficult to measure is that it is a normative notion. This differentiates it from other complex notions (and *Ballung* concepts) like, for example, intelligence. RRI expresses what is desirable, not what is factually the case. What seems particularly relevant here is the so-called naturalistic fallacy, i.e. the impossibility to understand normative notions fully in descriptive terms. According to the philosopher G. E. Moore, descriptive definitions of normative terms are always vulnerable to what he calls the open-question argument (Moore, 1903). If we provide a descriptive definition of a normative term like good, we can always ask the question: "It meets these and these descriptive requirements, but is it good?" Similarly, if we understand RRI fully in descriptive terms (or attributes), we can ask the question: "But is the innovation (process) really responsible?" It follows that if we are to measure a normative concept like RRI, at least some of the attributes need to be normative or involve normative judgments.

14.6 Strategic behavior

Performance measurement will almost always influence behavior, and in many cases it will invite strategic behavior (De Bruijn, 2007). This is most obviously the case if the performance measurement is directly connected to rewards or punishments. But even without such a direct connection, it is likely that performance measurement will have (behavioral) effects because (almost) nobody wants to perform poorly.

The fact that performance measurement affects behavior is in itself not bad; it may even be desirable or intended. After all, one reason to do RRI performance measurement may be to increase RRI performance. Nevertheless, it is important to realize that if we measure the RRI performance of an innovator, we do not just carry out a measurement but make an intervention. This intervention will have effects – desirable ones, but potentially also undesirable ones.

Moreover, the fact that there is not just measurement but intervention may affect the quality of the measurement. For one, it may affect the reliability of the measurement. This is most obviously the case if RRI performance measurement depends on self-assessment. Similarly, if the innovator is also the standard setter, there may be an inclination to choose RRI performance measures on which the innovator scores relatively well. In such cases, strategic behavior may potentially affect the chosen RRI construct, which may affect validity.

If the innovator is also the RRI assessor and/or the standard setter, there is a danger of *window dressing*. Window dressing is the phenomenon where an agent pretends to meet certain ethical standards (and makes efforts to show that) while in reality these standards are not met, or at least not to the extent pretended. Certain combinations of actor roles increase the risk of window dressing; in particular there is a risk of window dressing if the innovator is also the RRI assessor and/or is also the standard setter (Figure 14.1). This not to say that if actors combine these roles, it will always or necessary result in window dressing. An innovator may be genuine and even self-critical in an RRI self-assessment; it may potentially even be more critical than an outsider. Conversely, window dressing may also occur with an external assessor and standard setter. Even if the innovator, assessor and regulator are different persons or organizations they may still be closely aligned.[5] There may be collaborations between these actors, or economic or political dependencies, or they may share mutual ideological commitments.

Such collaborations, dependencies and shared commitments are indeed quite common in today's innovations systems, as for example underlined by such notions as the triple-helix model of innovations, that assumes close collaboration between universities, innovating companies and governments (Etzkowitz & Leydesdorff, 1997). Still, *ceteris paribus*, the risk of window dressing seems lower when the different roles outlined in Section 14.2 are separated rather than combined. However, as Table 14.1 testifies, most current RRI assessments are based on self-assessment and thus combine at least the role of innovator, or the one being assessed, with that of RRI assessor.

Also, when innovators cannot influence the construct chosen or the measurement, RRI measurement and assessment are likely to have behavioral effects. More specifically, innovators may make efforts to score better on the measured RRI attributes. This is in itself of course not undesirable, and in many cases, it is even desirable. However, in some cases, it may have detrimental effects.[6] This is particularly the case if the chosen RRI construct is not completely valid. In that case, an increase in performance in terms of the chosen RRI attributes may not signal a real increase in RRI performance. For example, certain RRI aspects may not be included in the measured attributes because they are difficult to measure. It is perfectly conceivable that increased performance in the measured RRI attributes goes hand in hand with decreased performance in aspects of RRI that are not measured, so that an increase in measured RRI performance does not signal an increase in the "real" RRI performance of the innovator.

Such effects may be particularly apparent if RRI performance measurements come with strong incentives. The effect will be, in general, that what is incentivized is the particular construct of RRI that is being measured. As long as construct validity is high, this is not a big problem. But since, as we have seen, RRI is a complex notion, it is not unlikely that what is being incentivized is actually a particular, somewhat narrow, interpretation of RRI. While strong incentives make it more likely that the particular construct of RRI that is measured is achieved, they also seem to make it likely that other aspects of RRI – that are not being measured and incentivized – are ignored. The reason for this is that resources are limited, so that an increased performance in some respects is likely to come with less attention for, or even decreased performance in, other aspects.

Incentivizing RRI may be problematic for other reasons as well. One such reason is the phenomenon known as "crowding out" (see e.g. Gneezy, Meier, & Rey-Biel, 2011). Crowding out occurs if an intrinsic motivation to achieve some good (in our case RRI) is replaced by external incentives, so that the initial intrinsic motivation decreases, or even disappears. For example, if sustainable behavior is incentivized with financial incentives, people may start to behave sustainably because of the financial gains rather than because they believe it to be good to behave sustainably. This may "crowd out" their intrinsic motivation to do good. Paradoxically, the effect may be that they start to behave less sustainably (cf. Gneezy & Rustichini, 2000). There are different possible explanations for this phenomenon (Gneezy et al., 2011). One possible explanation is that financial incentives may erode social norms. Another possible explanation is that by introducing (financial) incentives, people start to see sustainable behavior as something that can be traded for financial gains (and losses) rather than as something done for intrinsic moral reasons. As a consequence of this change in perspective, they might be willing to bear the financial costs that come with unsustainable behavior, for example because it brings them comfort, while – previously – the unsustainable behavior might have come with a moral guilt, that they wanted to avoid. So, incentivizing may sometimes make moral goods tradable, so that they lose their special moral status.

Crowding out may also be a concern when contemplating assessment and incentivizing schemes for RRI. There may be many motives for an innovator to engage in RRI, but the desire "to do good" is certainly one of them. On the other hand, the currently limited uptake of RRI suggests that moral motivations alone may not be enough, and that some form of incentivizing with connected RRI assessment or measurement schemes may be required, also to overcome some of the barriers for RRI uptake (Gurzawska et al., 2017). Nevertheless, an awareness of the risk of crowding out would be helpful in shaping assessment and incentivizing schemes for RRI.

14.7 Discussion

We have seen that measuring and assessing RRI performance are not straightforward. RRI is a complex and multidimensional notion. Moreover, RRI is what has been called a *Ballung* concept, a concept without clear borders, and such concepts are notoriously difficult to operationalize and measure. These measurement problems are further aggregated by the fact that there is not an agreed definition of RRI and that RRI is a normative notion, not just a descriptive one.

Navigating the various pitfalls of RRI assessment is not straightforward. Choices or directions that avoid certain pitfalls may well increase the likelihood of other pitfalls. Two tensions stand out in particular. One is that between reliability and validity. In order to make RRI measurement and assessment more reliable, one might want to aim for RRI attributes that are (more) objectively measurable. However, since RRI is a complex and normative notion, a focus on only attributes that can be objectively measured is likely to decrease construct validity as it will leave out aspects of RRI that are less tangible or more contextual, but are not less important. The other tension is that between avoiding window dressing and avoiding crowding out. The risk of window dressing can be reduced by a clearer division of roles and also by focusing more on the quantitative, easily measurable aspects of RRI. Both, however, may well increase the risk of crowding out. A sharper division of roles may make RRI an external obligation or constraint, rather than something that is aimed at for intrinsic reasons. Also a focus on objectively measurable numbers seems to increase the risk of crowding out, certainly if it is coupled with predictable rewards and punishments.

How are we to move forward given these pitfalls and dilemmas? Without suggesting that all dilemmas can be avoided, I think that major headway can be made by distinguishing more clearly between different rationales for doing RRI assessment. The rationales I have in mind are learning, accountability and incentivizing. I will explain these in more detail below and will argue that these three rationales are very hard – if not impossible – to combine in one RRI assessment method. As a consequence, in developing and deploying RRI assessment we are best advised to deliberately focus on one of these rationales, rather than trying to combine all three in one RRI assessment approach or tool. This does not rule out the use of *different* RRI assessment approaches (or tools) for different rationales, but it makes it advisable not to try to combine different rationales in one RRI assessment approach (or tool).

Learning is aimed at improving RRI performance through gaining new insights and learning new skills. For example, monitoring the efforts of certain RRI actions can provide insights in what RRI actions are (cost) effective and so help to improve RRI performance. But learning may take others forms as well. It may lead to a greater awareness and sensitivity of RRI issues. Learning may also relate to the ability to work with different stakeholders and the ability to connect research and innovation to the values, expectations and needs of society (cf. Klaassen et al., Chapter 9). Also what has been called second-order learning (Argyris & Schön, 1978) is relevant. While first-order learning is about learning how to better achieve given goals, second-order learning is learning about what goals to achieve and puts in question existing value and belief systems. If RRI assessment is to support second-order learning, it should probably leave room for changing goals and perspectives rather than being based on pregiven targets.

Accountability may take different forms, like for example showing that resources for RRI have been spent well, or that the organization meets a certain minimum level of RRI performance or that it can explain its choices in innovation to stakeholders. In all cases, accountability typically involves an external agent to which the innovator is accountable. Moreover, accountability in most cases assumes clear standards or expectations against which the innovator is held accountable.

Incentivizing is, like learning, aimed at improving the RRI performance of the innovator, but whereas in learning the (implicit) assumption is that the innovator is intrinsically motivated to improve RRI performance and thus wants to learn, the incentivizing rationale assumes that an (external) incentive scheme is required to motivate the innovator to do RRI and to improve RRI performance. This external perspective is somewhat similar to that in accountability, but whereas accountability is mainly backward-looking (accounting for what one has done in the past), incentivizing is forward-looking (i.e. incentivizing some future behavior).

Although learning, accountability and incentivizing may all three be proper rationales for doing RRI assessment, they seem hard to combine because there are fundamental tensions between them:

- *Learning versus accountability*: learning typically requires openness for failures and deviations; accountability makes it much harder to recognize these. In terms of RRI assessment, learning is often served by self-assessment, by subjective items that require judgment and by context-specificity, while accountability typically requires independent assessment, objectively measurable indicators and comparability (between contexts).
- *Learning versus incentivizing*: learning assumes an intrinsic motivation to do RRI and to improve; incentivizing assumes external motivation and lack of improvement without incentives. Incentivizing may lead to crowding out of intrinsic motivation and thus may diminish and undermine learning.
- *Incentivizing versus accountability*: the contrast is perhaps less stark than in the other two cases, but there are still potential tensions. Accountability requires objectively measurable indicators and comparability (between contexts); it will often lead to incentivizing what can be measured (and compared) rather than the underlying aim. Accountability may well lead to window dressing rather than a real improvement in RRI performance

Table 14.2 provides a summary of the differences between the three potential rationales for RRI assessment, including their main differences and tensions. Within each rationale, more specific aims may be formulated, like the aims mentioned in Table 14.1. It is important to note, however, that seemingly the same aim may serve different rationales. For example, an aim like monitoring RRI performance, as mentioned in Table 14.1, may be relevant in all three rationales. However, what is a good way to monitor RRI and what are appropriate RRI assessment approaches and tools heavily depend on the underlying rationale. If the underlying rationale is learning, some form of self-assessment and subjective items that require judgment, as for example in approaches 6 and 7 in Table 14.1 would be perfectly appropriate. However, for accountability these would be inappropriate methods, and approaches like 2 and 3 in Table 14.1 would be much more suitable.

The important lesson that can be drawn from this is that in developing and applying methods for RRI assessment one should not only be clear about what the aim of a specific method is, but should also be clear about the underlying rationale (learning, accountability or incentivizing). If one is vague about the underlying rationales or makes an attempt to combine different rationales, it is much more likely that an RRI assessment method is developed or applied that is not fit for purpose.

Table 14.2 Rationales for Responsible Research and Innovation (RRI) assessment

Rationale	*Main aim of RRI assessment*	*Assumed motivation for RRI*	*Self-assessment*	*Objectively measurable indicators*	*Context*
Learning	Improve RRI performance through learning	Intrinsic	Possible and even desirable	Not necessary	Need for context-specificity
Accountability	Show compliance and reliability to outside world	Can be both	Undesirable	Preferable if not required	Need for (some) comparability between contexts
Incentivizing	Improve RRI performance through external incentives	Extrinsic	Possible but usually undesirable	Preferable	Need for (some) comparability between contexts

14.8 Conclusions

The increased attention on RRI has led to proposals for assessing the RRI performance of innovators. Several methods and tools for RRI assessment are now available. I have discussed and illustrated several pitfalls of such methods. Some of these pitfalls are more specific to RRI, like the lack of a uniform definition and the normative character of RRI; others are more general, like the risks of window dressing and that of crowding out intrinsic motivation. Some pitfalls also have a dilemmatic character, in the sense that they relate to different requirements for RRI assessment that are difficult, if not impossible, to combine. To navigate these dilemmas, I have argued that it is best to start from the underlying rationales for doing RRI assessment. I have distinguished three such rationales – learning, accountability and incentivizing – and I have argued that these three rationales are by and large at tension with each other. The conclusion is that if one wants to develop and deploy RRI assessment one should make an explicit choice of one of the rationales rather than trying to serve all three at once. This is even the case if one's assessment method is geared towards an aim, like monitoring RRI performance, that is seemingly important for all three rationales.

Acknowledgment

This chapter has been written as part of the PRISMA project that received funding from the European Union's Horizon 2020 research and innovation program under grant agreement no. 710059.

Notes

1 This is similar to the definition of direct stakeholders in value-sensitive design; see e.g. Friedman, Kahn, and Borning (2006).
2 In particular the aim of assessment was not always explicitly mentioned, as well as who was the (assumed) assessor. In most cases, both could reasonably be derived from the further description or context.
3 These methods are typically aimed more at learning than at assessing RRI levels (and accountability). I will return to this issue in more detail below.
4 I would like to thank one of the reviewers for drawing my attention to this point.
5 I would like to thank one of the reviewers for pointing this out.
6 The point is similar to what is known as Goodhart's law in economics, which has been paraphrased by Strathern (1997: 308) as "When a measure becomes a target, it ceases to be a good measure."

References

Argyris, C., & Schön, D. A. (1978). *Organizational learning*. Reading, MA: Addison-Wesley.

Bradburn, N. M., Cartwright, N. L., & Fuller, J. (2017). A theory of measurement. In L. McClimans (Ed.), *Measuremnet in medicine. Philosophical essays on assessmemt and evaluation* (pp. 73–88). London: Rowman & Littlefield.

Carmines, E. G., & Zeller, R. A. (1979). *Reliability and validity assessment.* Beverly Hills, CA: Sage Publications.

De Bruijn, J. A. (2007). *Managing performance in the public sector* (second edition). London: Routledge.

Dreyer, M., Chefneux, L., Goldberg, A., von Heimburg, J., Patrignani, N., Schofield, M., & Shilling, C. (2017). Responsible innovation: A complementary view from industry with proposals for bridging different perspectives. *Sustainability, 9*(10), 1719.

EPSRC. (2019). Anticipate, reflect, engage and act (AREA). Retrieved from https://epsrc.ukri.org/research/framework/area/

Etzkowitz, H., & Leydesdorff, L. A. (1997). *Universities and the global knowledge economy: A triple helix of university–industry–government relations.* London: Pinter.

European Commission. (2012). Responsible Research and Innovation. Europe's ability to respond to societal challenges. Retrieved from http://ec.europa.eu/research/science-society/document_library/pdf_06/responsible-research-and-innovation-leaflet_en.pdf

European Commission. (2014). Rome Declaration on Responsible Research and Innovation in Europe. Retrieved from https://ec.europa.eu/research/swafs/pdf/rome_declaration_RRI_final_21_November.pdf

Flipse, S. M., Dam, K. H. V., Stragier, J., Vrielink, T. J. C. O., & Sanden, M. C. A. V. d. (2015). Operationalizing responsible research and innovation in industry through decision support in innovation practice. *Journal on Chain and Network Science, 15*(2), 135–146. doi:10.3920/JCNS2015.x004

Friedman, B., Kahn, P. H. J., & Borning, A. (2006). Value sensitive design and information systems. In P. Zhang & D. Galletta (Eds.), *Human–computer interaction in management information systems: Foundations* (Vol. 5, pp. 348–372). Armonk, NY: M.E. Sharpe.

Gneezy, U., Meier, S., & Rey-Biel, P. (2011). When and why incentives (don't) work to modify behavior. *Journal of Economic Perspectives, 25*(4), 191–210. doi:10.1257/jep.25.4.191

Gneezy, U., & Rustichini, A. (2000). A fine is a price. *The Journal of Legal Studies, 29*(1), 1–17. doi:10.1086/468061

Gurzawska, A., Mäkinen, M., & Brey, P. (2017). Implementation of Responsible Research and Innovation (RRI) practices in industry: Providing the right incentives. *Sustainability, 9*(10), 1759.

Keeney, R. L. (1992). *Value-focused thinking: A path to creative decisionmaking.* Cambridge, MA: Harvard University Press.

Kroes, P., & van de Poel, I. (2015). Design for values and the definition, specification, and operationalization of values. In J. van den Hoven, P. E. Vermaas, & I. van de Poel (Eds.), *Handbook of ethics, values, and technological design: sources, theory, values and application domains* (pp. 151–178). Dordrecht: Springer Netherlands.

Lubberink, R., Blok, V., van Ophem, J., & Omta, O. (2017). Lessons for responsible innovation in the business context: A systematic literature review of responsible, social and sustainable innovation practices. *Sustainability, 9*(9), 721.

Moore, G. E. (1903). *Principia ethica.* Cambridge: Cambridge University Press.

Oftedal, G. (2014). The role of philosophy of science in Responsible Research and Innovation (RRI): The case of nanomedicine. *Life Sciences, Society and Policy, 10*(1), 5. doi:10.1186/s40504-014-0005-8

Owen, R., Stilgoe, J., Macnaghten, P., Gorman, M., Fisher, E., & Guston, D. (2013). A framework for responsible innovation. In R. Owen, J. Bessant, & M. Heintz (Eds.), *Responsible innovation* (pp. 27–50). Chichester: Wiley.

Ravn, T., Nielsen, M. W., & Mejlgaard, N. (2015). *Metrics and Indicators of Responsible Research and Innovation.* Retrieved from www.rri-tools.eu/documents/10184/47609/MORRI-D3.2/aa871252-6b2c-42ae-a8d8-a8c442d1d557

Stahl, B. C., Obach, M., Yaghmaei, E., Ikonen, V., Chatfield, K., & Brem, A. (2017). The Responsible Research and Innovation (RRI) maturity model: Linking theory and practice. *Sustainability, 9*(6), 1036.

Star, S. L., & Griesemer, J. R. (1989). Institutional ecology. Translations and boundary objects: Amateurs and professionals in Berkeley's Museum of Vertebrate Zoology, 1907–39. *Social Studies of Science, 19*(3), 387–420.

Stilgoe, J., Owen, R., & Macnaghten, P. (2013). Developing a framework for responsible innovation. *Research Policy, 42*(9), 1568–1580. doi: http://dx.doi.org/10.1016/j.respol.2013.05.008

Strand, R., Spaapen, J., Bauer, M. W., Hogan, E., Revuelta, G., Stagl, S., … Pereira, Â. G. (2015). *Indicators for Promoting and Monitoring Responsible Research and Innovation. Report from the Expert Group on Policy Indicators for Responsible Research and Innovation.* Retrieved from: http://ec.europa.eu/research/swafs/pdf/pub_rri/rri_indicators_final_version.pdf

Strathern, M. (1997). 'Improving ratings': Audit in the British university system. *European Review, 5*(3), 305–321. doi:10.1002/(SICI)1234-981X(199707)5:3<305::AID-EURO184>3.0.CO;2–4

van de Poel, I., Asveld, L., Flipse, S., Klaassen, P., Scholten, V., & Yaghmaei, E. (2017). Company strategies for Responsible Research and Innovation (RRI): A conceptual model. *Sustainability, 9*(11), 2045.

Van den Hoven, J. (2013). Value sensitive design and responsible innovation. In R. Owen, J. Bessant, & M. Heintz (Eds.), *Responsible innovation* (pp. 75–84). Chichester: Wiley.

Von Schomberg, R. (2019). Why responsible innovation? In R. Von Schomberg & H. Jonathan (Eds.), *International handbook on responsible innovation* (pp. 12–33). Cheltenham (UK): Edward Elgar.

Wickson, F., & Carew, A. L. (2014). Quality criteria and indicators for responsible research and innovation: Learning from transdisciplinarity. *Journal of Responsible Innovation*, 1(3), 254–273. doi:10.1080/23299460.2014.963004

Index

Note: Locators in *italics* refer to figures and those in **bold** to tables.